TMU International Symposium on New Quantum Phases Emerging from Novel Crystal Structure 2015

Journal of Physics: Conference Series Volume 683

Tokyo, Japan
24 - 25 September 2015

ISBN: 978-1-5108-2093-7
ISSN: 1742-6588

TMU International Symposium on New Quantum Phases Emerging from Novel Crystal Structure 2015

Journal of Physics: Conference Series Volume 683

Tokyo, Japan
24 - 25 September 2015

IOP
Institute of Physics

Printed from e-media with permission by:

Curran Associates, Inc.
57 Morehouse Lane
Red Hook, NY 12571

Some format issues inherent in the e-media version may also appear in this print version.

Printed by Curran Associates, Inc. (2016)

For permission requests, please contact the Institute of Physics
at the address below.

Institute of Physics
Dirac House, Temple Back
Bristol BS1 6BE UK

Phone: 44 1 17 929 7481
Fax: 44 1 17 920 0979

techtracking@iop.org

Additional copies of this publication are available from:

Curran Associates, Inc.
57 Morehouse Lane
Red Hook, NY 12571 USA
Phone: 845-758-0400
Fax: 845-758-2634
Email: curran@proceedings.com
Web: www.proceedings.com

Journal of Physics: Conference Series **683** (2016) 011001 doi:10.1088/1742-6596/683/1/011001

TMU International Symposium on "New Quantum Phases Emerging from Novel Crystal Structure"

Preface

Papers

BiS_2 systems

Quantum criticality

Frustration systems

Other materials

Theory

Journal of Physics: Conference Series **683** (2016) 011002 doi:10.1088/1742-6596/683/1/011002

Preface

This volume of Journal of Physics: Conference Series contains both invited and contributed papers presented at the International Symposium on "New Quantum Phases Emerging from Novel Crystal Structure", which was held from 24–25 September 2015 at the Minami-Osawa Campus of Tokyo Metropolitan University (TMU).

The Graduate School of Science and Engineering of TMU is now promoting a research project on "New Quantum Phases Emerging from Novel Crystal Structure" with the support of the university. This is the cooperative project involving the electrical and electronic engineering and physics departments to discover new quantum phases in strongly correlated electron systems on novel crystal structures, with geometrically characteristic properties such as cage, layered, and geometrical frustrated structures.

In this international symposium, we have mainly picked up BiS_2-based layered superconductors, cage-structure materials such as 1-2-20 and filled skutterudites, geometrically frustrated systems such as pyrochlore compounds, and noncentrosymmetric materials. Topics on other materials with exotic crystal structure have been also discussed.

I believe that this symposium provides a good opportunity to present recent research results on magnetism and superconductivity in such materials, and to discuss future directions of research on strongly correlated electron systems with novel crystal structure.

I would like to give thanks, on behalf of the organizing committee, to all participants of the TMU International Symposium and all members of the Advisory Committee, who have contributed to the success of this symposium. I further thank the TMU Research Organization for the financial support of this symposium.

Takashi Hotta
Chairman, TMU International Symposium

Committees

Advisory Committee

H. Sato (Professor Emeritus, Tokyo Metropolitan University)

K. Ueda (Director, Physical Society of Japan)

Organizing Committee

Chairman: T. Hotta (Department of Physics, Tokyo Metropolitan University)

Y. Aoki (Department of Physics, Tokyo Metropolitan University)

K. Hattori (Department of Physics, Tokyo Metropolitan University)

R. Higashinaka (Department of Physics, Tokyo Metropolitan University)

H. Kadowaki (Department of Physics, Tokyo Metropolitan University)

T. D. Matsuda (Department of Physics, Tokyo Metropolitan University)

O. Miura (Department of Electrical and Electronic Engineering, Tokyo Metropolitan University)

Y. Mizuguchi (Department of Electrical and Electronic Engineering, Tokyo Metropolitan University)

A. Shudo (Department of Physics, Tokyo Metropolitan University)

H. Takatsu (Department of Energy and Hydrocarbon Chemistry, Kyoto University)

Publication Committee

T. Hotta, Y. Aoki, K. Hattori, T. D. Matsuda, and A. Shudo

Journal of Physics: Conference Series **683** (2016) 011004 doi:10.1088/1742-6596/683/1/011004

Invited Speakers

E. Bauer (Vienna University of Technology)
J. Custers (Charles University)
R. Higashinaka (Tokyo Metropolitan University)
K. Izawa (Tokyo Institute of Technology)
S. Kambe (Japan Atomic Energy Agency)
K. Kuroki (Osaka University)
H. Kusunose (Meiji University)
G. B. Martins (Oakland University)
K. Matsubayashi (University of Electro-Communications)
Y. Matsumoto (Institute for Solid State Physics, University of Tokyo)
Y. Mizuguchi (Tokyo Metropolitan University)
J. Nasu (Tokyo Institute of Technology)
T. Onimaru (Hiroshima University)
S. Onoda (Riken)
S. Ramakrishnan (Tata Institute of Fundamental Research)
H. Sagayama (High Energy Accelerator Research Organization)
N. L. Saini (Rome University)
H. Takatsu (Kyoto University)
T. Yokoya (Okayama University)

Peer review statement

All papers published in this volume of *Journal of Physics: Conference Series* have been peer reviewed through processes administered by the proceedings Editors. Reviews were conducted by expert referees to the professional and scientific standards expected of a proceedings journal published by IOP Publishing.

Superconductivity phase diagram of Se-substituted $CeO_{0.5}F_{0.5}Bi(S_{1-x}Se_x)_2$

Yoshikazu Mizuguchi, Takafumi Hiroi, Osuke Miura

Tokyo Metropolitan University, 1-1, Minami-osawa, Hachioji 192-0397, Japan

mizugu@tmu.ac.jp

Abstract. We investigated the effects of Se substitution on the lattice constants and superconducting properties of $CeO_{0.5}F_{0.5}Bi(S_{1-x}Se_x)_2$. With increasing Se concentration, the *a* lattice constant increased, while the *c* lattice constant did not show any significant increase between $x = 0.1$ and $x = 0.5$. Bulk superconductivity was observed in samples with $x = 0.2$–0.4, and the superconducting transition temperature was the highest at $x = 0.3$. The obtained superconductivity phase diagram was compared to those of $LaO_{0.5}F_{0.5}Bi(S_{1-x}Se_x)_2$ and $NdO_{0.5}F_{0.5}Bi(S_{1-x}Se_x)_2$.

1. Introduction

$BiCh_2$-based (Ch: S, Se) layered superconductors have been drawing considerable attention because of the superconductivity observed in layered structures composed of conducting $BiCh_2$ layers alternating with electrically insulating blocking layers, which resemble the structures of cuprate or Fe-based high-transition-temperature (T_c) superconductors [1,2]. Since the discovery of superconductivity in $Bi_4O_4S_3$ [3] and $LaO_{1-x}F_xBiS_2$ [4], several types of BiS_2-based superconductors have been discovered [5]. In addition, a $BiSe_2$-based superconductor ($LaO_{0.5}F_{0.5}BiSe_2$) was also synthesized [6]. However, the superconductivity mechanisms of $BiCh_2$-based superconductors have not been clarified yet.

Superconductivity in $BiCh_2$-based materials is essentially induced by electron carrier doping into Bi-6p orbitals; parent compounds, such as $LaOBiS_2$, are insulators with band gaps [7]. In addition, a recent synchrotron X-ray diffraction study indicated that crystal structure optimization is also required for the appearance of bulk superconductivity in $REOBiCh_2$-type (RE: rare earth) materials [8]. Specifically, bulk superconductivity states can be achieved when both the electron carrier concentration and the crystal structure are optimized. We have systematically studied the relationship between superconductivity and crystal structure by tuning the *chemical pressure*, which is lattice shrinkage (or expansion) not accompanied by carrier doping. For example, in $LaOBiS_2$, the chemical pressure can be tuned by substituting La^{3+} (1.16 Å) with Ce^{3+} (1.14 Å) or Nd^{3+} (1.12 Å) [9] or by substituting S^{2-} (1.84 Å) with Se^{2-} (1.98 Å). In $LaO_{0.5}F_{0.5}Bi(S_{1-x}Se_x)_2$, bulk superconductivity is induced with increasing Se concentration, and the highest T_c is observed when $x = 0.5$ [10]. In $NdO_{0.5}F_{0.5}Bi(S_{1-x}Se_x)_2$, however, T_c decreases with Se substitution [11]. Having considered the fact that these Se substitutions in $REO_{0.5}F_{0.5}Bi(S_{1-x}Se_x)_2$ do not dope electron carriers, clarifying the universal relationship between the superconductivity and the crystal structure of Se-substituted $REO_{0.5}F_{0.5}Bi(S_{1-x}Se_x)_2$ should provide information useful for understanding the superconductivity mechanisms in $BiCh_2$-based superconductors. Herein, we present the effects of Se substitution on the crystal structure (lattice constants) and superconducting properties of $CeO_{0.5}F_{0.5}Bi(S_{1-x}Se_x)_2$.

2. Experimental details

Polycrystalline samples of $CeO_{0.5}F_{0.5}Bi(S_{1-x}Se_x)_2$ were prepared using the solid-state-reaction method. Initial powders of CeO_2 (99.99%), CeF_3 (99.9%), Ce_2S_3 (99.9%), Bi_2Se_3 (synthesized from Bi (99.999%) and Se (99.999%) grains), Bi, and S (99.99%) with the nominal compositions of the target phases ($x = 0.1$–0.5) were mixed, pelletized, and sealed into evacuated quartz tubes. The samples were heated for 15 h at 700 °C. The obtained (reacted) pellets were ground, pelletized, sealed into evacuated quartz tubes, and heated for an additional 15 h at 700 °C. The purities and lattice constants of the samples were examined using powder X-ray diffraction (XRD) with CuKα radiation and the θ–2θ method. Their superconducting properties were investigated by performing magnetic susceptibility (χ) measurements using a superconducting quantum interference device (SQUID) magnetometer with an applied field of about 5 Oe after zero-field cooling down to 2 K.

3. Results and discussion

Figure 1(a) shows the XRD patterns for $CeO_{0.5}F_{0.5}Bi(S_{1-x}Se_x)_2$. The target phase of $CeO_{0.5}F_{0.5}Bi(S_{1-x}Se_x)_2$ was obtained for $x \leq 0.5$. For $x = 0.1$–0.3, almost all of the XRD peaks could be indexed using the space group of *P4/nmm* [2] except for the small peaks due to impure CeF_3, which are indicated with circles (●). For $x = 0.4$ and 0.5, peaks resulting from impure Bi_2Se_3 are evident. Based on these results, we assume that the solubility limit of S and Se in the $CeO_{0.5}F_{0.5}Bi(S_{1-x}Se_x)_2$ structure exists near $x = 0.5$. The lattice constants were calculated from the peak positions of the (200) and (004) reflections. The calculated values of the a and c lattice constants are plotted in Figs. 1(b) and 1(c), respectively, as functions of the nominal x value. The a lattice constant increases monotonically with increasing Se concentration, while the c lattice constant firstly increases (from $x = 0$ [9] to $x = 0.1$) but does not significantly change between $x = 0.1$ and $x = 0.5$. These lattice constant evolutions suggest that the substituted Se ions occupy in-plane Ch sites (see Fig. 1(d)). This tendency is similar to that observed in $LaO_{0.5}F_{0.5}Bi(S_{1-x}Se_x)_2$ [10]. The anisotropic elongations of the a- and c-axes should affect the structure of the $BiCh_5$ pyramid, which is an important factor in the superconductivity of $BiCh_2$-based materials [12,13]. To analyze the crystal structure parameters of $CeO_{0.5}F_{0.5}Bi(S_{1-x}Se_x)_2$, further synchrotron XRD, X-ray absorption spectroscopy, and/or single-crystal structural analyses are necessary.

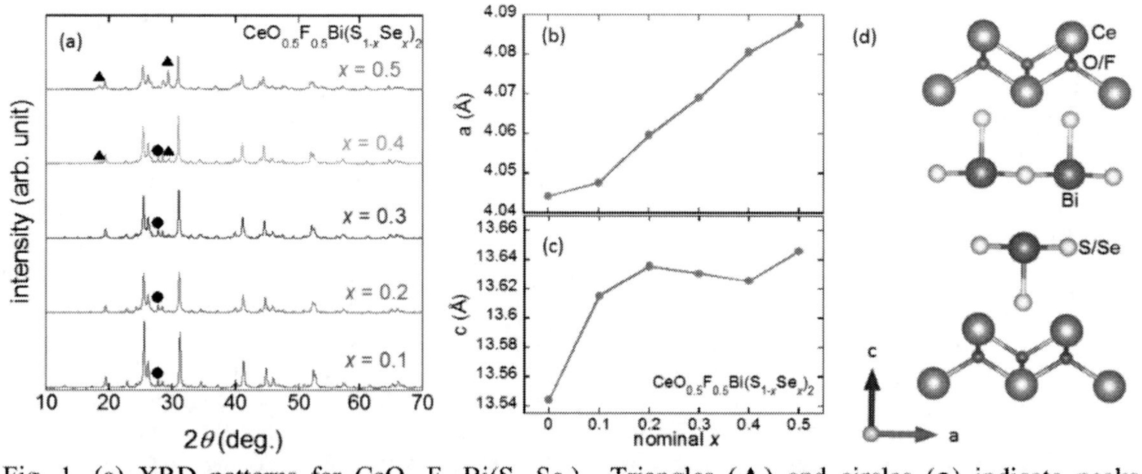

Fig. 1. (a) XRD patterns for $CeO_{0.5}F_{0.5}Bi(S_{1-x}Se_x)_2$. Triangles (▲) and circles (●) indicate peaks resulting from impure Bi_2Se_3 and CeF_3, respectively. (b) Nominal x dependence of a lattice constant of $CeO_{0.5}F_{0.5}Bi(S_{1-x}Se_x)_2$. (c) Nominal x dependence of c lattice constant of $CeO_{0.5}F_{0.5}Bi(S_{1-x}Se_x)_2$. (d) Schematic image of crystal structure of $CeO_{0.5}F_{0.5}Bi(S_{1-x}Se_x)_2$. Image was drawn using VESTA [14].

Figure 2(a) shows the temperature dependences of the magnetic susceptibilities ($4\pi\chi$) of $CeO_{0.5}F_{0.5}Bi(S_{1-x}Se_x)_2$ with different x values. Superconducting (diamagnetic) signals are not apparent for $x \leq 0.1$. Large diamagnetic signals are evident for $x = 0.2$–0.4, indicating the emergence of bulk superconductivity states in these samples. For $x = 0.5$, the superconducting transition is not observable. Using the estimated values of T_c, we established a superconductivity phase diagram for $CeO_{0.5}F_{0.5}Bi(S_{1-x}Se_x)_2$ (Fig. 2(b)). The phase diagram exhibits dome-shaped behavior upon Se substitution, which is similar to that in the phase diagram of $LaO_{0.5}F_{0.5}Bi(S_{1-x}Se_x)_2$ [10]. However, the x value at the peak of the $CeO_{0.5}F_{0.5}Bi(S_{1-x}Se_x)_2$ diagram ($x = 0.3$) is significantly different from that of the $LaO_{0.5}F_{0.5}Bi(S_{1-x}Se_x)_2$ diagram ($x = 0.5$). Furthermore, the phase diagram for $NdO_{0.5}F_{0.5}Bi(S_{1-x}Se_x)_2$ does not exhibit dome-shaped behavior; instead, T_c monotonically decreases with increasing x [11]. To compare these $REO_{0.5}F_{0.5}Bi(S_{1-x}Se_x)_2$ superconductivity phase diagrams, all three are plotted in Fig. 3.

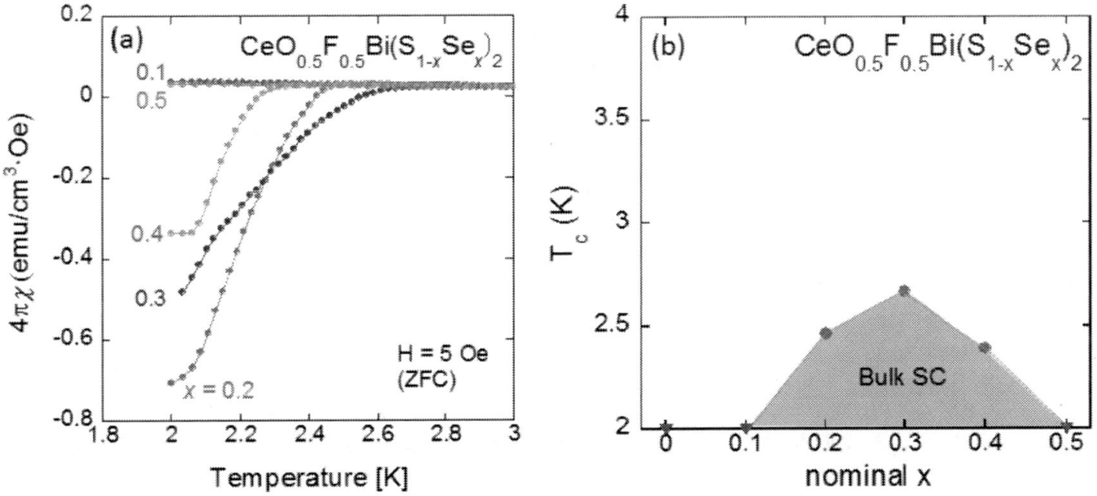

Fig. 2. (a) Temperature dependences of magnetic susceptibilities ($4\pi\chi$) for $CeO_{0.5}F_{0.5}Bi(S_{1-x}Se_x)_2$ with different x values. (b) Superconductivity phase diagram for $CeO_{0.5}F_{0.5}Bi(S_{1-x}Se_x)_2$. Bulk SC denotes region showing bulk superconductivity. Inverted triangles (▼) indicate samples showing no superconductivity above 2 K (measurement limit of our SQUID system).

Figure 3 presents the phase diagrams for $REO_{0.5}F_{0.5}Bi(S_{1-x}Se_x)_2$ (RE = La, Ce, and Nd) as functions of the nominal x value. For $LaO_{0.5}F_{0.5}Bi(S_{1-x}Se_x)_2$, bulk superconductivity is apparent for $x \geq 0.2$, while filamentary (weak) superconducting signals are observable at $x = 0$ and 0.1. T_c increases as x increases to 0.5, and it decreases above $x = 0.5$. For $CeO_{0.5}F_{0.5}Bi(S_{1-x}Se_x)_2$, bulk superconductivity is apparent at $x = 0.2$, and the highest T_c occurs at $x = 0.3$. With further Se substitution, superconductivity is suppressed. For $NdO_{0.5}F_{0.5}Bi(S_{1-x}Se_x)_2$, the T_c value of $NdO_{0.5}F_{0.5}BiS_2$ ($x = 0$) is the highest ($T_c \sim 5$ K) among all of the $REO_{0.5}F_{0.5}Bi(S_{1-x}Se_x)_2$ samples in this study, but it decreases with Se substitution. Substituting S with Se in $NdO_{0.5}F_{0.5}Bi(S_{1-x}Se_x)_2$ degrades its superconductivity. These differences in the effects of Se substitution on superconductivity should be related to the differences in the REO blocking layers in these materials, which are related to their chemical pressures. As mentioned in the introduction, the ionic radii of RE^{3+} are different: assuming a coordination number of 8, La^{3+}, Ce^{3+}, and Nd^{3+} have ionic radii of 1.16 Å, 1.14 Å, and 1.12 Å, respectively. The peak superconductivities for RE = La, Ce, and Nd occur at $x = 0.5$, 0.3, and 0 (or less than 0), respectively. Therefore, the x values at the peaks seem to be correlated with the sizes of the blocking layers in these materials.

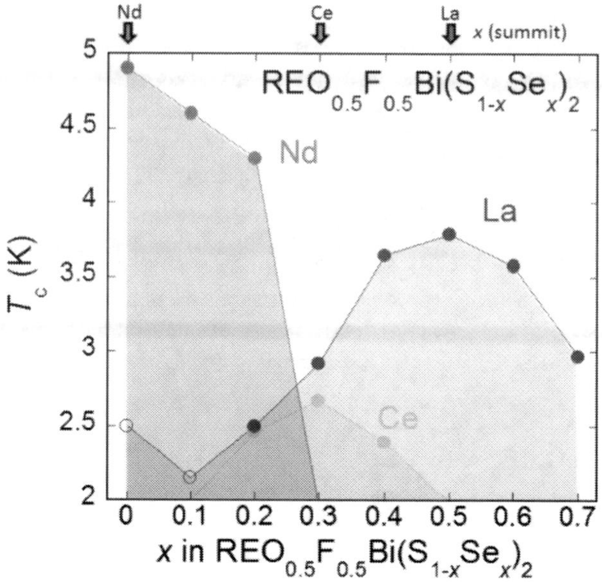

Fig. 3. Superconductivity phase diagrams for $REO_{0.5}F_{0.5}Bi(S_{1-x}Se_x)_2$ (RE = La, Ce, and Nd). Open blue circles for $LaO_{0.5}F_{0.5}Bi(S_{1-x}Se_x)_2$ indicate samples showing filamentary superconductivity.

4. Conclusion

We studied the effects of Se substitution on the crystal structure and superconductivity of $CeO_{0.5}F_{0.5}Bi(S_{1-x}Se_x)_2$. With increasing Se concentration, the *a* lattice constant increased, while the *c* lattice constant did not show any significant increase between $x = 0.1$ and $x = 0.5$. Bulk superconductivity was observed for the samples with $x = 0.2$–0.4, and the highest T_c value was attained at $x = 0.3$. Comparing the phase diagrams for $REO_{0.5}F_{0.5}Bi(S_{1-x}Se_x)_2$, we suggested that the *x* values yielding the peak superconductivities were related to the sizes of their REO blocking layers (RE^{3+} ionic radii).

Acknowledgements

This work was partly supported by Grant-in-Aid for Young Scientist (A) (25707031), Grant-in-Aid for challenging Exploratory Research (26600077), and Grant-in-Aid for Scientific Research on Innovative Areas (15H05886).

References

[1] J. B. Bednorz, and K. Müller 1986 Z. Physik B Condens. Matter **64** 189-193.
[2] Y. Kamihara *et al.* 2008 J. Am. Chem. Soc. **130** 3296-3297.
[3] Y. Mizuguchi *et al.* 2012 Phys. Rev. B **86** 220510(1-5).
[4] Y. Mizuguchi *et al.* 2012 J. Phys. Soc. Jpn. **81** 114725(1-5).
[5] Y. Mizuguchi 2015 J. Phys. Chem. Solid **84** 34-48.
[6] A.K. Maziopa *et al.* 2014 J. Phys.: Condens. Matter **26** 215702(1-5).
[7] H. Usui, K. Suzuki, and K. Kuroki 2012 Phys. Rev. B **86** 220501(1-5).
[8] Y. Mizuguchi *et al.* 2015 Sci. Rep. **5** 14968(1-8).
[9] J. Kajitani, *et al.* 2015 J. Phys. Soc. Jpn. **84** 044712(1-6).

[10] T. Hiroi *et al.* 2015 J. Phys. Soc. Jpn **84** 024723(1-4).
[11] T. Hiroi *et al.* 2014 J Supercond. Nov. Magn. **28** 1149-1153.
[12] T. Sugimoto *et al.* 2014 Phys. Rev. B **89** 201117(1-5).
[13] E. Paris *et al.* 2014 J. Phys.: Condens. Matter **26** 435701(1-6).
[14] K. Momma, and F. Izumi 2008 J. Appl. Crystallogr. **41** 653-658.

Comparative ARPES studies of $LaO_xF_{1-x}BiS_2$ ($x = 0.23$ and 0.46)

K Terashima[1,2], T Wakita[1,2], M Sunagawa[1], H Fujiwara[1], T Nagayama[1], K Ono[3], H Kumigashira[3], M Nagao[4], S Watauchi[4], I Tanaka[4], H Okazaki[5], Y Takano[5], Y Mizuguchi[6], H Usui[7], K Kuroki[7], Y Muraoka[1,2], and T Yokoya[1,2]

[1] Research Laboratory for Surface Science and Graduate School of Natural Science and Technology, Okayama University, Okayama, 700-8530, Japan
[2] Research Center of New Functional Materials for Energy Production, Storage, and Transport, Okayama University, Okayama, 700-8530, Japan
[3] High Energy Accelerator Research Organization (KEK), Photon Factory, Tsukuba, Ibaraki 305-0801, Japan
[4] Center for Crystal Science and Technology, Yamanashi University, Kofu 400-8510, Japan
[5] National Institute for Materials Science, 1-2-1, Sengen, Tsukuba, 305-0047, Japan
[6] Department of Electrical and Electronic Engineering, Tokyo Metropolitan University, 1-1, Minami-osawa, Hachioji, 192-0397, Japan
[7] Department of Physics, Osaka University, 1-1 Machikaneyama Toyonaka, Osaka 560-0043, Japan.

E-mail: yokoya@cc.okayama-u.ac.jp

Abstract. BiS_2-based layered superconductors are new superconductors, having been paid special attention due to the layered crystal structure and the characteristic electronic structure that can lead to exotic mechanism of superconductivity as well as cooperation of superconductivity and ordered phases. In this article, we report valence band dispersions and Fermi surface topology of $LaO_{1-x}F_xBiS_2$ (x = 0.23, T_c < 2 K) studied by angle-resolved photoemission spectroscopy, and compared them with those of $LaO_{1-x}F_xBiS_2$ (x = 0.46, T_c = 3.1 K). We discuss experimental x dependent electronic structures and its relation to T_c. We also point to anomalous spectroscopic features in $LaO_{1-x}F_xBiS_2$ (x = 0.46) that can be explained by band structure calculations considering distorted lattice, suggesting role of dynamical or static lattice distortion to the physical properties.

1. Introduction

The new family of layered superconductors possessing double BiS_2 layers has provoked much attention since the discovery of $Bi_4O_4S_3$ [1]. So far mostly studied compounds are $LnO_{1-x}F_xBiS_2$ (Ln: Lanthanides), where the highest superconducting critical temperature (T_c) of 10.6 K has been reported for $LaO_{1-x}F_xBiS_2$ [2,3]. The crystal structure consists of alternating stack of double BiS_2 layers and $La_2(O,F)_2$ spacer layers along the c axis. Some of the compounds exhibit magnetism (Ln = Ce) [4] or charge density of waves (Ln = Eu) above T_c [5], suggesting interplay between the superconductivity and ordered phases.

In $LaO_{1-x}F_xBiS_2$, the parent compound shows semiconducting temperature dependence on resistivity. As the fluorine concentration (x) increases, superconductivity appears at $x = 0.2$, and T_c increases till $x = 0.5$ and decreases with further increase of x [3]. Theoretically, band structure calculations have predicted interesting electronic structure and its relation to the mechanism of the superconductivity. In $LaO_{1-x}F_xBiS_2$, band structure calculations predicted that $LaOBiS_2$ is a band insulator having a band gap of ~ 0.8 eV [6-9]. While the valence band consists of S $3p$ (in plane and out of plane) and O $2p$ orbitals, the conduction band is derived from Bi $6p_{x,y}$ and in-plane S $3p$. Since the O substitution of F introduces electron carriers, the chemical potential of $LaO_{0.5}F_{0.5}BiS_2$ is expected to locate within the Bi $6p_{x,y}$ and in-plain S $3p$ hybridized band. This makes the system metallic. Importantly, calculated Fermi surface has a topological change around $x \sim 0.5$ from isolated electron pockets centered around X(R) in the Brillouin zone for smaller x to connected hole-like sheets around Γ(Z) and M(A) for higher x with a strong nesting with a vector (π, π) [along the Γ(Z)-M(A) direction]. The nested Fermi surface is originated from quasi-one-dimensional nature of the electronic structure [10] and leads to CDW instability [6,7], which can explain the pressure dependence of T_c in $LaO_{0.5}F_{0.5}BiS_2$ [11]. Possibilities of exotic pairing mechanisms due to electron-electron correlation originating from the nested Fermi surface have also been examined [10,12-14,15]. Therefore, to understand the mechanism of superconductivity, it is highly desirable to experimentally investigate the normal state electronic structure.

Experimentally, photoemission spectroscopy (PES) has played an important role to elucidate the electronic structure of $LnO_{1-x}F_xBiS_2$. From valence band soft x-ray PES of polycrystalline $LaO_{1-x}F_xBiS_2$, experimental electronic structure and its x dependence, namely higher binding energy shift of the valence band as well as appearance of new states near the Fermi level (E_F) having dominant Bi $6p$ character, were found to be consistent with the predictions of band structure calculations in general [16].

Angle-resolved photoemission spectroscopy (ARPES) of $NdO_{1-x}F_xBiS_2$ observed small electron-like pocket around the X(R) point in the Brillouin zone [17,18]. Large difference between the carrier concentrations determined from the observed area of the electron pocket and those estimated from nominal concentration were noticed [17,18]. ARPES studies of $CeO_{1-x}F_xBiS_2$ [19] and $LaO_{1-x}F_xBiSe_2$ (x = 0.18) [20] have also reported Fermi surface topology for smaller carrier concentrations. In contrast to the other ARPES studies, our vacuum ultraviolet and soft x-ray ARPES of $LaO_{1-x}F_xBiS_2$ ($x = 0.46$) showed Fermi surface topology for higher carrier concentration [21]. The observed ARPES intensity map showed electron like Fermi surface sheets and square-like intensity distribution, latter of which suggests that in the compound E_F lies close to the position where topological change occurs. This means that E_F is close to van Hove singularity leading to high density of states, different from ARPES of $NdO_{1-x}F_xBiS_2$. ARPES of $LaO_{1-x}F_xBiS_2$ observes the Fermi surface topology consistent with that expected from the nominal concentration determined from EDX. This indicates that $LaO_{1-x}F_xBiS_2$ is a suitable system for studying x dependent changes in band structures and Fermi surface topology.

Previous PES studies have also reported anomalous features that cannot be explained by band structure calculations. Soft x-ray PES reported reduced intensity for the new states at E_F [16]. Anomalous dispersive features of the electron pocket, which exhibit marked temperature dependence, was observed for ARPES data of $NdO_{1-x}F_xBiS_2$ and ascribed to strong electron-phonon coupling effects [17]. From the ARPES studies of $LaO_{1-x}F_xBiS_2$, we reported anomalous features around 0.4 eV binding energy in momentum regions away from the electron pocket [21]. Further investigation of the anomalous features can give some hint to understand the underlying physics.

In this paper, we report ARPES study of $LaO_{1-x}F_xBiS_2$ ($x = 0.23$) and compared it with our previous ARPES study of $LaO_{1-x}F_xBiS_2$ ($x = 0.46$) [21], in order to reveal x-dependent changes in band dispersion and Fermi surface topology. Comparative ARPES studies provide evidence for correlation between carrier concentration and T_c in $LaO_{1-x}F_xBiS_2$. We also discuss a possible origin of the anomalous spectral features observed in $LaO_{1-x}F_xBiS_2$ ($x = 0.46$), suggesting that strong electron-phonon coupling and CDW instability can give influence on the physical properties.

2. Experimental

$LaO_{1-x}F_xBiS_2$ (x = 0.23) single crystals were grown using a CsCl/KCl flux method. The ratio of the components was estimated by electron probe microanalysis. T_cs were determined from the temperature where the resistivity became zero [T_c^{zero} < 2 K (x = 0.23) and T_c^{zero} = 3.1 K (x = 0.46)].

ARPES measurements were performed at BL-28A in Photon Factory with a VG Scienta electron analyzer, and the energy of a circularly polarized light was set at 70 eV. The energy resolution was set to 30 meV for hν = 70 eV, and the angular resolution of ARPS was set to 0.2°. The binding energies of samples were deduced by referencing E_F of gold electronically contacted with samples. Samples were cleaved and measured *in situ* under a vacuum better than 3 × 10^{-8} Pa at T = 10 K, which was above the transition temperature of the sample. The samples were stable in the vacuum and showed no sign of degradation during the measurement period of 72 h.

3. Results and discussion

3.1. Valence band dispersions

Figures 1 (a) and (b) show experimental band dispersions obtained from $LaO_{1-x}F_xBiS_2$ single crystals with x = 0.23 and 0.46, respectively, measured along Γ(Z)-X(R) in the Brillouin zone. The darker regions correspond to bands. Theoretical band dispersions are superimposed on the plot of x = 0.46 with dotted orange lines, for comparison. In both data, many bands are closely located in the region between 2 eV and 6 eV binding energies, which are derived from mainly S 3p and O 2p orbitals. The dispersions of two samples appear to be similar with small differences in detail. The location in energy and width of the band are nearly the same, especially at Γ (Z). In contrast, the energy position of the

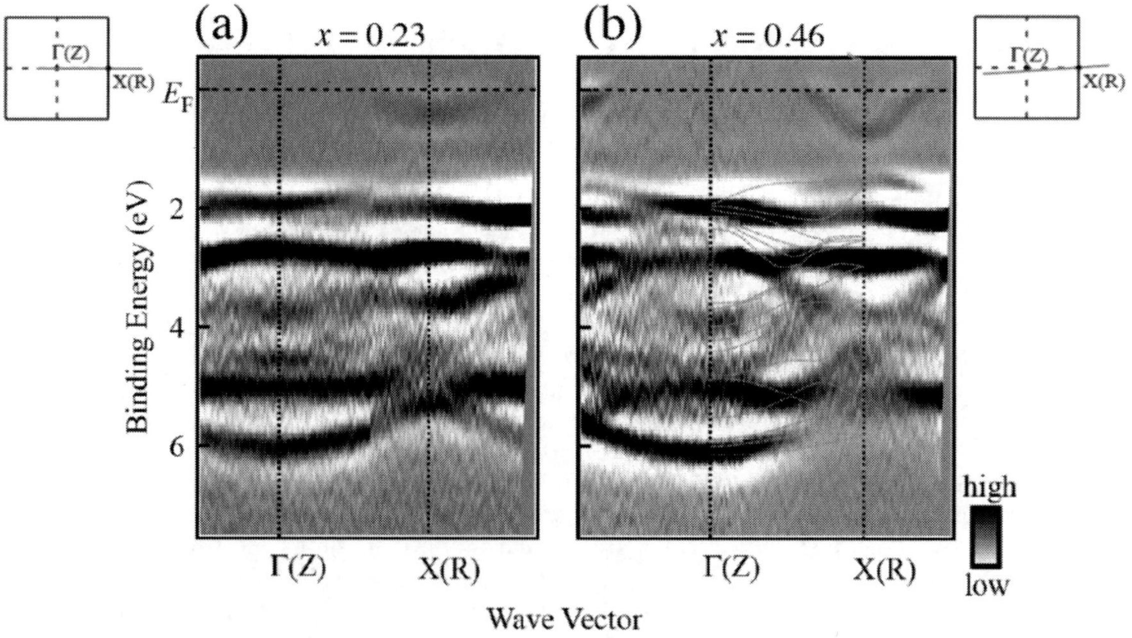

Figure 1. The valence band ARPES result of $LaO_{1-x}F_xBiS_2$ single crystal with x = 0.23, compared with that with x = 0.46 [21]. The results are made by plotting minus second derivative of near E_F ARPES intensity as functions of binding energy and wave vector. Darker parts correspond to experimentally observed bands. Orange lines are theoretical band dispersions calculated from band structure calculations with SOC. The measured direction is indicated by a line in each schematic Brillouin zone.

Journal of Physics: Conference Series **683** (2016) 012002 doi:10.1088/1742-6596/683/1/012002

bottom of the electron pocket is largely different. This indicates that the change of the electronic structure as a function of x is not rigid-band like. This is consistent with the band structure calculations, which predicted that the band dispersions of $LaO_{1-x}F_xBiS_2$ are sensitive to both lattice structure and fluorine doping [22]. Nevertheless, the change in the energy position of the bottom of the electron pocket is in line with electron doping with increasing of x and therefore the change of Fermi surface topology is expected.

Before examining Fermi surface, it is worth noticing the importance of spin-orbit coupling (SOC) of Bi 6p to explain the band structure of $LaO_{1-x}F_xBiS_2$. The comparison between the experimental and theoretical band dispersions in x = 0.46 sample shows some correspondence. For example, the experimental band dispersion of the electron pocket is well explained by the calculations. For the valence band, the experimental band dispersions around Γ(Z) at 2, 5, and 6 eV as well as that around X(R) at 1.5 eV well follow the calculated dispersions. The band structure calculations without SOC predicted a larger gap between the top of the valence band and the bottom of the electron pocket and also predicted another smaller electron pocket at X(R), which is not observed in the present data. These findings tell that the electronic structure with SOC better explains the observed band dispersions, indicating the importance of SOC for the electronic structure of $LaO_{1-x}F_xBiS_2$. Role of SOC has been examined from the splitting of the electron pocket [17].

3.2. Fermi surface

In order to look at the x dependent Fermi surface topology, we plotted the near E_F ARPES intensity map of x = 0.23 as functions of k_x and k_y and compared it with that of x = 0.46, as shown in Figs. 2(a) and (b). The darker parts correspond to the Fermi surface or the momentum regions where the band is close to E_F. In both the data, Fermi surface around X(R) are clearly observed, whose electron like character are evident from the band dispersions shown in each inset. While the shape of the electron pocket in x = 0.23 sample appears to be rather circular, that in x = 0.46 sample is rectangular. The area of the electron pocket in x = 0.46 sample appears to be approximately two times larger than that in x = 0.23 sample, indicating x dependent increase of carrier concentration. Interestingly, in x = 0.46 sample, square-like intensity distribution is also found centered at the Γ(Z) point. These features in x = 0.46 are consistently explained by the calculated Fermi surface, which is superimposed in the figure with orange curves. The rather smaller intensity around the mid point of the corner of the square-like

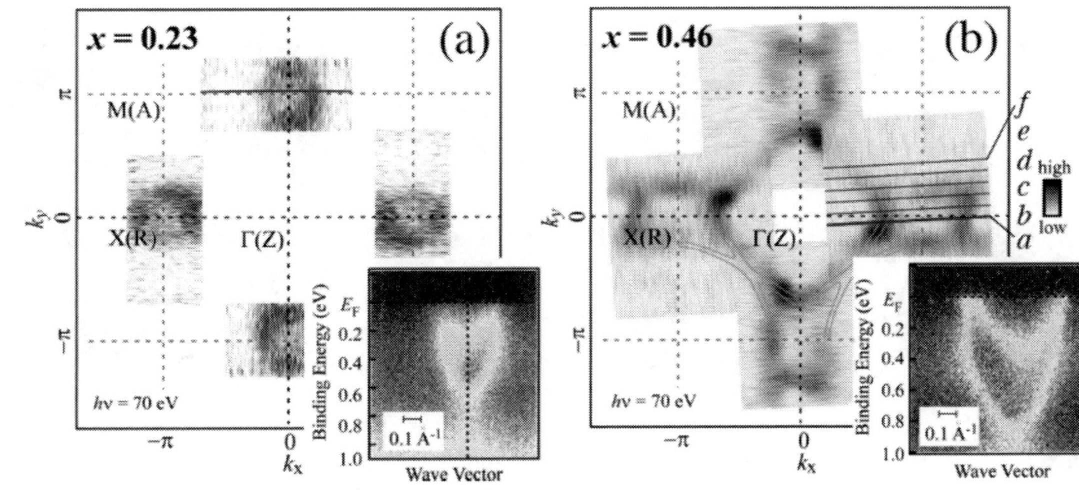

Figure 2. Near E_F ARPES intensity map of x = 0.23, together with that of x = 0.46 [21], as functions of k_x and k_y. Inset to each Fermi surface map is band dispersions for the electron pocket.

Journal of Physics: Conference Series **683** (2016) 012002 doi:10.1088/1742-6596/683/1/012002

intensity distribution around Γ(Z) will be discussed in the next subsection.

In $LaO_{1-x}F_xBiS_2$, T_c shows x dependence: the larger x concentration is, the higher T_c is. The present ARPES provides direct evidence that T_c is directly related to the carrier concentration, at least for $LaO_{1-x}F_xBiS_2$. On the other hand, very small carrier concentration estimated from previous ARPES of $NdO_{1-x}F_xBiS_2$ with T_c of 4.5 K suggests existence of other parameters for determining T_c in $LnO_{1-x}F_xBiS_2$.

3.3. Anomalous features observed in ARPES

So far we reported experimental band structure and its consistency with band structure calculations. We have also found strong deviations of the experimental electronic structure from theoretical studies. One of the results is related to the anomalous spectral intensity found in the compounds. In the insets of Fig. 2, the ARPES intensity tends to be smaller for lower binding energy. A related finding for the electron pocket was reported for ARPES of $NdO_{1-x}F_xBiS_2$ [17], where the feature was discussed in terms of strong coupling of electrons with lattice. Anomalous features in the other momentum regions may give further understanding for the electron-phonon coupling. In the upper row of Fig. 3, we show ARPES intensity maps along cuts *a-f*. We found that, though the band dispersions along cuts *a-c* are reasonably well explained by the calculated bands, those along cuts *d-f* cannot be done very well. In cuts *d-f*, large intensity is left around 0.4 eV binding energy, though the calculated bands are located close to E_F. Actually, the structure around 0.4 eV can be observed every cut, as shown in the second derivative maps in the lower row. One can also notice that the intensity variation of the 0.4 eV structure shows a clear momentum dependence (cuts *d-f*), which follows existence of the bands near E_F. These results seem to suggest that the structure around 0.4 eV is made up of two components: momentum dependent one and momentum independent one.

STS/STM studies reported a broad structure centered around 0.4 eV with an energy gap at E_F [23], indicating insulating nature of the surface states. The momentum independent part of ARPES data, which is a localized state, may be something to do with the insulating surface electronic states observed in STS/STM. For the momentum dependent component, we would like to point out striking similarity with the band structure calculations for a distorted lattice [12,13]. According to the band structure calculations, $LaO_{1-x}F_xBiS_2$ with $x \sim 0.5$ has instability toward lattice distortion with a wave vector of (π, π), due to the strong nesting and large electron-photon coupling. The band structure calculations for the distorted lattice predicted the momentum dependent modification of bands. The band dispersions around the region where the van Hove singularity is located are largely affected by

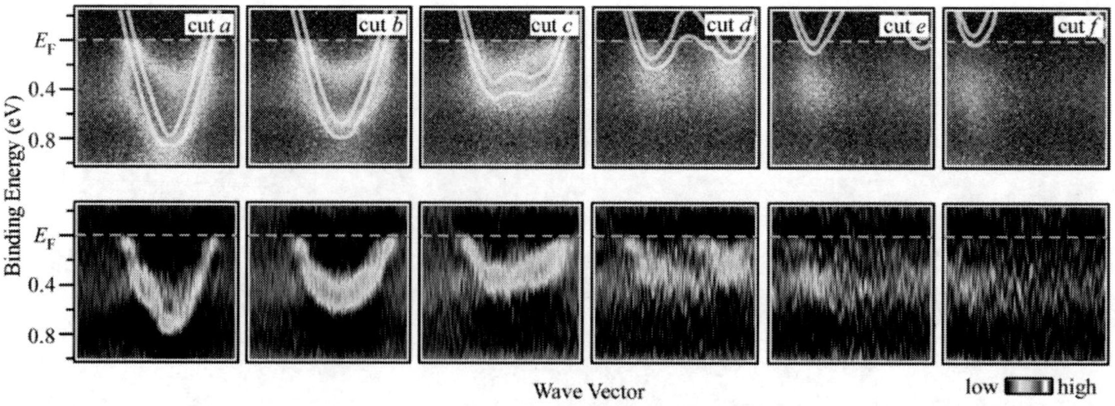

Figure 3. ARPES intensity (upper row) maps [21] and their minus second derivative (lower row) maps as functions of binding energy and wave vector measured along 6 cuts ($a-f$) shown in Fig. 2 (b). The white curves in the intensity maps are results of band calculations for $LaO_{0.54}F_{0.46}BiS_2$ with SOC.

doi:10.1088/1742-6596/683/1/012002

the distortion. The momentum region corresponds to that around the midpoint of the square-like intensity distribution in Fig. 2(b). More specifically, the band forming the van Hove singularity, which is located slightly above E_F in the undistorted lattice, can slip into occupied part in the distorted lattice. Theoretical studies predicted either static or dynamical type of distortion [6, 7]. If it is dynamical, we may be observing the sum of the snap shots of electronic structure affected by the distortion, as the time scale of PES is extremely short, an order of 10^{-15} second, compared to the lattice motion. We may be observing electronic structures from both the undistorted and statically distorted regions simultaneously. This may be able to explain the relatively small intensity at E_F around the midpoint of the square-like intensity distribution and existence of the momentum dependent 0.4 eV structure. The similarity between the experimental observation and the theoretical prediction for distorted lattice in $LaO_{1-x}F_xBiS_2$ ($x = 0.46$) suggests that the strong electron-phonon coupling near the instability may be important to understand the physical properties.

4. Summary

The experimental band dispersions and Fermi surface topology of $LaO_{1-x}F_xBiS_2$ ($x = 0.23$) were observed by ARPES and compared with those of $x = 0.46$. The non rigid band like x dependence of $LaO_{1-x}F_xBiS_2$ was observed from comparative valence band ARPES data. Increase of the area of the electron-like Fermi surface around X(R), which indicates increase of carrier concentration, was observed and found to correlate with T_c. This indicates that the carrier concentration is a parameter directly related to T_c in $LaO_{1-x}F_xBiS_2$. The anomalous features observed around the region at the midpoint of the square-like intensity distribution can be explained by band structure calculations for distorted lattice, which suggests the importance of dynamical or static lattice distortion to consider the physical properties of $LaO_{1-x}F_xBiS_2$ ($x = 0.46$).

Acknowledgments

We thank Dr. T. Yildirim for valuable discussion. ARPES experiments at Photon Factory were performed under proposal numbers 2013G703. This work was partially supported by JSPS KAKENHI Grant Numbers 25800205 and 15H03691. This work was also partially supported by the Program for Promoting the Enhancement of Research Universities from MEXT.

References

[1] Mizuguchi Y 2015 *J. Phys. Chem. Solid* **84** 34
[2] Mizuguchi Y, Demura S, Deguchi K, Takano Y, Fujihisa H, Gotoh Y, Izawa H and Miura O 2012 *J. Phys. Soc. Jpn.* **81** 114725
[3] Deguchi K, Mizuguchi Y, Demura S, Hara H, Watanabe T, Denholme S J and Takano Y 2013 *Europhys. Lett.* **101** 17004
[4] Demura S, Deguchi K, Mizuguchi Y, Sato K, Honjyo R, Yamashita A, Yamaki T, Hara H, Watanabe T, Denholme S J, Fujioka M, Okazaki H, Ozaki T, Miura O, Yamaguchi T, Takeya H and Takano Y 2015 *J. Phys. Soc. Jpn.* **84** 024709
[5] Zhai H-F, Tang Z-T, Jiang H, Xu K, Zhang K, Zhang P, Bao J-K, Sun Y-L, Jiao W-H, Nowik I, Felner I, Li Y-K, Xu X-F, Tao Q, Feng C-M, Xu Z-A and Cao G-H 2014 *Phys. Rev. B* **90** 064518
[6] Wan X, Ding H-C, Savrasov S Y and Duan C-G 2013 *Phys. Rev. B* **87** 115124
[7] Yildirim T 2013 *Phys. Rev. B* **87** 020506
[8] Li B, Xing Z W and Huang G Q 2013 *Europhys. Lett.* **101** 47002
[9] Shein I R and Ivanovskii A L 2013 *JETP Letters* **96** 769
[10] Usui H, Suzuki K and Kuroki K 2012 *Phys. Rev. B* **86** 220501(R)
[11] Kotegawa H, Tomita Y, Tou H, Izawa H, Mizuguchi Y, Miura O, Deguchi K and Takano Y 2012 *J. Phys. Soc. Jpn.* **81** 103702
[12] Zhou T and Wang Z D 2013 *J. Supercond. Nov. Magn.* **26** 2735
[13] Liang Y, Wu X, Tsai W F and Hu J 2014 *Front. Phys.* **9** 194

[14] Martins G B, Moreo A and Dagotto E 2013 *Phys. Rev. B* **87** 081102
[15] Agatsuma T and Hotta T 2015 To appear in *J. Magn. Magn. Mater.*
[16] Nagira S, Sonoyama J, Wakita T, Sunagawa M, Izumi Y, Muro T, Kumigashira H, Oshima M, Deguchi K, Okazaki H, Takano Y, Miura O, Mizuguchi Y, Suzuki K, Usui H, Kuroki K, Okada K, Muraoka Y and Yokoya T 2014 *J. Phys. Soc. Jpn.* **83** 033703
[17] Zeng L K, Wang X B, Ma J, Richard P, Nie S M, Weng H M, Wang N L, Wang Z, Qian T, Ding H 2014 *Phys. Rev. B* **90** 054512
[18] Ye Z R, Yang H F, Shen D W, Jiang J, Niu X H, Feng D L, Du Y P, Wan X G, Liu J Z, Zhu X Y, Wen H H and Jiang M H 2014 *Phys. Rev. B* **90** 045116
[19] Saini N L, Ootsuki D, Paris E, Joseph B, Barinov A, Tanaka M, Takano Y, Mizokawa T 2014 *Phys. Rev. B* **90** 214517
[20] Sugimoto T, Ootsuki D, Morice C, Artacho E, Saxena S S, Schwier E F, Zheng M, Kojima Y, Iwasawa H, Shimada K, Arita M, Namatame H, Taniguchi M, Takahashi M, Saini N L, Asano T, Higashinaka R, Matsuda T D, Aoki Y, and Mizokawa T 2015 *Phys. Rev. B* **92** 041113(R)
[21] Terashima K, Sonoyama J, Wakita T, Sunagawa M, Ono K, Kumigashira H, Muro T, Nagao M, Watauchi S, Tanaka I, Okazaki H, Takano Y, Miura O, Mizuguchi Y, Usui H, Suzuki K, Kuroki K, Muraoka Y and Yokoya T 2014 *Phys. Rev. B* **90** 220512(R)
[22] Suzuki K, Usui H and Kuroki K 2013 *Physics Procedia* **45** 21
[23] Machida T, Fujisawa Y, Nagao M, Demura S, Deguchi K, Mizuguchi Y, Takano Y and Sakata H 2014 *J. Phys. Soc. Jpn.* **83** 113701

Bulk sensitive angle-resolved photoelectron spectroscopy on $Nd(O,F)BiS_2$

K Terashima[1,2], J Sonoyama[1], M Sunagawa[1], H Fujiwara[1], T Nagayama[1], T Muro[3], M Nagao[4], S Watauchi[4], I Tanaka[4], H Okazaki[5], Y Takano[5], Y Mizuguchi[6], H Usui[7], K Suzuki[7], K Kuroki[7], T Wakita[1,2], Y Muraoka[1,2], and T Yokoya[1,2]

[1] Research Laboratory for Surface Science and Graduate School of Natural Science and Technology, Okayama University, Okayama, 700-8530, Japan
[2] Research Center of New Functional Materials for Energy Production, Storage, and Transport, Okayama University, Okayama, 700-8530, Japan
[3] Japan Synchrotron Radiation Research Institute (JASRI)/SPring-8, Sayo, Hyogo 679-5198, Japan
[4] Center for Crystal Science and Technology, Yamanashi University, Kofu 400-8510, Japan
[5] National Institute for Materials Science, 1-2-1, Sengen, Tsukuba, 305-0047, Japan
[6] Department of Electrical and Electronic Engineering, Tokyo Metropolitan University, 1-1, Minami-osawa, Hachioji, 192-0397, Japan
[7] Department of Physics, Osaka University, 1-1 Machikaneyama Toyonaka, Osaka 560-0043, Japan.

k-terashima@cc.okayama-u.ac.jp

Abstract. Bulk electronic structure of novel layered superconductor $Nd(O,F)BiS_2$ was studied by using soft x-ray angle-resolved photoelectron spectroscopy (ARPES). Electron-like Fermi surface centered at the X(R) point was observed, consistent with earlier ARPES reports on surface-sensitive VUV light source. Based on the comparison of the electronic structure between $Nd(O,F)BiS_2$ and $La(O,F)BiS_2$, we discuss possible important factors for the superconductivity in this series of material.

1. Introduction

$BiCh_2$-based superconductors (Ch: Chalcogen) are newly discovered materials [1] which possess similar layered structure with high-transition temperature (T_c) cuprates and iron-based superconductors. Their typical chemical formula is $REOBiCh_2$ where RE is rare earth atom, and it consists of alternate stacking layers of RE_2O_2 and $BiCh_2$. Except for the RE = Eu case [2], the parent compounds with stoichiometry show insulating behavior, and a partial substitution of O with F changes the system from insulator to metal and superconductivity emerges with maximum T_c of 10.6 K [1]. This substitution is regarded as effective carrier doping into Bi-Ch hybridized bands [3]. Such a crystal structure of block layers and conductive layers is suggestive of a possibility that one can design a new $BiCh_2$-based

Journal of Physics: Conference Series **683** (2016) 012003 doi:10.1088/1742-6596/683/1/012003

material with higher- T_c using a combination of ingredients, just as in cuprates and iron-based superconductors.

Interestingly, the T_c in this series of superconductor shows RE-atom dependence, that is, the smaller the ion-radius of RE-atom is, the higher T_c tends to be obtained. More specifically, single crystals of $La(O,F)BiS_2$ show maximum T_c of ~ 3.1 K [4] while that of $Nd(O,F)BiS_2$ is ~5.1 K [5]. It has been pointed out that the evaluated value of the in-plane chemical pressure effect correlates well with T_c [6]. There, the microscopic reason for the observed correlation has been argued in relation with electronic structures. Therefore, a comparative study of the electronic structure in a series of $RE(O,F)BiS_2$ superconductors are highly valuable.

Angle-resolved photoelectron spectroscopy (ARPES) can directly observe the experimental band dispersions thus has been served as a strong tool for studying the electronic structure in solids. In earlier ARPES results on $Nd(O,F)BiS_2$ [7,8], it has been reported that only small electron-like Fermi surface near the X(R) point exists. This is in contrast to the theoretical prediction that the Bi-S hybridized band holds van hove singularity near ($\pi/2$, $\pi/2$) which is expected to tough Fermi energy (E_F) by electron doping [3]. The carrier number estimated from Luttinger volume deviates far from the F content value estimated by electron-probe microanalysis (EPMA) [7,8]. Recent scanning tunneling microscope (STM) [9] reported that the electronic structure of topmost surface of $Nd(O,F)BiS_2$ obtained by cleaving does not match the density of states predicted by first principles calculations [10], suggesting that the surface electronic states can be strongly modulated from that of bulk system. Since previous ARPES studies have been performed at surface-sensitive settings, a re-examination of the electronic structure by bulk-sensitive method has been highly desired.

In this paper, we use soft x-ray ARPES to revisit the bulk electronic structure of $Nd(O,F)BiS_2$. While the surface component was observed in the Bi 4*f* core level spectra, the obtained Fermi surface as well as the overall electronic structure near E_F well correspond to those reported earlier by surface-sensitive ARPES. We discuss the possible explanation for this phenomena and compare the results with $La(O,F)BiS_2$ [11].

2. Experimental

The single crystal of $Nd(O,F)BiS_2$ was synthesized by flux method [12]. The nominal F composition was 0.3, while EPMA analysis showed that the value was ~0.29. The soft x-ray ARPES was performed at BL25SU of SPring-8 using SES200 analyzer. The used photon energy was 880 eV for both of the core level and the valence band measurements. The energy resolution was set at approximately 0.2 eV. E_F was determined by the Fermi cut-off energy of Au that was electronically contacted with the sample. The clean surface for the measurement was obtained by in-situ cleaving, and ARPES measurement was performed in an ultrahigh vacuum of ~ 10^{-8} Pa.

3. Results and discussion

3.1. Core level spectra

Figure 1(a) shows the core level spectrum of $Nd(O,F)BiS_2$ in a wide energy range. Multiple peaks are observed such as Nd 4*d* (~120 eV), O 1*s* (~543 eV), F 1*s* (~697 eV), Bi 4*f* (~160 eV), and S 2*s* (~230 eV) , that are consistent with the constituents of the sample. We did not find C 1s component around 280 eV, suggesting the high quality of the crystal. In figure 1(b), we show Bi 4*f* and S 2*p* core level spectra and their detection-angle dependence in an expanded energy range. The spectra are normalized by total intensity integrated from 157 to 167 eV. It is clear from the figure that both of Bi $4f_{7/2}$ and $4f_{5/2}$ core levels have several components. Such multiple peaks are unexpected in bulk since Bi atoms occupy only one chemically-equivalent site in the unit cell. The peak intensity at higher binding energy side (~159 and ~164 eV) becomes stronger as the detection-angle (surface-sensitivity) is increased. This indicates the higher binding energy peaks are mainly originating from surface component and the rest would be the bulk component. The observation of surface Bi states is consistent with earlier STM results [9]. Next we discuss the electronic structure in the near-E_F region.

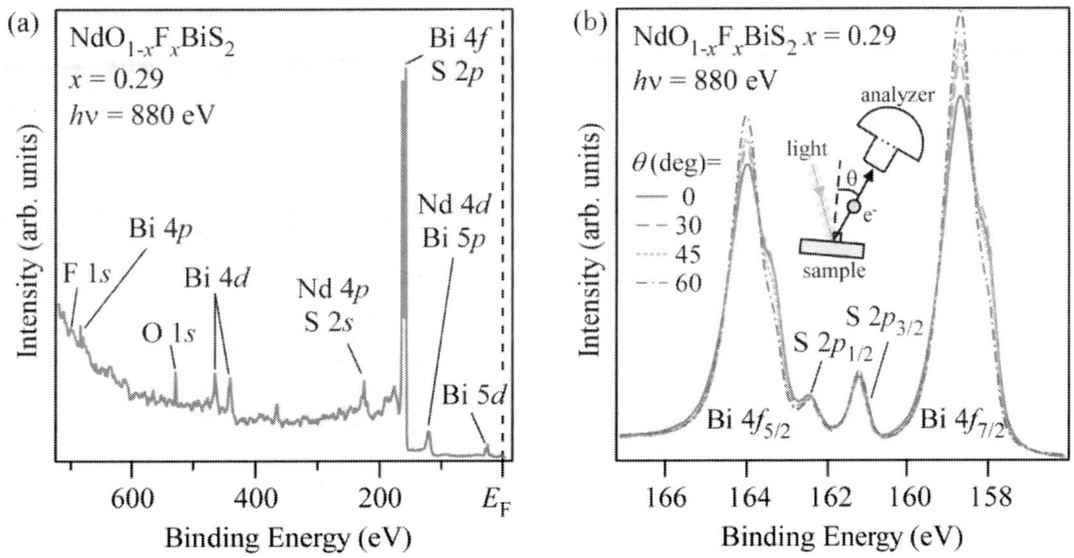

Figure 1. (a) Core level spectrum of Nd(O,F)BiS_2 in a wide energy range. (b) Detection-angle (θ) dependence of Bi 4*f* core level spectrum.

3.2. Near E_F region

We show in Figure 2(a) the ARPES intensity plot near E_F ($E_F \pm 0.05$ eV) as a function of two-dimensional wave vectors to map out the experimental Fermi surface of Nd(O,F)BiS_2. We have observed rectangular-like Fermi surface centered at the X(R) point. In upper panel of Fig. 2(b), we show *E-k* plot along X(R) to X(R) direction depicted as a blue line in Fig. 2(a). Near the X(R) point, the observed band shows downward dispersion, which indicates the sign of the carrier is electron. By combining those observations, we can conclude that bulk electronic structure of Nd(O,F)BiS_2 holds electron-like Fermi surface topology. Then one might ask that where is the surface component in electronic structure? In order to answer this question, we show in lower panel of Fig. 2(b) the second derivative of *E-k* mapping. In the figure, not only the dispersive feature near the X(R) point but also a nondispersive component at ~ 0.2 eV is seen. Such a nondispersive character and the energy position of the band well correspond to the local density of states observed on the surface of Nd(O,F)BiS_2 in STM measurements. Therefore, we attribute this nondispersive structure to be surface state.

Next we compare the electronic structure of Nd(O,F)BiS_2 with that of La(O,F)BiS_2 [11]. The Bi 4*f* core level spectrum of La(O,F)BiS_2 also shows a surface component. In the near E_F region, a nondispersive component is observed also in La(O,F)BiS_2 although its energy is slightly different (~0.4 eV for nearly optimal doped La(O,F)BiS_2 case). So it is likely that the electronic structure of BiS_2 plane is near to instability toward charge localization. On the other hand, the Fermi surface topology in the superconducting phase may not be the same between $LaO_{0.54}F_{0.46}BiS_2$ (T_c = 3.1 K) and $NdO_{0.71}F_{0.29}BiS_2$ (T_c = 5.1 K). It has been reported that the electronic structure of $LaO_{0.54}F_{0.46}BiS_2$ is in close proximity to Lifshitz transition [11]. Thus, in RE = La case, the evolution of superconductivity seems to be related to the occupation of van Hove singularity predicted near (π/2, π/2) [3]. In this study, we have clarified that the bulk Fermi surface of $NdO_{0.71}F_{0.29}BiS_2$ is electronlike and the Fermi energy is far from the one where the van Hove singularity touches E_F. If we assume that the electronic structures of both samples are similar, the present study also indicates that even though the density of states at E_F is lower, T_c of Nd(O,F)BiS_2 is higher. In this series of material, there has to be other factors than the carrier density that effectively determine T_c. An in-plane chemical pressure effect has been proposed to be the one. If so, its influence should be observed in the electronic structure as discussed in iron-based superconductors. A more comprehensive study with high energy resolution for both of La(O,F)BiS_2 and Nd(O,F)BiS_2 has to be performed.

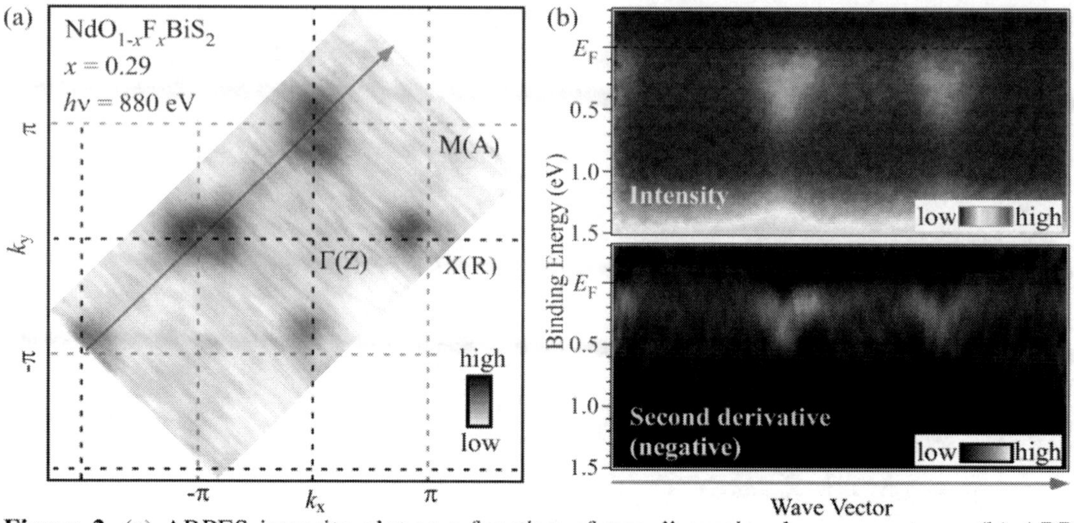

Figure 2. (a) ARPES intensity plot as a function of two-dimensional wave vectors. (b) ARPES intensity plot (upper panel) and it's negative second derivative (lower panel) as a function of wave vector and binding energy taken along the direction shown as an arrow in (a).

4. Summary

We have performed soft x-ray ARPES on Nd(O,F)BiS_2 to study its bulk electronic structure. The core level spectrum of Bi 4*f* shows detection-angle dependence, indicative of the presence of surface electronic states of Bi. In near E_F region, we have observed both of dispersive electron-like band forming Fermi surface centered at the X(R) point and nondispersive structure at ~0.2 eV. Therefore, the bulk electronic structure is essentially consistent with the one reported earlier by surface-sensitive PES regarding the Fermi surface topology, where the lower carrier number in Nd(O.F)BiS_2 produces higher T_c than La(O,F)BiS_2 that points towards possible unconventional mechanism of superconductivity in this series of material.

Acknowledgment

ARPES experiments at SPring-8 were performed under proposal numbers 2013A1324. This work was partially supported by JSPS KAKENHI Grant Numbers 25800205 and 15H03691. This work was also partially supported by the Program for Promoting the Enhancement of Research Universities from MEXT.

References

[1] Mizuguchi Y, Demura S, Deguchi K, Takano Y, Fujihisa H, Gotoh Y, Izawa H and Miura O 2012 *J. Phys. Soc. Jpn.* **81**, 114725

[2] Zhai H F, Tang Z T, Jiang H, Xu K, Zhang K, Zhang P, Bao J K, Sun Y L, Jiao W H, Nowik I, Felner I, Li Y K, Xu X F, Tao Q, Feng C M, Xu Z A and Cao G H 2014 *Phys. Rev. B* **90**, 064518

[3] Usui H, Suzuki K and Kuroki K 2012 *Phys. Rev. B* **86**, 220501(R)

[4] Nagao M, Miura A, Demura S, Deguchi J, Watauchi S, Takei T, Takano Y, Kumada N and Tanaka I 2014 *Solid State Commun.* **178**, 33

[5] Xing J, Li S, Ding X, Yang H and Wen H H 2012 *Phys. Rev. B* **86**, 214518

[6] Mizuguchi Y, Miura A, Kajitani J, Hiroi H, Miura O, Tadanaga K, Kumada N, Magome E, Moriyoshi C and Kuroiwa Y 2015 *Sci. Rep.* **5**, 14968

[7] Ye Z R, Yang H F, Shen D W, Jiang J, Niu X H, Feng D L, Du Y P, Wan X G, Liu J Z, Zhu X Y, Wen H H and Jiang M H 2014 *Phys. Rev. B* **90**, 045116

[8] Zeng L K, Wang X B, Ma J, Richard P, Nie S M, Weng H M, Wang N L, Wang Z, Qian T and Ding H 2014 *Phys. Rev. B* **90**, 054512

[9] Machida T, Fujisawa Y, Nagao M, Demura S, Deguchi K, Mizuguchi Y, Takano Y and Sakata H 2014 *J. Phys. Soc. Jpn.* **83**, 113701

[10] Yildirim T 2013 *Phys. Rev. B* **87**, 020506(R)

[11] Terashima K, Sonoyama J, Wakita T, Sunagawa M, Ono K, Kumigashira H, Muro T, Nagao M, Watauchi S, Tanaka I, Okazaki H, Takano Y, Miura O, Mizuguchi Y, Usui H, Suzuki K, Kuroki K, Muraoka Y and Yokoya T 2014 *Phys. Rev. B* **90** 220512(R)

[12] Nagao M, Demura S, Deguchi K, Miura A, Watauchi S, Takei T, Takano Y, Kumada N and Tanaka I 2013 *J. Phys. Soc. Jpn.* **82** 113701

Journal of Physics: Conference Series **683** (2016) 012004 doi:10.1088/1742-6596/683/1/012004

One-dimensional model for BiS_2 superconductivity: analyzing the pressure effect over T_c

M A Griffith[1], K Foyevtsova[2,3], M A Continentino[1] and G B Martins[4]

[1]Centro Brasileiro de Pesquisas Físicas, Rua Dr. Xavier Sigaud 150, Urca, 22290-180 Rio de Janeiro, RJ, Brazil

[2]Department of Physics & Astronomy, University of British Columbia, Vancouver, British Columbia V6T 1Z1, Canada

[3]Quantum Matter Institute, University of British Columbia, Vancouver, British Columbia V6T 1Z4, Canada

[4]Department of Physics, Oakland University, Rochester, MI 48309, USA

E-mail: `martins@oakland.edu`

Abstract. The recently synthesized compound $SrFBiS_2$ [Lei H et al. 2013 *Inorg. Chem.* **52** 106851] has been shown to become superconducting [Lin X et al. 2013 *Phys. Rev. B* **87** 020504] through the substitution of Sr^{2+} by La^{3+}. In addition, application of moderate hydrostatic pressure in $Sr_{0.5}RE_{0.5}FBiS_2$ (RE=Ce, Nd, Pr, and Sm) [Jha R et al. 2015 *J. Appl. Phys.* **117** 013901] has shown a dramatic increase in the superconducting transition temperature T_c. In this work, using a recently proposed one-dimensional model for BiS_2-based superconductors [Griffith M A et al. 2015 submitted to *Phys. Rev. B*], the authors analyze the effect of pressure over T_c.

1. Introduction

After the discovery of the cuprates in 1986 [1], the search for new layered superconducting materials has attracted much attention, with important discoveries occurring in the last 15 years. For example, it was discovered in 2001 that MgB_2 has $T_c = 39$ K [2] and in 2008 superconductivity (SC) in the iron pnictides was reported [3]. Both MgB_2 and the iron pnictides have highlighted the importance of multiband SC [4], to the point that the recent literature on cuprates devoted to multiband models has substantially increased [5]. An unrelated development has been the explosion of research in topological superconductors [6], due to proposals to 'engineer' Majorana fermion quasiparticles through midgap excitations of a chiral p-wave superconductor. This has led to renewed interest in the Ruthenate compound Sr_2RuO_4, discovered in 1994 [7], which is one of the few candidates to realizing p-wave-type SC [8; 9], another candidate being the organic superconductor $(TMTSF)_2PF_6$. It should also be emphasized that, as was the case for intermetallics with A15 structure (like Nb_3Sn or V_3Si) [10], the Ruthenates display 'hidden' quasi-one-dimensional (quasi-1d) SC [11] (while organic superconductors are explicitly 1d). Finally, we also mention SC in doped semiconductors, studied since before the 60s [12], with the interest greatly increasing after the discovery of SC in Boron-doped Diamond with $T_c = 4$ K [13].

It is then interesting to notice that one of the latest families of superconductors to be discovered, those containing BiS_2 planes, [14] presents many of the aspects mentioned above:

a layered structure with intercalated charge reservoirs [15], like in the cuprates and in the iron superconductors; a possible double superconducting gap [16], as in MgB_2 [2]; its minimal model contains two bands [17], and Fermi surface nesting effects seem to be important [18] (as in the iron pnictides [3]); because it contains a heavy element (Bismuth), spin-orbit effects are enhanced and some proposals linking BiS_2 superconductivity to spin-triplet pairing and a weak topological superconducting state have been made [19]; based on first-principles electronic structure calculations, it has been pointed out the 'subtle' 1d character of its band structure [17], which has been recently experimentally confirmed through polarization-dependent Angular Resolved Photoemission Spectroscopy (ARPES) measurements [20]; finally, a few members of the BiS_2 family have semiconducting parent compounds that become metallic/superconducting with electron doping and/or application of moderate hydrostatic pressure, which also can lead to sizable increase in T_c [21].

In this work, to advance the understanding of SC in BiS_2, where there is no consensus yet if SC is of the conventional or unconventional type [22], we concentrate in these last two aspects: one-dimensionality of the electronic structure and the pronounced effects pressure has over the superconducting phase. To model that, the authors take the following approach: i) adopt a 1d three-orbital model for BiS_2, adding the Cooper-pairing by hand, ii) solve the gap equations at the mean-field level, iii) study the dependence of the superconducting gap with the variation of the hopping terms, whose magnitude one expects to increase under applied pressure iv) decide on the acceptance or not of specific pair-scattering terms based on semi-quantitative agreement with experiments. Regarding this last point, we look specifically in what range of electron-filling a superconducting dome is obtained (see Fig. 4) and how SC varies with hopping parameters. To make the connection with BiS_2 more explicit, and thus obtain semi-quantitative agreement with experiments, all the parameter values of the single-particle Hamiltonian were obtained through first-principles Density Functional Theory (DFT) calculations for a two-dimensional (2d) five-band model (see Table 1). It should be noted that, although our calculations are done specifically for $SrFBiS_2$, this marked increase of T_c with pressure has been observed in many other members of the BiS_2 family of superconductors [23].

We can summarize our results as follows: Taking into account a three-orbital model, where Sulfur contributes with orbitals s and p, and Bismuth with a p orbital (see Fig. 1), we considered all possible pair-scattering terms (intra and interband, restricted to pairs formed by same-band electrons), individually and in conjunction, and solved the resulting gap equations at the mean-field level. We obtain that i) no single-band pair-scattering process, acting isolatedly, can describe the experiments (as specifically defined above), unless an unrealistic pair-coupling is assumed ($g > 0.1$ eV); this seems to indicate that multi-gap SC is a natural consequence of our model ii) two different types of multi-gap SC (see detailed description below) are in semi-quantitative agreement with experiments iii) the gap dependence with hopping (see Fig. 5) indicates a qualitative difference between the two hoppings considered in our model. These important results establish an appropriate effective 1d model to simulate the properties of BiS_2. We expect that our work will motivate other groups to investigate other similar purely 1d effective models (an abbreviated version of these results has been submitted elsewhere [24]).

2. Model

We consider a linear chain with a unit cell consisting of two sites denoted a and b, see Fig. 1. The a sites (Sulfur) have orbitals s and p, while b sites (Bismuth) have just one p orbital. In second quantization notation, the annihilation operator for an s orbital in unit cell n is denoted as c_n, and those for Sulfur and Bismuth p orbitals are denoted $p_{a,n}$ and $p_{b,n}$, respectively. The non-interacting part of the Hamiltonian can then be written as

$$\mathcal{H} = \sum_n \{(\epsilon_{s,n} + \mu) n_{s,n} + \sum_{i=a,b} (\epsilon_{p_i,n} + \mu) n_{p_i,n} + t_{pp}[p^\dagger_{a,n} p_{b,n} + p^\dagger_{b,n} p_{a,n+1} + \text{h.c.}]$$

Journal of Physics: Conference Series **683** (2016) 012004 doi:10.1088/1742-6596/683/1/012004

Table 1. Partial list of tight-binding parameters (in eV) for the 2d five-orbital model. The same parameters are used for the 1d three-orbital model depicted in Fig. 1. The chemical potential corresponds to 1/8-filling of the p_b orbital in the 1d model.

$\epsilon_{s,n}$	$\epsilon_{p_a,n}$	$\epsilon_{p_b,n}$	t_{sp}	t_{pp}	μ
−11.2840	−1.2691	0.1635	−0.9952	−0.8155	0.5007

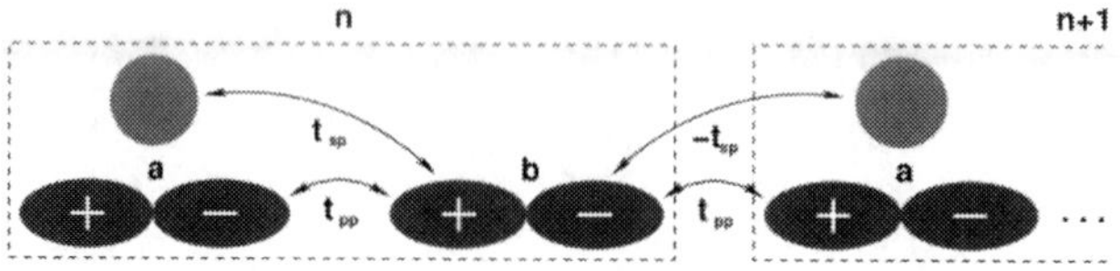

Figure 1. 1d model for BiS_2. t_{pp} and t_{sp} are the hopping terms considered in our model. The dashed box indicates the unit cell with atoms a (Sulfur, with orbitals s and p) and b (Bismuth, with just one p orbital). Notice the alternating signs of the t_{sp} hoppings [25].

$$+ \quad t_{sp}[c_n^\dagger p_{b,n} - p_{b,n}^\dagger c_{n+1} + \text{h.c.}]\} \tag{1}$$

where $\epsilon_{s,n}$ and $\epsilon_{p_i,n}$ describe the energy levels of orbitals s and p (for site $i = a, b$) at unit cell n, respectively; $n_{s,n} = c_n^\dagger c_n$ and $n_{p_i,n} = p_{i,n}^\dagger p_{i,n}$ are the number operators, and μ is the chemical potential. The hopping parameters are indicated in Fig. 1 and the values used in this work (along with orbital energies and chemical potential) are listed in Table 1 in eV units. Note that these parameter values were obtained through a full DFT calculation. The hoppings kept for the 1d model here studied were all the nearest neighbor hoppings in excess of 0.5 eV.

An early *2d* minimal model for BiS_2 contains two orbitals: Bismuth p_x and p_y orbitals [17]. Therefore, before deriving the self-consistent gap equations, the inclusion of the Sulfur p and s orbitals should be justified, mainly the latter one, which lies deep below the Fermi energy (see DFT parameter values in Table 1). Figure 2 shows the density of states (DOS) obtained for a *2d* model of BiS_2 involving five orbitals: two Bismuth p orbitals (p_x and p_y), two Sulfur p orbitals (p_x and p_y), and one Sulfur s orbital. From the examination of the DOS one can conclude that, at the Fermi energy $E_F = 0$ (which, in this plot, is between 1/8- and 1/4-filling, for the 2d model), the participation of the Sulfur s orbital [dot-dashed (green) curve], is quite relevant, even more than that of the Sulfur p orbitals [dashed (blue) curve]. In addition, it is easy to recognize the characteristic 1d DOS profile for the Sulfur p orbital at the top of the valence band and for the Bismuth p and Sulfur s orbitals at the bottom of the conduction band (compare with the DOS for the actual 1d model, shown in Fig. 3). This, coupled to the above mentioned polarized ARPES results indicating the one-dimensionality of the electronic structure of BiS_2, justifies our model. We now proceed to the derivation of the self-consistent gap equations.

3. Self-consistent gap equations at zero temperature

After taking a Fourier transform of the non-interacting part, and introducing pair-scattering terms between the electrons, the total Hamiltonian can be written as

$$\mathcal{H}(k) = \sum_k \{(\epsilon_{as} + \mu)c_k^\dagger c_k + \sum_{i=a,b} (\epsilon_{p_i} + \mu)p_{ik}^\dagger p_{ik} + 2t_{pp}\cos(k)[p_{ak}^\dagger p_{bk} + \text{h.c.}]$$

Journal of Physics: Conference Series **683** (2016) 012004 doi:10.1088/1742-6596/683/1/012004

$$+ \quad 2it_{sp}\sin(k)[c_k^\dagger p_{bk} - \text{h.c.}] - \sum_{i,j,k,k'} g_{ij}[\gamma_{i,k+}^\dagger \gamma_{i,\bar{k}-}^\dagger \gamma_{j,\bar{k}'-}\gamma_{j,k'+} + \text{h.c.}]\} \tag{2}$$

where, in the last line, $\gamma_{i/j,k\sigma}$ ($\sigma = \pm$ and $\bar{k}$ indicates $-k$) stands for either one of $c_{k\sigma}$, $p_{a,k\sigma}$, or $p_{b,k\sigma}$. Note that it is implicit in the form of the expression for the pair-scattering term that we are only considering singlet Cooper pairs composed of electrons from the same band, as pairing of different-band electrons tends to promote pair-density-wave (inhomogeneous) superconducting ground states [26]. Already anticipating results that will be discussed below (see Fig. 3), we describe how to obtain the gap equations when an specific set of pair-scattering processes are taken in account. Considering terms involving *intraband* scattering in the s and p_b bands and *interband* scattering between the s and p_b bands, the last term of eq. (2) (which we denote as $\mathcal{H}^{SC}$) can be written as

$$\begin{aligned}\mathcal{H}^{SC} &= -\sum_{k,k'}\{g_{ss}[c_{k+}^\dagger c_{\bar{k}-}^\dagger c_{\bar{k}'-}c_{k'+}] + g_{p_bp_b}[p_{b,k+}^\dagger p_{b,\bar{k}-}^\dagger p_{p,\bar{k}'-}p_{b,k'+}] \\ &+ g_{sp_b}[c_{k+}^\dagger c_{\bar{k}-}^\dagger p_{b,\bar{k}'-}p_{b,k'+} + p_{b,k+}^\dagger p_{b,\bar{k}-}^\dagger c_{\bar{k}'-}c_{k'+}]\}.\end{aligned} \tag{3}$$

Note that, based on experimental evidence [23], we consider the pairing couplings g_{ij} as being k-independent, *i.e.*, we assume s-wave pairing functions. Applying a mean-field decoupling eq. (3) becomes

$$\begin{aligned}\mathcal{H}^{SC} &= -\sum_{k}\{\Delta_{ss}(c_{\bar{k}-}c_{k+} + c_{k+}^\dagger c_{\bar{k}-}^\dagger) + \Delta_{p_bp_b}(p_{p,\bar{k}-}p_{b,k+} + p_{b,k+}^\dagger p_{b,\bar{k}-}^\dagger) \\ &+ \Delta'_{ss}(p_{b,\bar{k}-}p_{b,k+} + p_{b,k+}^\dagger p_{b,\bar{k}-}^\dagger) + \Delta'_{p_bp_b}(c_{k+}^\dagger c_{\bar{k}-}^\dagger + c_{\bar{k}-}c_{k+})\},\end{aligned} \tag{4}$$

which can be rewritten as

$$\mathcal{H}^{SC} = -\sum_{k}[\Delta_1(c_{\bar{k}-}c_{k+} + c_{k+}^\dagger c_{\bar{k}-}^\dagger) + \Delta_2(p_{b,\bar{k}-}p_{b,k+} + p_{b,k+}^\dagger p_{b,\bar{k}-}^\dagger)], \tag{5}$$

with the following definitions

$$\Delta_1 = \Delta_{ss} + \Delta'_{p_bp_b} = g_{ss}\sum_k\langle c_{\bar{k}-}c_{k+}\rangle + g_{sp_b}\sum_k\langle p_{b,\bar{k}-}p_{b,k+}\rangle \tag{6}$$

and

$$\Delta_2 = \Delta'_{ss} + \Delta_{p_bp_b} = g_{sp_b}\sum_k\langle c_{\bar{k}-}c_{k+}\rangle + g_{p_bp_b}\sum_k\langle p_{b,\bar{k}-}p_{b,k+}\rangle \tag{7}$$

where $<>$ indicates an average over the ground state. If we consider the following relation, $g_{ss} = g_{p_bp_b} = g_{sp_b} = g$, then it follows that $\Delta_{ss} = \Delta'_{ss}$ and $\Delta_{p_bp_b} = \Delta'_{p_bp_b}$ (therefore $\Delta_1 = \Delta_2 = \Delta$), and we obtain the gap equation as

$$\Delta = g\sum_k(\langle c_{\bar{k}-}c_{k+}\rangle + \langle p_{b,\bar{k}-}p_{b,k+}\rangle). \tag{8}$$

We want to derive a self-consistent equation for Δ and then analyze the effect of variations in the hopping parameters over it. In order to determine the correlations $\langle c_{\bar{k}-}c_{k+}\rangle$ and $\langle p_{b,\bar{k}-}p_{b,k+}\rangle$ in the gap equation, we need to calculate the anomalous Green's functions $\langle\langle c_{k+}; c_{\bar{k}-}\rangle\rangle$ and $\langle\langle p_{b,k+}; p_{b,\bar{k}-}\rangle\rangle$. These calculations are long and tedious, and thus are presented in the appendix. After writing the equation of motion for the propagators $\langle\langle p_{b,k+}; p_{b,\bar{k}-}\rangle\rangle$ and $\langle\langle c_{k+}; c_{\bar{k}-}\rangle\rangle$, and through lengthy algebraic manipulations, we arrive at expressions for Δ_1 and Δ_2

$$\Delta_{1;2} = -g_{ss;sp_b}\frac{k_F}{2}\int_{-1}^{1}d\tilde{k}\sum_{j=1}^{3}\frac{|D_{ss}(-\omega_{\tilde{k},j})|}{\omega_{\tilde{k},j}r_j} - g_{sp_b;p_bp_b}\frac{k_F}{2}\int_{-1}^{1}d\tilde{k}\sum_{j=1}^{3}\frac{|D_{p_bp_b}(-\omega_{\tilde{k},j})|}{\omega_{\tilde{k},j}r_j} \tag{9}$$

which, after the simplifying step mentioned above, results in $\Delta_1 = \Delta_2 = \Delta$ (the terms under the two integrals are fully developed in the appendix).

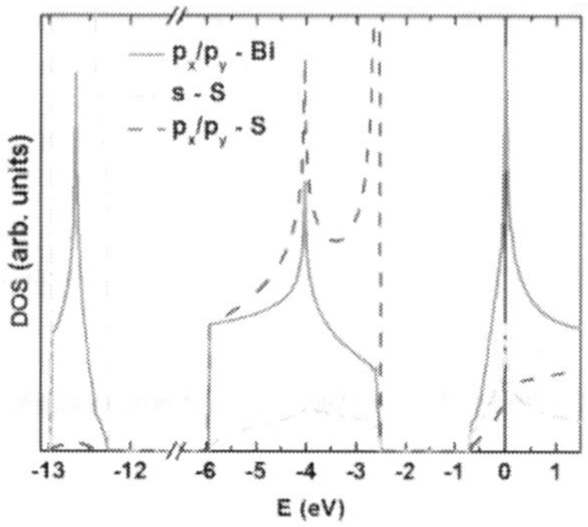

Figure 2. Density of states obtained through DFT for the five-orbital model of the $SrFBiS_2$ compound. Note the importance of the Sulfur s-orbitals [dot-dashed (green) line] at the Fermi energy ($E_F = 0$).

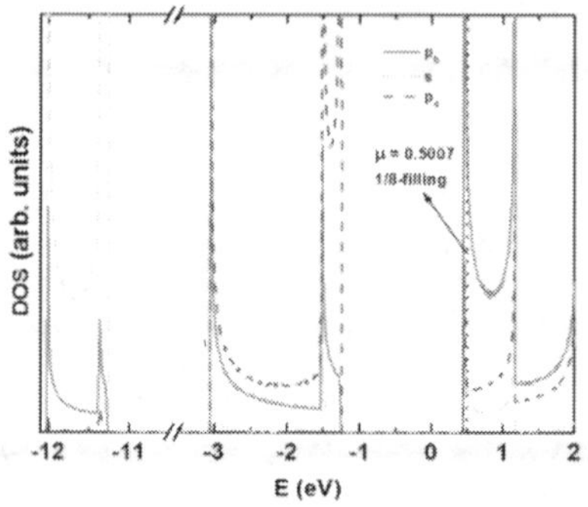

Figure 3. DOS for the 1d model using the same parameters as for the 2d DOS calculation. Note that, as in the case of the 2d DOS, the strong hybridization between the s and p_b orbitals brings a considerable amount of the s-orbital spectral weight to the Fermi energy, which for 1/8-filling ($\mu = 0.5007$) is indicated by the vertical dotted line.

4. Results

As mentioned in the Introduction, our strategy was to solve the gap equations at the mean-field level (hopping amplitudes fixed at the values obtained by DFT), and look for solutions at least qualitatively compatible with experiments, i.e., for chemical potential values around 1/8-filling and for coupling strengths g that are not unrealistically large. Taking in account the pair-scattering terms in eq. (4) and following the derivations up to eq. (9), we obtain the gap function Δ, which has a dependence with μ as shown in Fig. 4, for three different values of coupling g. It is interesting to note that the value of μ around which the three domes are centered corresponds to an electron filling close to that where SC has been found for most members of the BiS2 family, i.e., 1/8-filling [27]. This is an important result, as μ was *not* fixed from the start. It is taken as a free parameter, whose value, obtained self-consistently, was used to determine which gap equations (for specific pair-scattering terms) produced acceptable results. Indeed, if the value of μ for which SC was found is too far removed from 1/8-filling, that gap equation (and the pair-scattering term generating it) is rejected.

If one takes the maximum value obtained for Δ in Fig. 4 for $g = 25$ meV [(red) circles], $\Delta \approx 4$ meV, and uses the BCS relation $2\Delta/k_BT_c = 3.52$, one obtains $T_c \approx 26$ K. A maximum $T_c \approx 11$ K has been measured for $LaO_{1-x}F_xBiS_2$ at 1/8-filling ($x = 0.5$) [23], indicating that our results, for a realistic value of g, produce a T_c qualitatively similar to experiments. A comment should be made on the horizontal width of the dome for the (red) solid circles curve in Fig. 4. At the base of the dome, the electron filling varies roughly from 0.25 to 0.26 electrons per p_b-orbital (Bismuth). Although there is still some controversy about the actual filling around which SC occurs [27], a few of the published T_c vs. doping results indicate a broader dome. We believe that the narrower dome we obtain is an artifact of the 1d model. Indeed, the DOS close to 1/8-filling for our 1d model (see Fig. 3) has a very pronounced van Hove singularity, therefore a very strong variation of DOS with the chemical potential. This strong dependence, for smaller values of g (as the ones plotted in the main panel in Fig. 4), seems to result in a superconducting phase that is very sensitive to the chemical potential, leading to a narrow dome. In the inset to

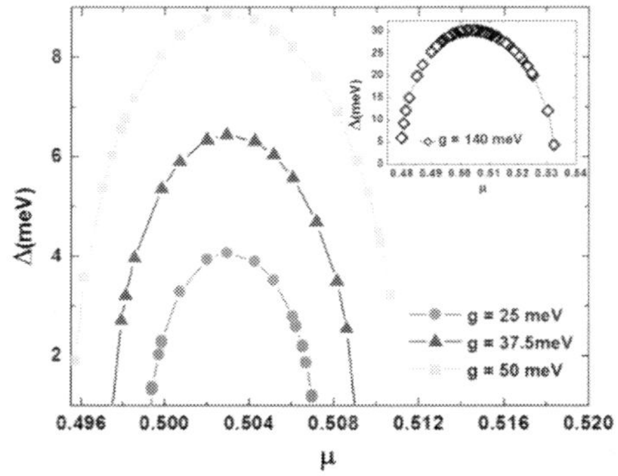

Figure 4. Pairing interaction Δ as a function of the chemical potential μ for different values of the coupling constant g.

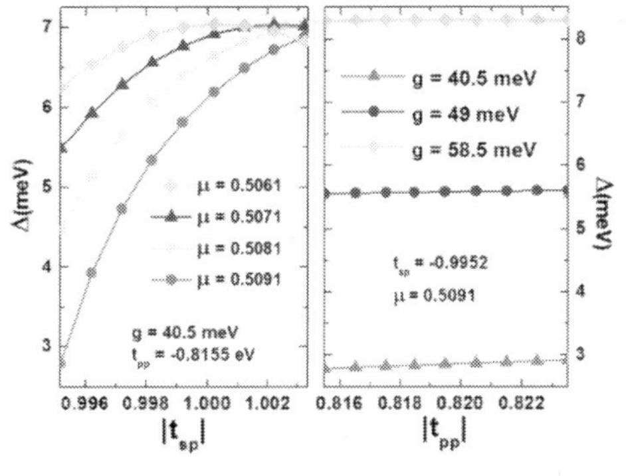

Figure 5. Left panel: Results showing the dependence of Δ with t_{sp} for a few values of the chemical potential μ in the dome region in Fig. 4 for $g = 40.5$ meV and $t_{pp} = -0.8155eV$. Right panel: Same as left panel, but now varying t_{pp} and keeping t_{sp} fixed.

Fig. 4, we show results for a larger $g = 140$ meV value. In it, we see a much broader variation in electron filling, from 0.2 to 0.31 [(black) open diamonds curve]. The actual system is quasi-1d, implying that once a three-dimensional superconducting state stabilizes, it will be less sensitive to variations in the chemical potential. To have the same effect in a purely 1d model we have to increase the pairing coupling, as shown in the inset to Fig. 4.

It is reasonable to expect that applying hydrostatic pressure in a crystal lattice will enhance the overlap between the orbitals and therefore increase the magnitude of the hopping terms. Making the assumption that this increase is similar to the change in lattice parameter, which for an applied pressure of 2 GPa will amount to a change of $\approx 1\%$ [28], we solve the gap equations for increasing values (in magnitude) of t_{sp} and plot the results in the left panel in Fig. 5 for some μ values in the dome region in Fig. 4 (for $g = 40.5$ meV). For a variation of $|t_{sp}| \approx 0.5\%$ the value of Δ roughly doubles, which is in semi-quantitative agreement with experimental results for T_c obtained for $LaO_{0.5}F_{0.5}BiS_2$ and $CeO_{0.5}F_{0.5}BiS_2$ [29]. A similar calculation for the variation in t_{pp}, shown in the right panel in Fig. 5, shows no changes in Δ, up to the same percent variation as for t_{sp}. This seems to be consistent with previous results [25] showing that antisymmetric hybridization is very effective in increasing the SC gap amplitude. To test this hypothesis, extensive calculations are underway where the condition $g_{ss} = g_{p_b p_b} = g_{sp_b} = g$ is relaxed [30].

There is another choice of pair-scattering terms in eq. (4) which produces results (not shown) very similar to the ones just described. One just needs to replace s by p_a in eq. (4). As already mentioned, these two were the only situations where the results obtained were compatible with the criteria described above for acceptance of the gap equation results. For all the other possibilities, either the coupling parameter g was unrealistically large or the electron-filling was too far removed from 1/8-filling.

5. Conclusions

Motivated by recent experiments in superconducting members of the BiS_2 family of compounds showing its 'hidden' 1d electronic structure and the strong effect that pressure has over its superconducting state, we propose an effective 1d model where the kinetic energy part of the

Journal of Physics: Conference Series **683** (2016) 012004 doi:10.1088/1742-6596/683/1/012004

Hamiltonian is obtained through DFT calculations for the 2d model for BiS_2. Supported by the DOS results shown in Fig. 2, we add the Sulfur p- and s-orbital to the p-orbital of Bismuth. Despite being several eV below the other two orbitals, the s-orbital undergoes strong hybridization with the Bismuth p-orbital and has a sizable contribution to the DOS at the Fermi energy, justifying its inclusion in the model (see Figs. 1 and 2). Pair scattering terms are then added and treated at the mean-field level. We solve the gap equations and systematically probe what combination of pair-scattering terms produce results in qualitative agreement with the experiments, i.e., approximate location of the superconducting phase in a T vs. doping phase diagram, realistic coupling constant values, and dependence with hopping parameters (simulating application of hydrostatic pressure). We find that single-gap SC does not produce acceptable results. This is quite relevant, as there is experimental evidence that BiS_2 presents two gaps [23]. We find that if we consider s- and p_b-type pairs, and allow for intra and interband scattering we obtain results in semi-quantitative agreement with experiments. The same is true if we choose p_a- and p_b-type pairs, and also allow for intra and interband scattering. The interesting point here is that the t_{sp} hopping is the one that, in both cases, enhances SC when its magnitude increases, whereas the effect on Δ of increasing t_{pp} is marginal. This last point reinforces the need for considering the Sulfur s orbital explicitly. We argue that the anti-symmetric character of the t_{sp} hopping (as stressed in previous work by one of the authors [25]) may explain its enhanced effect in the superconducting state.

Appendix

In the gap equations for Δ_1 and Δ_2 (eqs. (6) and (7) in the main text), the correlation functions are related to the Green's functions (propagators, from now on) $\langle\langle c_{k+}; c_{\bar{k}-}\rangle\rangle$ and $\langle\langle p_{bk+}; p_{b\bar{k}-}\rangle\rangle$ through the equation

$$\langle \gamma_{\bar{k}-}\gamma_{k+}\rangle = i\int_{-\infty}^{+\infty} d\omega f(\omega)[\langle\langle \gamma_{k+};\gamma_{\bar{k}-}\rangle\rangle_{\omega+i\eta} - \langle\langle \gamma_{k+};\gamma_{\bar{k}-}\rangle\rangle_{\omega-i\eta}], \tag{A.1}$$

where γ stands for the annihilation operators c or p_b, and $\eta \to 0$. In order to calculate the propagators we will write their equations of motion (taking from now on $\hbar = 1$)

$$\omega\langle\langle c_{k+}; c_{\bar{k}-}\rangle\rangle = \frac{1}{2\pi}\langle\{c_{k+}, c_{\bar{k}-}\}\rangle + \langle\langle [c_{k+}, H]; c_{\bar{k}-}\rangle\rangle \tag{A.2}$$

and

$$\omega\langle\langle p_{bk+}; p_{b\bar{k}-}\rangle\rangle = \frac{1}{2\pi}\langle\{p_{bk+}, p_{b\bar{k}-}\}\rangle + \langle\langle [p_{bk+}, H]; p_{b\bar{k}-}\rangle\rangle, \tag{A.3}$$

where $\mathcal{H}(k)$ is the Hamiltonian for the system (eq. (2) in the main text) and $\{,\}$ and $[,]$ indicate an anticommutator and a commutator, respectively.

Let us develop further the equation of motion for the first propagator ($\langle\langle c_{k+}; c_{\bar{k}-}\rangle\rangle$). Making use of standard relations for fermion creation and annihilation operators, we obtain

$$\omega_s^{(-)}\langle\langle c_{k+}; c_{\bar{k}-}\rangle\rangle - 2it_{sp}\sin(k)\langle\langle p_{bk+}; c_{\bar{k}-}\rangle\rangle + \Delta_1\langle\langle c^\dagger_{\bar{k}-}; c_{\bar{k}-}\rangle\rangle = 0, \tag{A.4}$$

where $\omega_s^{(-)}$ stands for $\omega - \epsilon_s + \mu$.

In the process above, two new propagators were created, viz., $\langle\langle p_{bk+}; c_{\bar{k}-}\rangle\rangle$ and $\langle\langle c^\dagger_{\bar{k}-}; c_{\bar{k}-}\rangle\rangle$. Therefore, in order to close the system of equations for the propagators, we also need the equation of motion for $\langle\langle c^\dagger_{\bar{k}-}; c_{\bar{k}-}\rangle\rangle$, $\langle\langle p_{bk+}; c_{\bar{k}-}\rangle\rangle$, $\langle\langle p^\dagger_{b\bar{k}-}; c_{\bar{k}-}\rangle\rangle$, $\langle\langle p_{ak+}; c_{\bar{k}-}\rangle\rangle$, and $\langle\langle p^\dagger_{a\bar{k}-}; c_{\bar{k}-}\rangle\rangle$.

This procedure generates a system of equations given by

$$D\cdot\begin{pmatrix}\langle\langle c^{\dagger}_{\bar{k}-};c_{\bar{k}-}\rangle\rangle\\ \langle\langle p_{ak+};c_{\bar{k}-}\rangle\rangle\\ \langle\langle p^{\dagger}_{a\bar{k}-};c_{\bar{k}-}\rangle\rangle\\ \langle\langle c_{k+};c_{\bar{k}-}\rangle\rangle\\ \langle\langle p_{bk+};c_{\bar{k}-}\rangle\rangle\\ \langle\langle p^{\dagger}_{b\bar{k}-};c_{\bar{k}-}\rangle\rangle\end{pmatrix}=\begin{pmatrix}\frac{1}{2\pi}\\0\\0\\0\\0\\0\end{pmatrix},\tag{A.5}$$

where

$$D=\begin{bmatrix}\omega_s^{(-)}&0&0&\Delta_1&-\bar{t}_{sp_b}&0\\0&\omega_{p_a}^{(+)}&0&0&0&\bar{t}_{p_ap_b}\\0&0&\omega_{p_b}^{(-)}&0&-\bar{t}_{p_ap_b}&0\\\Delta_1&0&0&\omega_s^{(+)}&0&\bar{t}_{sp_b}\\\bar{t}_{sp_b}&0&-\bar{t}_{p_ap_b}&0&\omega_{p_b}^{(+)}&\Delta_2\\0&\bar{t}_{p_ap_b}&0&-\bar{t}_{sp_b}&\Delta_2&\omega_{p_b}^{(-)}\end{bmatrix},\tag{A.6}$$

and we defined $\omega_q^{(\pm)}=\omega\pm\epsilon_q\mp\mu$, for $q=s,\ p_a$, or p_b; $\bar{t}_{sp_b}=2it_{sp_b}\sin(k)$; and $\bar{t}_{p_ap_b}=2t_{p_ap_b}\cos(k)$.

Using Cramer's method to solve the system of equations in (5), we have that

$$\langle\langle c_{k+};c_{\bar{k}-}\rangle\rangle=\frac{|D_{ss}|}{|D|},\tag{A.7}$$

where the matrix D_{ss} is obtained by replacing the 4th column in matrix D by the column-matrix defined in the right side of eq. (5) (note that $|D|$ means the determinant of matrix D). Repeating the same procedure for $\langle\langle p_{bk+};p_{b\bar{k}-}\rangle\rangle$ we obtain

$$D\cdot\begin{pmatrix}\langle\langle c^{\dagger}_{\bar{k}-};p_{b\bar{k}-}\rangle\rangle\\ \langle\langle p_{ak+};p_{b\bar{k}-}\rangle\rangle\\ \langle\langle p^{\dagger}_{a\bar{k}-};p_{b\bar{k}-}\rangle\rangle\\ \langle\langle c_{k+};p_{b\bar{k}-}\rangle\rangle\\ \langle\langle p_{bk+};p_{b\bar{k}-}\rangle\rangle\\ \langle\langle p^{\dagger}_{b\bar{k}-};p_{b\bar{k}-}\rangle\rangle\end{pmatrix}=\begin{pmatrix}0\\0\\0\\0\\0\\\frac{1}{2\pi}\end{pmatrix},\tag{A.8}$$

obtaining, as above,

$$\langle\langle p_{bk+};p_{b\bar{k}-}\rangle\rangle=\frac{|D_{p_bp_b}|}{|D|},\tag{A.9}$$

where $D_{p_bp_b}$ is obtained, as above, by replacing the 5th column in matrix D by the column-matrix defined in the right side of eq. (8).

In eqs. (7) and (9), $|D|$ is a sixth degree biquadratic polynomial that can be written as

$$|D|=\sum_{n=0}^{3}B_{2n}\omega^{2n}=\prod_{n=1}^{6}(\omega-\omega_n)=\prod_{n=1}^{3}(\omega^2-\omega_n^2)\tag{A.10}$$

where the last equation is obtained by noting that $\omega_1=-\omega_4$, $\omega_2=-\omega_5$, and $\omega_3=-\omega_6$. The coefficients B_{2n} are functions of the parameters of the Hamiltonian. The ω_n are the zeros of $|D|$ and represent the energy excitations of the system.

To finally determine the gap equations, it is appropriate to make use of the following identity

$$\frac{1}{|D|} = \sum_{n=0}^{3} \frac{1}{r_n} \left[\frac{1}{2\omega_n} \left(\frac{1}{\omega - \omega_n} - \frac{1}{\omega + \omega_n} \right) \right], \tag{A.11}$$

where

$$r_1 = (\omega_1^2 - \omega_2^2)(\omega_1^2 - \omega_3^2);\ r_2 = (\omega_2^2 - \omega_1^2)(\omega_2^2 - \omega_3^2);\ r_3 = (\omega_3^2 - \omega_1^2)(\omega_3^2 - \omega_2^2).$$

Substituting eq. (11) into eqs. (7) and (9), and after using eq. (1), we have

$$\langle \gamma_{\bar{k}-} \gamma_{k+} \rangle = i \sum_{j=1}^{3} \int_{-\infty}^{+\infty} d\omega f(\omega) \frac{D_{\gamma\gamma}(\omega)}{2\omega_j r_j} \times (C_+(\omega) - C_-(\omega)), \tag{A.12}$$

where we have defined the following quantities

$$C_\pm(\omega) = \lim_{\eta \to 0^+} \left(\frac{1}{\omega \pm \omega_j - i\eta} - \frac{1}{\omega \pm \omega_j + i\eta} \right). \tag{A.13}$$

Now, using the fact that

$$\delta(x) = \frac{1}{2\pi i} \lim_{\eta \to 0^+} \left(\frac{1}{x - i\eta} - \frac{1}{x + i\eta} \right), \tag{A.14}$$

we finally get

$$\langle \gamma_{\bar{k}-} \gamma_{k+} \rangle = \pi \sum_{j=1}^{3} \frac{|D_{\gamma\gamma}(\omega_j)| f_{FD}(\omega_j)}{\omega_j r_j} - \pi \sum_{j=1}^{3} \frac{|D_{\gamma\gamma}(-\omega_j)| f_{FD}(-\omega_j)}{\omega_j r_j}, \tag{A.15}$$

where $f_{FD}(\omega_j) = 1/[\exp(\omega_j/k_B T) + 1]$.

Substituting eq. (15) into the equations for Δ_1 and Δ_2 in the previous page, and taking the limit $T \to 0$, we can write the following self-consistent equations

$$\Delta_1 = \Delta_{ss} + \Delta'_{p_b p_b} = -g_{ss} \sum_k \pi \sum_{j=1}^{3} \frac{|D_{ss}(-\omega_j)|}{\omega_j r_j} - g_{sp_b} \sum_k \pi \sum_{j=1}^{3} \frac{|D_{p_b p_b}(-\omega_j)|}{\omega_j r_j} \tag{A.16}$$

and

$$\Delta_2 = \Delta'_{ss} + \Delta_{p_b p_b} = -g_{sp_b} \sum_k \pi \sum_{j=1}^{3} \frac{|D_{ss}(-\omega_j)|}{\omega_j r_j} - g_{p_b p_b} \sum_k \pi \sum_{j=1}^{3} \frac{|D_{p_b p_b}(-\omega_j)|}{\omega_j r_j}. \tag{A.17}$$

In the thermodynamic limit ($L \to \infty$), we can replace the sum by an integral [in the interval $-k_F \leq k \leq k_F$] by using the standard relation

$$\frac{2\pi}{L} \sum_k \Delta_k = \frac{1}{2\pi} \int_{-k_F}^{k_F} dk, \tag{A.18}$$

where, k_F is the Fermi wave vector and L is the length of the one-dimensional system. For practical purposes, we change the integration variable ($\tilde{k} = k/k_F$), and finally obtain

$$\Delta_1 = -g_{ss} \frac{k_F}{2} \int_{-1}^{1} d\tilde{k} \sum_{j=1}^{3} \frac{|D_{ss}(-\omega_{\tilde{k},j})|}{\omega_{\tilde{k},j} r_j} - g_{sp_b} \frac{k_F}{2} \int_{-1}^{1} d\tilde{k} \sum_{j=1}^{3} \frac{|D_{p_b p_b}(-\omega_{\tilde{k},j})|}{\omega_{\tilde{k},j} r_j}, \tag{A.19}$$

and

$$\Delta_2 = -g_{sp_b}\frac{k_F}{2}\pi\int_{-1}^{1}d\tilde{k}\sum_{j=1}^{3}\frac{|D_{ss}(-\omega_{\tilde{k},j})|}{\omega_{\tilde{k},j}r_j} - g_{p_bp_b}\frac{k_F}{2}\int_{-1}^{1}d\tilde{k}\sum_{j=1}^{3}\frac{|D_{p_bp_b}(-\omega_{\tilde{k},j})|}{\omega_{\tilde{k},j}r_j}. \quad (A.20)$$

In a one-dimensional system, the Fermi wave vector $k_F = \frac{\rho\pi}{2}$, where $\rho = \frac{N}{L}$ is the density of electrons in the material and N is the total number of electrons. Note that, when $g_{ss} = g_{p_bp_b} = g_{sp_b}$, we have $\Delta_1 = \Delta_2$.

References

[1] Bednorz J G and Müller K A 1986 *Zeit. Phys. B-Cond. Matt.* **64** 189–193

[2] Nagamatsu J, Nakagawa N, Muranaka T, Zenitani Y and Akimitsu J 2001 *Nature* **410** 63–64

[3] Kamihara Y, Watanabe T, Hirano M and Hosono H 2008 *J. Am. Chem. Soc.* **130** 3296–3297

[4] Lin S Z 2014 *J. Phys. Condens. Matter* **26** 493202

[5] White S R and Scalapino D 2015 *arXiv:1503.01533v1*

[6] Beenakker C 2013 *Annu. Rev. Condens. Matter Phys.* **4** 113–136

[7] Maeno Y, Hashimoto H, Yoshida K, Nishizaki S, Fujita T, Bednorz J G and Lichtenberg F 1994 Superconductivity in a layered perovskite without copper

[8] Mackenzie A P 2003 *Rev. Mod. Phys.* **75** 657–712

[9] Kallin C 2012 *Rept. Prog. Phys.* **75** 042501

[10] Bok J and Bouvier J 2012 *J. Supercond. Nov. Magn.* **25** 657–667

[11] Raghu S, Kapitulnik A and Kivelson S A 2010 *Phys. Rev. Lett.* **105** 1–4

[12] Cohen M L 1964 *Rev. Mod. Phys.* **36** 240–243

[13] Sidorov V and Ekimov E 2010 *Nature* **19** 351–357

[14] Mizuguchi Y, Demura S, Deguchi K, Takano Y, Fujihisa H, Gotoh Y, Izawa H and Miura O 2012 *J. Phys. Soc. Japan* **81** 114725

[15] Mizuguchi Y 2014 *J. Phys. Chem. Solids* 1–15

[16] Liu J, Fang D, Wang Z, Xing J, Du Z, Li S, Zhu X, Yang H and Wen H H 2014 *Europhys. Lett.* **106** 67002

[17] Usui H, Suzuki K and Kuroki K 2012 *Phys. Rev. B* **86** 220501

[18] Martins G B, Moreo A and Dagotto E 2013 *Phys. Rev. B* **87** 081102

[19] Yang Y, Wang W S, Xiang Y Y, Li Z Z and Wang Q H 2013 *Phys. Rev. B* **88** 094519

[20] Sugimoto T, Ootsuki D, Morice C, Artacho E, Saxena S S, Schwier E F, Zheng M, Kojima Y, Iwasawa H, Shimada K, Arita M, Namatame H, Taniguchi M, Takahashi M, Saini N L, Asano T, Higashinaka R, Matsuda T D, Aoki Y and Mizokawa T 2015 *Phys. Rev. B* **92** 041113

[21] Wolowiec C T, White B D, Jeon I, Yazici D, Huang K and Maple M B 2013 *J. Phys. Condens. Matter* **25** 422201

[22] Hirsch J E, Maple M B and Marsiglio F 2015 *arXiv:1504.03318*

[23] Yazici D, Jeon I, White B and Maple M 2015 *Physica C Supercond* **514** 218–236

[24] Griffith M A, Foyevtsova K, Continentino M A and Martins G B, submitted to *Phys. Rev. B*

[25] Continentino M A, Padilha I T and Caldas H 2014 *J. Stat. Mech. Theory Exp.* **2014** P07015

[26] Caldas H and Continentino M A 2012 *Phys. Rev. B* **86** 144503

[27] Most of the studies reporting T_c vs. x results were conducted on polycrystalline samples and the maximum T_c was obtained for $x = 0.5$, corresponding to 0.5 electron per Bismuth, i.e., 1/8-filling, as each Bismuth has two active orbitals (p_x and p_y). Recently, a study on the electronic structure of $NdO_{1-x}F_xBiS_2$ single crystals was reported and it was found that the charge carrier doping is smaller than that expected from the nominal x value. This was attributed to Bismuth deficiency [31]. The results we obtain here agree with the nominal concentation (1/8-filling for maximum T_c), which corresponds to 0.25 electron per Bismuth ($\mu = 0.5007$) in the 1d model (with one p orbital per Bismuth).

[28] Tomita T, Ebata M, Soeda H, Takahashi H, Fujihisa H, Gotoh Y, Mizuguchi Y, Izawa H, Miura O, Demura S, Deguchi K and Takano Y 2014 *J. Phys. Soc. Japan* **83** 063704

[29] Wolowiec C T, Yazici D, White B D, Huang K and Maple M B 2013 *Phys. Rev. B* **88** 064503

[30] M. A. Grifitth *et. al.*, to be submitted.

[31] Ye Z R, Yang H F, Shen D W, Jiang J, Niu X H, Feng D L, Du Y P, Wan X G, Liu J Z, Zhu X Y, Wen H H and Jiang M H 2014 *Phys. Rev. B* **90** 045116

Quantum Critical Behavior and Superconductivity in new multi-site Cerium Heavy Fermion Compound Ce_3PtIn_{11}

J. Custers[1], **M. Diviš**[1], **and M. Kratochvílová**[1]

[1] Charles University in Prague, Faculty of Mathematics and Physics, Ke Karlovu 5, 121 16 Prague 2, Czech Republic.

E-mail: `custers@mag.mff.cuni.cz`

Abstract. We discuss the results of thermodynamic and transport measurements at ambient and under hydrostatic pressure of the recently discovered heavy fermion superconductor Ce_3PtIn_{11}. At $p = 0$ superconductivity emerges in the complex antiferromagnetic state below $T_c = 0.32$ K. Both phenomena coexist in a wide range of the $p - T$ phase diagram. The critical pressure where $T_N \to 0$ and T_c is maximum equals $p_c = 1.3$ GPa. Here, the resistivity follows $\rho(T) \propto T^n$ with $n = 0.90 \pm 0.05$ suggesting a local moment-type of quantum critical point. The presence of a maximum structure in the susceptibility and the unusual pressure dependence of T_c hint to an understanding of properties of Ce_3PtIn_{11} in the view of distinct different Kondo screened sublattices. Therefore, we analyze the magnetic entropy S_{mag} within the spin-1/2 Kondo model while accounting for the fact that Ce_3PtIn_{11} unit cell possesses two inequivalent Ce-sites, Ce1 and Ce2. We employ an extreme case scenario of non-interacting sublattices where the Ce1-sublattice is paramagnetic and only Ce2-sublattice orders magnetically and discussed its consequences.

1. Introduction

In the past decade, condensed matter research has come to a point of convergence. The appearance of superconductivity in several classes of strongly correlated electron systems such as cuprates, organics, heavy fermions and the more recently discovered iron pnictides cover a putative quantum critical point (QCP). Here, these systems undergo a phase transition at zero temperature and critical fluctuations associated with the order parameter of the phase transition enhance the interaction between electrons in a similar way as phonons do in conventional superconductors leading to the observed superconducting dome [1]. This is probably best explored in heavy fermion (HF) materials where many different non-phononic pairing mechanism associated with different types of QCPs have been proposed [2]. Most notably, in numerous HF materials, superconductivity has been reported to emerge at the border of a magnetically ordered state and a paramagnetic state indicative of magnetically mediated superconductivity [3].

The vast majority of the cerium HF compounds studied to date have been compounds which have only one crystallographic site for the cerium ions. The ground state of these materials depends on the competition between the magnetic inter-site interaction, i. e. , the indirect RKKY exchange interaction, and the demagnetizing on-site Kondo interaction. This competition is well illustrated by the Doniach diagram [4]; the ground state depends on the relative value of the

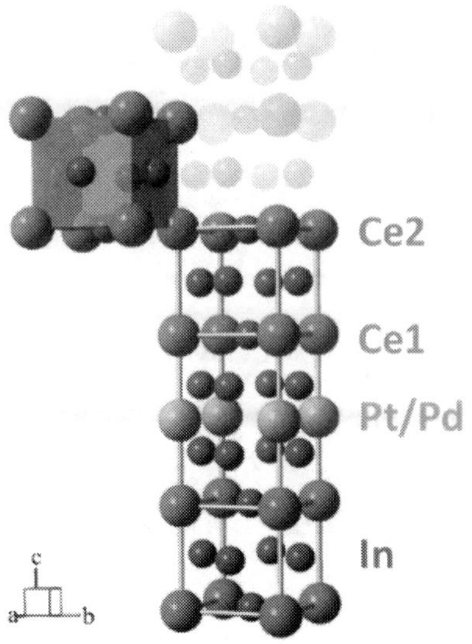

Figure 1. The crystallographic unit cell of Ce_3TIn_{11} (Ce: red spheres; Pd or Pt: green spheres; In: blue spheres). The two inequivalent Cerium sites Ce1 and Ce2 are marked, respectively. The $CeIn_3$ environment of the Ce2-site is shown explicitly. The lattice parameters yield $a = 4.6874(4)$ Å and $c = 16.8422(12)$ Å for the Pt compound and for $a = 4.6896(11)$ Å and $c = 16.891(3)$ Å for the Pd-analogue.

respective RKKY and Kondo energies, $k_BT_{RKKY} \sim |N_FJ_{cf}^2|$ and $k_BT_K \sim \exp[1/|N_FJ_{cf}|]$, with N_F being the conduction band density of states at the Fermi level and J_{cf} the coupling constant (hybridization) between $4f$(Ce) and conduction electrons. Because of the extended nature of the Ce $4f$ wave-function, J_{cf} depends strongly on the local environment of the Ce-ion. This has been capitalized on, for example, investigating quantum critical phenomena in these HF compounds, where by either applying hydrostatic pressure or chemical substitution a formerly magnetic ground state (small values of $|J_{cf}|$) undergoes a transition at a QCP into a non-magnetic ground state where the Kondo effect dominates (large values of $|J_{cf}|$).

In compounds with multiple inequivalent Ce sites, the local environment of these two– or more cerium sites will vary and consequently the respective Ce ion experiences different hybridization (different values for J_{cf}). This results in an interesting aspect; the ground state of such compounds might be a coexistence on microscopic scale of different electronic and magnetic states. For instance, it is quite conceivable that one Ce-site (Ce1) is magnetic, i. e., its sublattice orders magnetically, while the other Ce-site (Ce2) is non-magnetic as a results of dominant Kondo interaction. It was pointed out by Benlagra and co-workers [5] that in these compounds the inequivalent Ce sites (sublattices) do not act independently from each other. The onset of Kondo screening in one sublattice (Ce2-site in our example above) may boost or can reduce the onset of Kondo screening in the Ce1-sublattice. In particular with respect to quantum criticality and superconductivity this can have dramatic consequences as in a local moment scenario the Kondo scale breaks down at the QCP [6]. At present, the investigation of quantum critical behavior in these multi-site compounds is still rather an uncharted terrain. Two of the few examples in literature are Ce_7Ni_3 and $Ce_3Pd_{20}Si_6$. The first-mentioned compound crystallizes in the hexagonal Th_7Fe_3 structure (space group $P6_3mc$) in which the Ce-ions reside in three inequivalent sites, 1 Ce_I, 3 Ce_{II} and 3 Ce_{III}. The crystal structure is rather complex with the Ce_I surrounding being trigonal and the other two have monoclinic symmetry. Ce_7Ni_3 is a heavy-fermion antiferromagnet with $T_N = 1.8$ K. It was suggested by the entropy analysis of the specific heat and from magnetic studies that only the Ce_I sublattice orders. The Ce_{II} sublattice has been associated with heavy fermion (HF) behavior while the Ce_{III} one gives rise to intermediate valence contributions [7]. Under increasing pressure T_N decreases and vanishes near $p_c \approx 0.33$ GPa [8]. Non-Fermi-liquid (NFL) behavior appears at 0.4 GPa in both the specific heat and ac magnetic susceptibility, $C_{el}/T \propto \ln T$ and $\chi_{ac} \propto (1 - \alpha\sqrt{T})$, and has

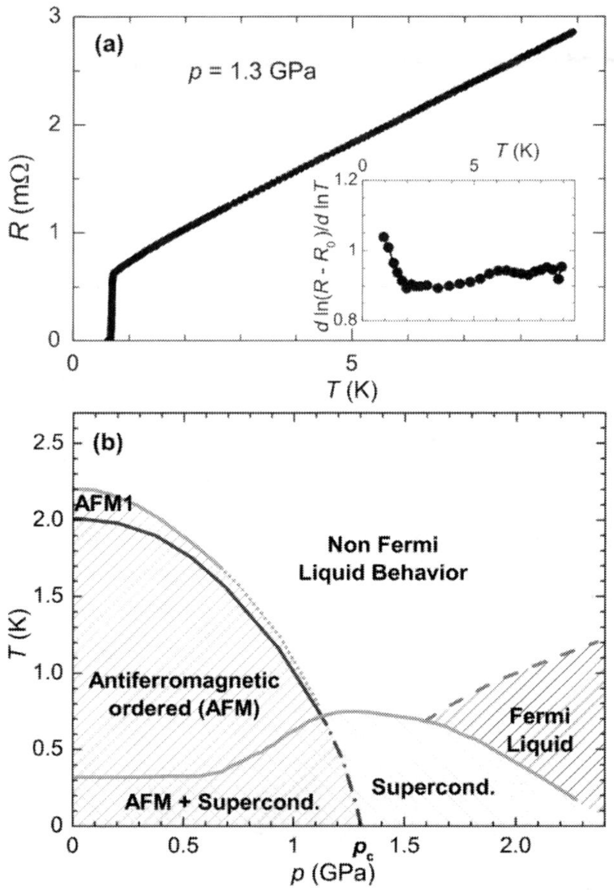

Figure 2. Summary of the properties of Ce_3PtIn_{11} (a) Observation of non-Fermi liquid behavior in the resistance R at the critical hydrostatic pressure of p_c = 1.3 GPa where $T_N \to 0$ [14]. Inset shows the corresponding resistance exponent in the normal state. (b) The $p - T$ phase diagram of Ce_3PtIn_{11} adapted from Ref. [14] showing the various ordered states.

been discussed in the frame of self consistent renormalization theory of spin fluctuations of heavy-fermion systems near an antiferromagnetic instability [9]. Above 0.62 GPa, the normal Fermi-liquid state is recovered. The latter compound, $Ce_3Pd_{20}Si_6$, is a so-called cage compound. $Ce_3Pd_{20}Si_6$ crystallizes in the cubic structure of space group $Fm\bar{3}m$. It harbors two different crystallographic Ce-sites, both with cubic symmetry but one with coordination 16 (tetrahedral $8c$: all Pd) and one with coordination 18 (octahedral $4a$: 12 Pd, 6 Si). At $T = 0.5$ K, $Ce_3Pd_{20}Si_6$ undergoes a quadrupolar phase transition and at $T_N = 0.31$ K a transition in an AFM ordered state [10, 11, 12]. The quadrupolar state has been linked to ordering of the $8c$ and the dipolar order to the $4a$ sublattice. Both ordered states coexist at low temperatures. The compound gained special attention displaying a field-induced QCP ($T_N \to 0$ for $B_c \approx 0.9$ T) of local moment (Kondo breakdown) type in a cubic system which separates the AFM and the AFQ state at $T = 0$ [13].

Until recently no superconductivity has been found in multi-site cerium (HF) systems. Our low-temperature experiments on newly discovered HF compounds Ce_3PdIn_{11} [15, 16] and Ce_3PtIn_{11} [16] revealed that both become superconducting [17, 14]. Ce_3PdIn_{11} and Ce_3PdIn_{11} belong to the Ce_nTIn_{3n+2} class of materials which comprises a numerous amount of compounds including $CeCoIn_5$, $CeRhIn_5$ and Ce_2RhIn_8. The compounds crystallize with space group P4/mmm (tetragonal structure) based on the AuPt3-type ($CeIn_3$block) and $PtHg_2$-type (TIn_2) units and possesses two non-equivalent Ce-sites labeled Ce1 (2 Ce-ions per f.u.) and Ce2 (1 Ce-ion per f.u.) as emphasized in Fig. 1. The advantage of both materials is that they are are less complex from a crystallographic point of view than the examples mentioned before. Ce2 resides the Wyckoff $1a$ position which has local C_{4v} symmetry. The ion experiences $CeIn_3$-like environment. The Ce1-site occupies the $2g$ position (D_{4h} symmetry). Its surrounding is identical to Ce-atoms in Ce_2PdIn_8 and Ce_2PtIn_8. At ambient pressure, both compounds shows remarkable properties: in the absence of magnetic field, Ce_3PtIn_{11} (Ce_3PdIn_{11}) undergoes two successive antiferromagnetic type of transitions at $T_1 = 2.2$ K and $T_N = 2$ K ($T_1 = 1.67$ K and $T_N = 1.53$ K in the case of Pd) and becomes superconducting below $T_c = 0.32$ K [14] (Pd: $T_c = 0.42$ K [17]).

1.1. Superconductivity in Ce_3PtIn_{11} and Ce_3PdIn_{11}

We briefly summarize the superconducting properties of both compounds. Details can be found in the references mentioned above. Ce_3PtIn_{11} and Ce_3PdIn_{11} take a unique place in the Ce-based HF family of materials. In both compounds, superconductivity emerges out of a magnetically ordered state at ambient pressure. This, in fact, makes the compounds be the first known full inversion symmetrical stoichiometric heavy fermion material based on Ce exhibiting, most likely, a coexistence of magnetism and superconductivity already at ambient pressure. The superconducting transition temperatures are $T_c = 0.32$ K and $T_c = 0.42$ K, respectively, and the parameter $\Delta C/(\gamma_n T_c)$ yields for both ≈ 0.7, being half of the expected value from BCS theory. The upper critical field for Ce_3PtIn_{11} equals $B_{c2} = 1.4$ T and for Ce_3PdIn_{11} $B_{c2} = 2.3$ T. The upper critical fields are orbital limited. The initial slope for the Pt-compound is $dB_{c2}/dT = -7.24$ T/K and for the Pd-analogue -8.6 T/K from which is evident that heavy quasiparticles, mass m^*, are involved in Cooper-pairing ($dB_{c2}/dT \propto (m^*)^2 T_c$). The superconducting phase in both has been linked to a nearby magnetic QCP.

1.2. Quantum critical behavior in Ce_3PtIn_{11}

So far quantum critical behavior has only been investigated on Ce_3PtIn_{11} [14]. The application of hydrostatic pressure (p) reduces the temperature of magnetic ordering transitions as expected for a Ce-compound (see Fig. 2b). The magnetic transitions intersect with the superconductivity at $p \approx 1.1$ GPa. Upon increasing p, a pronounced region of non-Fermi liquid behavior has been observed above T_c which is characterized by an almost T-linear dependence of the resistance (R) up to temperatures even higher than $T > 5$ K. It was pointed out that the NFL behavior can be seen as indication of the presence of strong quantum critical fluctuations originating from a magnetic QCP hidden inside the SC dome. The position of the QCP where $T_N \to 0$ was traced by the maximum in T_c [1] and is located at a critical pressure of $p_c \approx 1.3$ GPa ($T_c = 0.7$ K). A fit to $R(T)$ at p_c gives $R = R_0 + AT^n$ with $n = 0.90 \pm 0.05$. Alternatively, the exponent has been determined using $n = d\ln(R - R_0)/d\ln T$, i. e. , the temperature logarithmic derivative of the resistance measured relative to the residual low temperature value (Fig. 2a). From the value of exponent in the NFL-state it was suggested that the QCP might be of local moment type.

2. Results and analysis

this manuscript focusses on the analysis of the experimental data. The properties of Ce_3PdIn_{11} and Ce_3PtIn_{11} are very similar. In order to keep it transparent we will mainly discuss results of the latter compound.

2.1. Susceptibility at high temperatures

The temperature dependence of the susceptibility $\chi(T)$ for Ce_3PtIn_{11} and Ce_3PdIn_{11} $\chi(T)$ in field of $B = 1$ T applied perpendicular ($B \perp c$) and parallel c axis ($B \parallel c$) has been measured. Both compounds exhibit a weak magnetocrystalline anisotropy. The ratio $\chi^{\parallel c}/\chi^{\perp c}$ determined at $T = 3$ K yields 1.25 for the Pt-compound and 1.2 for the Pd-analogue [14, 17]. Figure 3a and b depicts the resulting inverse susceptibilities $1/\chi(T)$ for Ce_3PtIn_{11} and for comparison

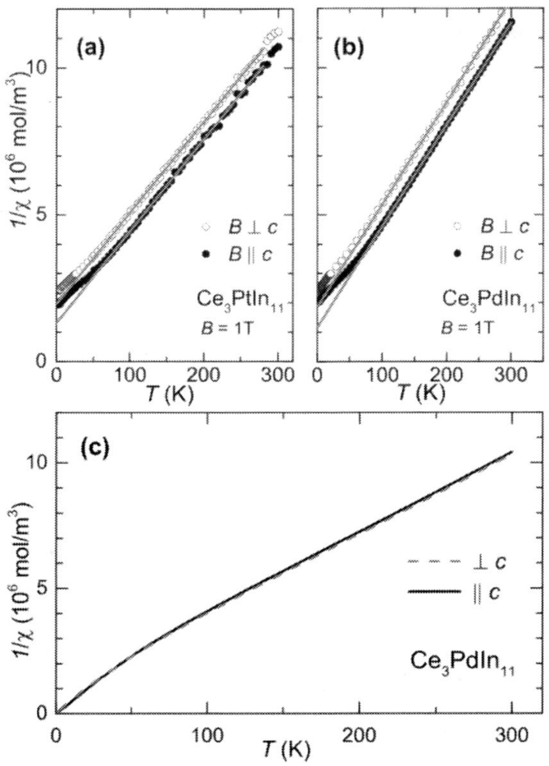

Figure 3. Temperature dependence of the inverse of the susceptibilities $1/\chi(T)$ (a) Ce_3PtIn_{11} and (b) Ce_3PdIn_{11} [17] for $B = 1$ T applied $\perp$ c (open) and $\parallel$ c (closed symbols). The solid (red) lines are Curie-Weiss fits to the data. (c) $1/\chi(T)$ of Ce_3PdIn_{11} from first-principles calculations for $\parallel$ c (solid black line) and $\perp$ c (dashed red line), respectively.

Ce_3PdIn_{11} [17], respectively. In higher temperatures $1/\chi(T)$ is fitted by a Curie-Weiss law (solid red lines) with values of the effective moment of $\mu_{\mathrm{eff}} = 2.60\mu_{\mathrm{B}}/\mathrm{Ce}$ for both directions in case of Ce_3PtIn_{11} and Weiss temperatures $\theta_p^{\perp c} = -64$ K and $\theta_p^{\parallel c} = -42$ K and $\mu_{\mathrm{eff}} = 2.43\mu_{\mathrm{B}}/\mathrm{Ce}$ and $\theta_p^{\perp c} = -49$ K and $\theta_p^{\parallel c} = -33.5$ K for the Pd-compound. The effective moments are in good agreement with expected Hund's rule value for free Ce^{3+}-ion (2.54 μ_{B}) which indicates that all 3 Ce-ions in the unit cell are in Ce^{3+}-state. This is further supported by theoretical calculations. Figure 3c shows the results of first-principles calculations performed for Ce_3PdIn_{11} based on density functional theory [18] from which we obtained the contribution to the susceptibility from each Ce-site separately when field applied along and perpendicular to c-axis. The high temperature $\chi(T)$ values are in qualitative agreement with experimental data and the figure reveals an almost isotropic behavior of $\chi(T)$. More specific, the susceptibility of the Ce1-site exhibits only a marginal magnetocrystalline anisotropy. Surprisingly, the Ce2-site shows a stronger anisotropy. $\chi^{\parallel c}(T)$ increases more upon decreasing temperature with respect to $\chi^{\perp c}(T)$. This increase in anisotropy is the origin of the crossing of $1/\chi^{\perp c}(T)$ and $1/\chi^{\parallel c}(T)$ in Fig. 3c. Note that in our experimental data such crossing has not been observed. A more detailed comparison with the experimental data, however, would require a correct description of the exchange interactions in the paramagnetic region $T > T_1$, including the exact crystal electric field level scheme of Ce_3PdIn_{11} which is unknown. We can assume that first principle calculations on Ce_3PtIn_{11} will give similar results. In the following we present results of Ce_3PtIn_{11} only.

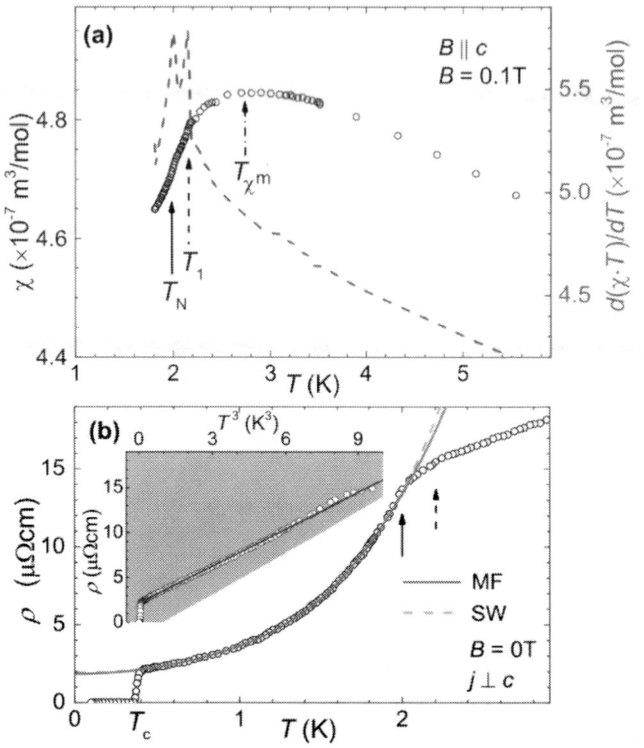

Figure 4. Low temperature susceptibility (χ) and resistivity (ρ) of Ce_3PtIn_{11} in greater detail. (a) Left ordinate: χ vs T in field $B = 0.1$ applied $\parallel$ c–axis. Right ordinate: derivative $d(\chi \cdot T)/dT$ vs T in accordance to Ref. [19]. (b) Resistivity in $B = 0$ T and $j \perp c$ (symbols). The lines show the mean field (MF) and spin-wave (SW) fit for $T_c < T < T_N$ (see text). Solid and dashed arrows mark T_N and T_1, respectively. The inset plots the $\rho(T)$ vs T^3. The solid blue line is a fit $\rho \propto T^3$.

2.2. Susceptibility and resistivity at low temperatures

Figure 4 displays the ambient pressure susceptibility and transport properties of Ce_3PtIn_{11} below 3 K. Two small anomalies can be identified in $\chi^{\parallel c}(T)$ (see Fig. 4a) which appear below a rounded maximum structure with vertex at $T_{\chi^m} = 2.8$ K. In accordance with work of Fisher [19] we attribute this maximum structure due to the presence of short-range magnetic correlations as precursor of a nearby AFM transition. In line with this work, the two anomalies which are inflexion points in $\chi^{\parallel c}(T)$ and appear as maximum in $d(\chi \cdot T)/dT$, hence correspond to T_1 and T_N (Fig. 4a right axis) yielding 2.15 K and 2 K. These values correspond well with those reported from specific heat being $T_1 = 2.2$ K and $T_N = 2$ K, respectively [14]. The antiferromagnetic-type of ordering at T_1 and T_N which we label AFM1 and AFM, is consistent with the observations in Fig. 5. Here we present C/T in field of $B = 3$ T applied along both crystallographic directions. Subtraction of phonon term has been omitted, because we are interested in the effect on the transitions only. For $B \parallel c$ both transitions temperatures shift to lower temperatures as expected in case of AFM ordering (Fig. 5a). Interestingly, along perpendicular direction the upper transition temperature T_1 remains unchanged while T_N decreased slightly being 1.97 K as seen in Fig. 5b. Equivalent has been observed in the Pd-analogue [17]. It suggests that the hard magnetic direction is perpendicular c-axis, which is in sharp contrast to other magnetic members of the Ce_nTIn_{3n+2} group of compounds, i. e. , $CeRhIn_5$ and Ce_2RhIn_8 [20]. The question arises whether or not the order-to-order transition at T_N is of first-order type, which then implies a symmetry change of the magnetic order parameter. At present we have no experimental evidence for this in form of, for instance, latent heat. It was pointed out in Ref. [14] that below T_N, C_{el}/T can be described by a second-order mean-field type expression (MF) or in equal matter by a simple spin-wave term (SW) for spin-waves with dispersion $\epsilon(q) \sim q$. The corresponding expected behavior of the resistivity is presented in Fig. 4b. The red solid line is a MF fit, $\rho(T) = \rho_0 + AT^2 + bT[1 + 2T/\Delta]e(-\Delta/T)$. The resulting parameters are

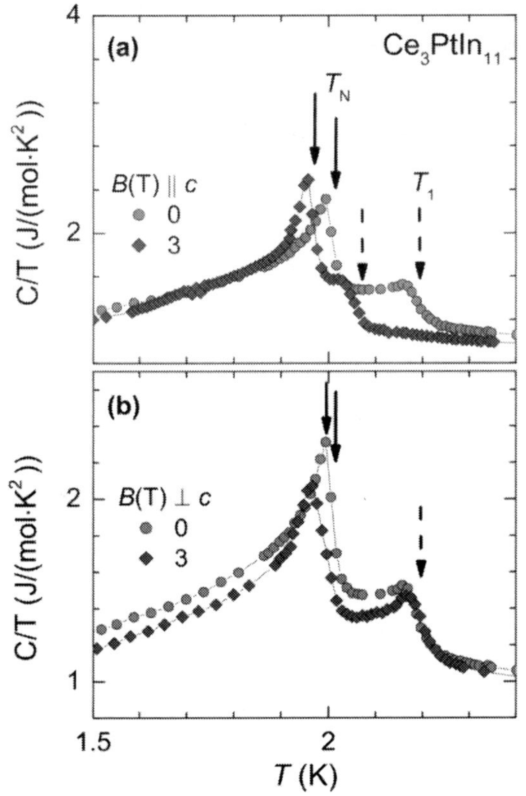

Figure 5. Comparison of the C/T vs T in zero field and in an applied field of 3 T for (a) $B \parallel c$-axis and (b) $B \perp c$-axis. Solid (dashed) arrows mark T_N (T_1).

$\rho_0 = 1.84\mu\Omega$cm, $A = 1.76\mu\Omega\text{cm}/\text{K}^2$, $b = 58.43\mu\Omega\text{cm}/\text{K}$ and the gap $\Delta = 7.35$ K. The dashed green line plots the respective SW expression, $\rho(T) = \rho_0 + AT^2 + BT^5$, where the second term is Fermi liquid (FL) contribution and the third term is due to scattering off magnons. We obtain $A = 1.57\mu\Omega\text{cm}/\text{K}^2$ and $B = 0.17\mu\Omega\text{cm}/\text{K}^5$. Note that for both fits the FL contribution is very similar. Moreover, the necessity of the addition T^2 Fermi liquid term suggests that two types of scattering mechanisms are present. We can insert A in the Kadowaki-Woods relation [21] to determine the Sommerfeld coefficient, $\gamma = \sqrt{A/10\mu\Omega\,\text{cm} \cdot \text{mol}^2 \cdot \text{K}^2 \cdot \text{J}^{-2}}$, and obtain a value of ≈ 0.4 J/(mol·K^2). It is still unclear whether this empirical relation holds for multi-site compounds. However, it is interesting to note that this value times 3 matches well with the normal state Sommerfeld coefficient from specific heat, $\gamma_n = 1.21$ J/(mol·K^2) [14]. Finally, a single T^3-behavior presented in the inset in Fig. 4c describes the resistivity data with similar quality, but such can only be explained by a more complex type of AFM ordering.

3. Discussion

The increase of T_c upon suppressing the magnetic transition temperatures T_1 and T_N as well as the observed NFL behavior the normal state in close vicinity around the critical pressure $p_c = 1.3$ GPa, strongly points to magnetic quantum fluctuations driving the SC state in Ce_3PtIn_{11}. The type of QCP is yet unclear. The exponent $n = 0.90 \pm 0.05$ in resisitvity at p_c suggest a local moment-type scenario. However, in another compound of the $Ce_nT_mIn_{3n+2m}$ layered class of materials, $CeRhIn_5$, 2 dimensional-type (2D) spin fluctuations have been reported [22]. Hence, it is not unreasonable to assume that 2D spin fluctuations are present in Ce_3PtIn_{11} too. In that case, the resistivity data can be also explained within an AFM spin-density wave type of QCP. It predicts $R(T) \propto T^{d/z}$, where the dynamical exponent $z = 2$ and d is the effective dimensionality of the critical spin-fluctuation being 2 here [23]. More intriguingly, the pressure independency of the superconducting state for $0 < p < 0.6$ GPa observed in Fig. 2b seems to contradict the relationship between SC and QCP. It can be taken as a clue that SC and magnetism arise from different electrons in this part of the phase diagram. A possible explanation arises from the two Ce-sites (3 Ce atoms/f.u.: 2Ce1 + 1Ce2) assuming one sublattice is responsible for magnetism and the other sublattice orders magnetically. Further support for this idea comes from Fig. 4b where the resisitvity has been fitted by the sum of a T^2–Fermi liquid expression and a T^5–magnon term (green dashed line). In the following, we quantitatively discuss possible implications of this idea with respect to the thermodynamic properties.

The electronic specific heat coefficient γ_n of Ce_3PtIn_{11} was reported to be 1.21 ± 0.05 J/(mol·K^2) when extrapolated to zero temperature [14]. This large γ_n characterizes Ce_3PtIn_{11} as heavy fermion compound. However, since there are two kinds of Ce-ions (Ce1 and Ce2) with different site symmetries and neighboring ions, it is quite likely that the two do not contribute equally to the density of states at the Fermi level, i. e. , $\gamma_n = 1.21$ J/(mol·K^2) is the sum of the individual contributions. The same argument accounts for the magnetic entropy S_{mag} which is depicted in the Fig. 6 (data taken from Ref. [14]). The value S_{mag} deliberated at T_1 equals $0.7R\ln 2$, that is one-fourth of the value expected for the ground state doublet of Ce_3PtIn_{11} (three times $R\ln 2$). Thus there is a considerable reduction of the magnetic entropy. We relate the deficiency of $S_{Ce}(T_1)$ to the presence of the Kondo effect which partially lifts the twofold degeneracy above T_1. In such case, it can be shown that $S_{mag}(T_1) = S_K(T_1/T_K)$, where $S_K(T_1/T_K)$ is the Kondo entropy at T_1. [24] The relation holds if the first exited level $\Delta_{CEF} \gg k_B T_1$ and $\Delta_{CEF} \gg k_B T_K$. From earlier discussion we conclude that the former condition is satisfied in Ce_3PtIn_{11} [14]. A value around $\Delta_{CEF} = 60$ K for the first excited state can be expected. Additionally, typical Kondo scales in the $Ce_nT_mIn_{3n+2m}$ class of materials ranges between 2 and 10 K suggesting that also the second condition is fulfilled. Further, using the Bethe ansatz for spin-1/2 Kondo model, Desgranges and Schotte calculated the specific heat and corresponding magnetic entropy. [25] By inserting the experimentally found S_{mag} into the above relation we obtain the ratio T_1/T_K. The estimated Kondo temperature for Ce_3PtIn_{11} hence yields 12 K, which is an average value, meaning we assume that the Kondo screening is equal for both Ce sites. Figure 6 compares S_{mag} (symbols) with the calculated values of the spin-1/2 Kondo model (S_{Ce}: green dashed line) [25]. The data agree fairly well suggesting that T_{K1} and T_{K2} are close. We observe that below T_1 the experimental data fall under the theoretical curve as expected (entropy is absorbed in the transition). For $T \gg T_1$ the S_{Ce} values exceed the calculated ones which is an indication for the presence of additional contributions to the specific heat, most likely originating from excited CEF levels.

In the extreme case of our scenario, the Ce1 and Ce2 sublattices do not interact and only one sublattice orders magnetically while the other sublattice is responsible for the observed HF behavior. Note that this is different from the discussion in Ref. [14] where the HF behavior is not allocated to a specific Ce-sublattice. Ideally, in this particular case the entropy well below the magnetic transition ($T \ll T_1$) is solely related to the second, paramagnetic sublattice. With this argument we can determine the paramagnetic sublattice Kondo temperature T_{K1}. We assume that the Ce2-sublattice orders magnetically, which is most plausible because of the $CeIn_3$-like environment, and the Ce1-sublattice is paramagnetic. The matching of the low-T S_{Ce} data to the spin-1/2 Kondo model reveals a Ce1-sublattice Kondo temperature of $T_{K1} \approx 15.4$ K (see Fig. 6 red solid line). With the same model the entropy $S_{Ce1}(T)$ of this Ce-sublattice above T_1 can be determined. The difference to the total entropy $\Delta S(T) = S_{mag}(T) - S_{Ce1}(T)$ originates from the Ce2-sublattice and hence allows to estimate its Kondo temperature T_{K2}. In order to avoid the influence of magnetic fluctuations from the magnetic transitions as well as contributions arising

Journal of Physics: Conference Series **683** (2016) 012005 doi:10.1088/1742-6596/683/1/012005

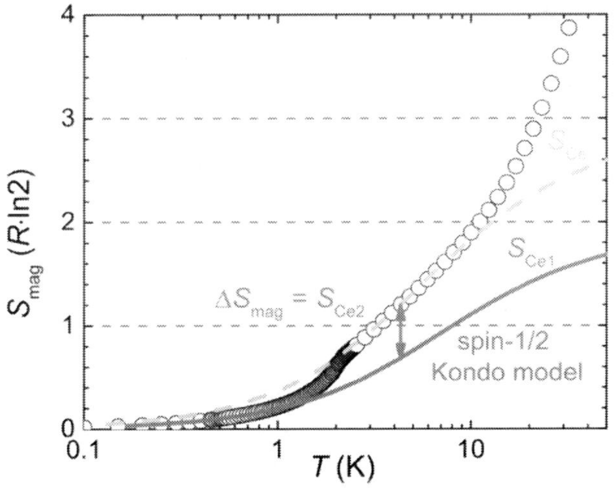

Figure 6. The magnetic entropy S_{mag} of Ce_3PtIn_{11} in units of $R \ln 2$ (symbols). The green dashed line marked S_{Ce} shows the the theoretical calculation of the spin-1/2 Kondo model [25] in case of $T_{\text{K}} = 12$ K and assuming all Ce-ions contribute equally to S_{mag}. The red curve (solid) depicts the contribution to the entropy arising form the Ce1-sublattice only (see text). The double arrow marks the difference $\Delta S_{\text{mag}} \equiv S_{\text{Ce2}} = S_{\text{mag}} - S_{\text{Ce1}}$ at $T = 4$ K which is the entropy associated with the Ce2-sublattice.

from CEF we determined T_{K2} from a single data point at 4 K. With $S(4\text{K})/R = 0.533$ we obtain $T_{\text{K2}} \approx 8$ K [25]. Note that this value is the lower limit of the Ce2-sublattice Kondo temperature. More realistic is a higher T_{K2} because some entropy of the Ce2-sublattice will be present below T_1. The calculation, however, shows that even in the very extreme case both sublattice Kondo temperatures are of the same order of magnitude and similar to the values obtained in Ref. [14] being $T_{\text{K1}} = 15$ K and $T_{\text{K2}} = 3$ K, respectively. It promotes the view that in Ce_3PdIn_{11} the two sublattice Kondo temperatures are mutually cooperative, meaning the set in of T_{K1} is expected to boost the lower Kondo scale T_{K2} [5].

We can further speculate about the implications of above scenario; with the Ce1-sublattice paramagnetic this would lead to a huge separation (16.891 Å) between adjacent magnetic Ce2-layers. Hence, one would expect a significant magnetically 2D behavior with for instance, a strongly reduced T_{N}. The observed maximum in $\chi(T)$ can be interpreted as a signature of this 2D-behavior [26]. Here, in a 2D-spin Heisenberg model for a spin-1/2 system the position of the maximum relates to $T_{\chi^{\text{m}}} \approx 0.93|J|$ which for $T_{\chi^{\text{m}}} = 2.8$ K gives $|J| = 3$ K. Any deviation from ideal 2D produces long-range order at $T_{\text{N}} \approx S(S+1)|J|/2 = 1.1$ K. This value is lower than the observed magnetic transition temperature but the model does not include Kondo interactions. These may suppress the fluctuations considerably shifting $T_{\chi^{\text{m}}}$ to lower temperatures. Assuming further that the fluctuations are stronger affected by the Kondo interaction than T_{N} which reduces for similar reasons could explain the discrepancy.

4. Summary and conclusion

A detailed analysis of the ambient pressure susceptibility, resistivity and the magnetic entropy of multi-site heavy fermion compound Ce_3PtIn_{11} has been presented. The compound crystalizes in a tetragonal structure and possesses two inequivalent Ce-sites (Ce1 and Ce2). Despite the layered structure $\chi(T)$ exhibits a weak magnetocrystalline anisotropy. First principle calculations attribute this anisotropy to the Ce2-site. We employ an extreme scenario of non-interacting sublattices with the Ce1-sublattice being paramagnetic and solely responsible for heavy fermion behavior and a magnetic Ce2-sublattice. Within this scenario we discuss the magnetic entropy and analyze the data using the spin-1/2 Kondo model of Desgranges and Schotte [25]. The corresponding sublattice Kondo temperatures $T_{\text{K1}} = 15.4$ K and $T_{\text{K1}} = 8$ K have been retrieved. Although the scenario is a strong simplification of the real property of Ce_3PtIn_{11}, it in principle explains the fact of two contributions to the resistivity, i. e., a FL and a magnon term as well as the possibility of existence of 2D magnetic fluctuations. The latter would be important to understand the nature of the QCP at $p_c = 1.3$ GPa. Inelastic neutron experiments are indispensable here. The observation of a pressure independent T_c in a large part of the $p - T$ phase diagram ($p < 0.6$ GPa) opens speculation that the ambient pressure SC state has different origin of Cooper-pairing than the SC-state around p_c. It complies, however, with our scenario. Here the heavy quasiparticles associated with the Ce1-sublattice form the Cooper-pairs. In fact, we may speculate that the gap-symmetry is s-wave. At higher pressures around p_c critical spin fluctuations linked to the suppression of magnetic order in the Ce2-sublattice enhance T_c and the SC state contains a mixture, or overlap, of distinct different SC order parameters resulting from different pairing mechanisms. Although this is still to be investigated, together with the unusual magnetic states and the question of the nature of the QCP makes Ce_3PtIn_{11} (and Ce_3PdIn_{11}) an exciting compound for future studies.

Acknowledgments

We are grateful to Christoph Geibel for helpful discussions. This work was supported by the Grant Agency of the Czech Science Foundation (Project P203/12/1201). Experiments were performed in MLTL (http://mltl.eu/) which is supported within the program of Czech Research Infrastructures (project no. LM2011025).

References

[1] Monthoux P, Pines D and Lonzarich G G 2007 *Nature* **450** 117
[2] Gegenwart P, Steglich F and Si Q 2007 *Nature Phys.* **4** 186
[3] Pfleiderer C 2009 *Rev. Mod. Phys.* **81** 1551
[4] Doniach S 1977 *Physica B + C* **91** 231
[5] Benlagra A, Fritz L and Vojta M 2011 *Phys. Rev. B* **84** 075126
[6] Si Q, Rabello S, Ingersent K and Smith J 2001 *Nature* **413** 804
[7] Sereni J G, Trovarelli O, Kappler J P, Paschke C, Trappmann T and v Löhneysen H 1994 *Physica B* **199 & 200** 567
[8] Umeo K, Kadomatsu H and Takabatake T 1997 *Phys. Rev. B* **55** 692(R)
[9] Moriya T and Takimoto T 1995 *J. Phys. Soc. Jpn.* **64** 960
[10] Strydom A M, Pikul A, Steglich F and Paschen S 2006 *J. Phys.: Conf. Series* **51** 239
[11] Goto T, Watanabe T, Tsuduku S, Kobayashi H, Nemoto Y, Yanagisawa T, Akatsu M, Ano G, Suzuki O, Takeda N, Dönni A and Kitazawa H 2009 *J. Phys. Soc. Jpn.* **78** 024716
[12] Mitamura H, Tayama T, Sakakibara T, Tsuduku S, Ano G, Ishiti I, Akatsu M, Nemoto Y, Goto T, Kikkawa A and Kitazawa H 2010 *J. Phys. Soc. Jpn.* **79** 074712
[13] Custers J, Lorenzer K A, Müller M, Prokofiev A, Sidorenko A, Winkler H, Strydom A M, Shimura Y, Sakakibara T, Yu R, Si Q and Paschen S 2012 *Nature. Mater.* **11** 189
[14] Prokleška J, Kratochvílová M, Uhlířová K, Sechovský V and Custers J 2015 *Phys. Rev. B* **92** 161114(R)
[15] Tursina A, Neterenko S, Seropegin Y, Noël H and Kaczorowski D 2013 *J. Solid State Chem.* **200** 7
[16] Kratochvílová M, Dušek M, Uhlířová K, Rudajevová A, Prokleška J, Vondráčková B, Custers J and Sechovský V 2014 *J. Cryst. Growth* **397** 47
[17] Kratochvílová M, Prokleška J, Uhlířová K, Tkáč V, Dušek M, Sechovský V and Custers J 2015 *Sci. Rep.* **5** 15904
[18] Diviš M, Rusz J, Michor H, Hilscher G, Blaha P and Schwarz K 2005 *J. Alloys. Compd.* **403** 29
[19] Fisher M E 1962 *Phil. Mag.* **7** 1731
[20] Cornelius A L, Pagliuso P G, Hundley M F and Sarrao J L 2001 *Phys. Rev. B* **64** 144411
[21] Kadowaki K and Woods S B 1986 *Solid State Commun.* **58** 507
[22] Bao W, Aeppli G, Lynn J W, Pagliuso P G, Sarrao J L, Hundley M F, Thompson J D and Fisk Z 2002 *Phys. Rev. B* **65** 100505(R)
[23] Millis A J 1993 *Phys. Rev. B* **48** 7183
[24] Mori H, Yashima H, N and Sato 1985 *J. Low Temp. Phys.* **58** 513
[25] Desgranges H and Schotte K 1982 *Phys. Lett.* **91A** 240
[26] Jongh L D and Miedema A 1974 *Adv. Phys.* **23** 1

Journal of Physics: Conference Series **683** (2016) 012006 doi:10.1088/1742-6596/683/1/012006

^{29}Si NMR spin-echo decay in $YbRh_2Si_2$

S. Kambe[1], H. Sakai[1], Y. Tokunaga[1], T. Hattori[1], G. Lapertot[2], T.D. Matsuda[3], G. Knebel[2], J. Flouquet[2], and R.E. Walstedt[4]

[1]Advanced Science Research Center, Japan Atomic Energy Agency, Tokai-mura, Ibaraki 319-1195, Japan

[2]SPSMS/CEA-Grenoble, 38000 Grenoble, France

[3]Division of Physics, Tokyo Metropolitan University, Hachioji-shi, Tokyo 192-0397, Japan

[4]Physics Department, The University of Michigan, Ann Arbor, MI 48109, USA

E-mail: kambe.shinsaku@jaea.go.jp

Abstract. ^{29}Si nuclear magnetic resonance (NMR) has been measured in a ^{29}Si-enriched single crystal sample of $YbRh_2Si_2$. The spin-echo decay for applied field $H \parallel, \perp$ the c-axes has been measured at 100 K. A clear spin-echo decay oscillation is observed for both cases, possibly reflecting the Ruderman-Kittel (RK) interaction. Since the observed oscillation frequency depends on the direction of applied magnetic field, anisotropic RK coupling and pseudo-dipolar (PD) interactions may not be negligible in this compound. The origin of spin-echo decay oscillations is discussed.

1. Introduction

The quantum critical phase transition (QCPT) is one of the important aspects of Kondo lattices at low temperatures. In $YbRh_2Si_2$ [1], the QCPT is not a case of the ordinary spin density wave (SDW) instability observed in Ce-based Kondo-lattices [2], but a candidate for a novel case of local criticality [3]. In $YbRh_2Si_2$, the weak antiferromagnetic transition below $T_N \sim$70 mK is easily tuned to $T = 0$ with a small applied magnetic field H [1]. At the critical magnetic field H_{cr} for $T_N \sim 0$ K, the T-dependence of the physical properties shows non-Fermi liquid (NFL) behavior owing to quantum critical fluctuations that occur on approaching the QCPT [4]. Having a tetragonal crystal structure (Fig. 1), $YbRh_2Si_2$ exhibits an anisotropic electronic state [5]. Thus, for applied field $H \perp$ and $\parallel$ the c-axis, the critical field is $H_{cr} \sim 0.06$ T and ~ 0.66 T, respectively, indicating that the effective energy scale for quantum criticality is ~ 10 times larger for $H \parallel$ the c-axis. In a previous study of the spin-lattice relaxation time T_1 as a function of temperature and field, we reported that coexisting, degenerate Fermi and non-Fermi liquid states appeared around the QCPT [6]. In a fashion similar to H_{cr}, the coexisting states showed rather anisotropic behavior [7, 8].

In the present study, we report data for the ^{29}Si nuclear spin-echo decay in $YbRh_2Si_2$. Here, we have chosen to apply H parallel and perpendicular to the c-axes in order to investigate the anisotropy of the electronic state. The observed spin-echo decay curves show a clear oscillation, which may be due to the Ruderman-Kittel (RK) interaction [9]. The anisotropic oscillation frequency indicates that the contribution of the pseudo-dipolar (PD) interaction [10] is not negligible, an effect that may also reflect the anisotropy of the electronic state.

The low natural abundance ($\sim$ 4.7%) of ^{29}Si ($I = 1/2$; $\gamma_n/(2\pi)$=845.77 Hz/Oe; γ_n: gyromagnetic ratio; $\omega_n = \gamma_n H$ corresponds to the NMR measurement frequency) has prevented

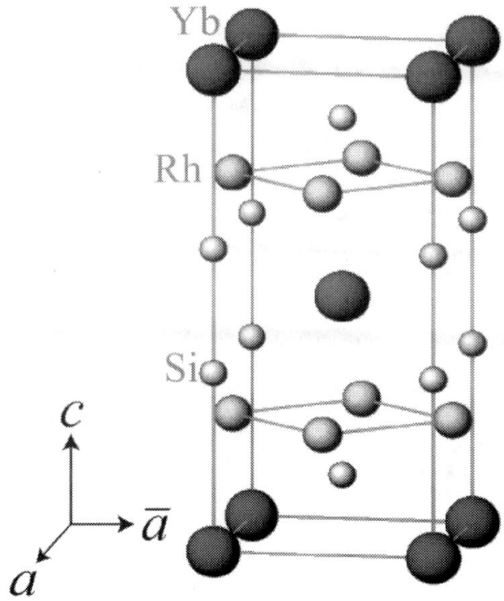

Figure 1. Tetragonal crystal structure ($I4/mmm$) of $YbRh_2Si_2$. The local symmetry of the Si site is tetragonal $4mm$.

highly accurate ^{29}Si NMR measurements on $YbRh_2Si_2$ up to now. For the present study a single crystal sample has been prepared with the ^{29}Si isotope enriched to 52 %, improving the NMR sensitivity by a factor $\sim$ 11.

2. Experimental

Single crystals of $YbRh_2Si_2$ were grown by the In-flux method. The starting materials were Yb, Rh, natural Si, 99.3 % enriched ^{29}Si and In. These materials were put in an alumina crucible and sealed in a quartz tube with the stoichiometric composition of Yb:Rh:Si:^{29}Si:In = 1:2:1:1:65. The ampoule was heated up to 1200 °C, maintained at this temperature for one day, and cooled to 900 °C at a rate of 1 °C/h, taking about 14 days in total. The excess flux was removed from the crystals by spinning the ampoule in a centrifuge.

High sample purity was confirmed by the small residual resistivity $\rho_0 \sim 0.99\,\mu\Omega$cm and the large RRR value $=\rho(300\text{K})/\rho(2\text{K}) = 104$. No magnetic impurity phase was detected in magnetic susceptibility measurements. The antiferromagnetic phase transition was observed near 80 mK at zero field.

A single crystal specimen ($6 \times 3 \times 0.1$ mm^3) was mounted in a ^{4}He cryostat with an NMR pickup coil. Using a standard $\pi/2-\pi$ pulse sequence, the spin-echo intensity $m(2\tau)$ so generated was measured as a function of the time τ between the pulses for the measurement of the spin-echo decay (see Fig. 2 a)). Here, a typical $\pi/2$ pulse width is 4 $\sim$ 5 μsec. Since the resonance linewidth is quite narrow (*e.g.* $\sim$ 9 kHz at 7.2 T), all nuclear spins in the spectrum were quite uniformly excited by the radio-frequency pulses used. The pulse repetition time $t_{rep}$was taken to be much longer than the previously determined T_1 [6, 7, 8]. Figure 2b) shows a calculated typical spin-echo decay curve without the oscillation term. (See below: This case is from Eqs. 2 and 4 with $J = 0$, *i.e.* $W(2\tau)/W(0) = 1$). Here we have adopted the values $1/T_{1L} = 15$ sec^{-1} and $1/T_{2G} = 470$ sec^{-1}, which are similar values to those in the $H \parallel c$ case shown below.

Journal of Physics: Conference Series **683** (2016) 012006 doi:10.1088/1742-6596/683/1/012006

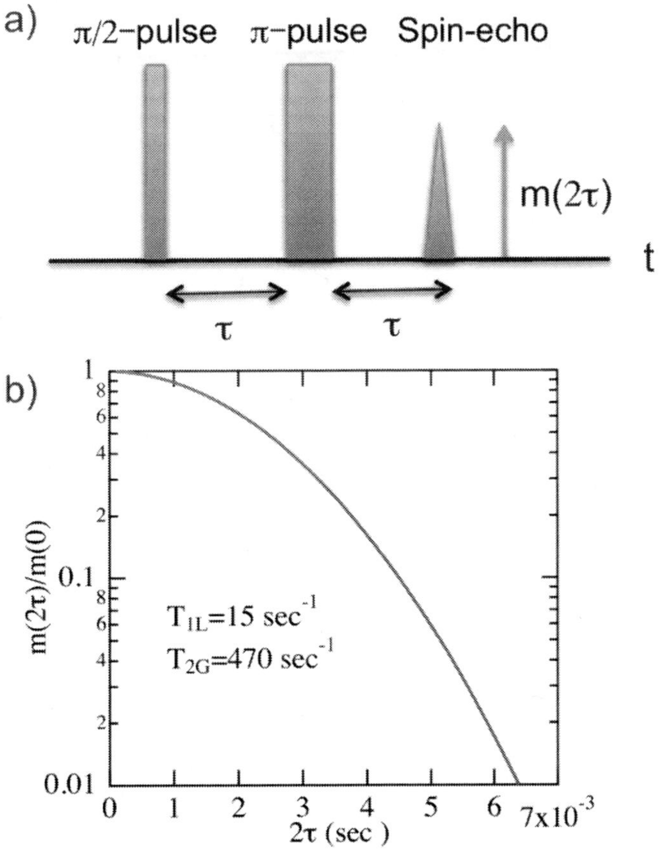

Figure 2. a) Schematic description of NMR spin-echo formation. In the present case, typical τ is considerably larger than the π-pulse width ($\sim 10^{-5}$ sec). b) Calculated spin-echo decay curve without oscillation term ($J = 0$ case for Eqs. 2 and 4). We adopt similar $1/T_{1L} = 15$ sec^{-1} and $1/T_{2G} = 470$ sec^{-1} values to these in the $H \parallel c$ case (see Table 1). In the real case for $H \parallel c$ with $J > 0$, the calculated curve is modulated with the $W(2\tau)/W(0)$ term as shown in Fig. 3 b).

In the actual case with $J > 0$, the spin-echo decay curve is modified by the oscillation factor $W(2\tau)/W(0)$.

3. Spin-echo oscillation due to RK interactions

It is well known that the RK interaction can induce oscillations in the spin-echo decay curve [9]. In the present analysis, the question of PD interaction effects is set aside, though it will be mentioned below. In metals the indirect RK interaction between nuclear spins $\mathbf{I}_i, \mathbf{I}_j$ occurs through second-order scattering of conduction electrons [11], taking the scalar form

$$J_{ij}\mathbf{I}_i \cdot \mathbf{I}_j. \tag{1}$$

The magnitude of J_{ij} can be estimated from the oscillation frequency of the spin–echo envelope [9]. In fact, this type of oscillation has been observed in Pt alloys [9]. Spin-echo oscillations due to RK interactions are predominant if the following conditions are satisfied: a) dipolar and PD

Table 1. Values of $1/T_{2G}$ and J obtained from fits, along with previously determined $1/T_1$ values [6, 7, 8], which are adopted as $1/T_{1L}$.

	$1/T_{1L}$ (sec^{-1})	$1/T_{2G}$ (sec^{-1})	$J/2\pi$(kHz)
$H \parallel c$	15	470 ± 10	1.07 ± 0.02
$H \perp c$	12.9	522 ± 10	0.50 ± 0.01

interactions can be neglected; and b) local inhomogeneous broadening is larger than the indirect RK interaction [9]. Condition a) is considered to be satisfied in the present case. The lattice sum of nuclear dipolar field at the Si site in $YbRh_2Si_2$ is estimated as ~ 0.1 Oe, which corresponds to $\sim 0.1 \times \gamma_n/(2\pi) \cong 85$ Hz. This is much smaller than the estimated RK interaction (see below). The PD interaction is not negligible, a point to be discussed later. Condition b) is satisfied, since the observed inhomogeneous linewidth ~9 kHz at 7.2 T is much larger than the estimated RK interaction.

An RK oscillation factor $W(2\tau)$ for the spin echo ($I = 1/2$ nuclei, $\pi/2 - \pi$ pulses) is analyzed here considering only the nearest–neighbor (nn) nuclear spins, *i.e.* $J_{ij} = J$. Then, we adopt the expression [12]

$$W(2\tau) = \sum_{r=0}^{N} A_r \cos^r(J\tau), \tag{2}$$

where A_r is the probability that a ^{29}Si nucleus has r nn ^{29}Si, and $N = 4$ is the total number of nn Si sites in $YbRh_2Si_2$.

If conditions a) and b) are satisfied, with ^{29}Si nuclei distributed randomly, A_r may be expressed by the binomial coefficient $\binom{N}{r}$ with concentration p of ^{29}Si nuclear species,

$$A_r = A_{rbn} \equiv \binom{N}{r} p^r (1-p)^{N-r}. \tag{3}$$

If in addition to the oscillation factor $W(2\tau)$ there is a spin-lattice relaxation time $T_{1L} \approx T_1$ and a Gaussian component of spin-spin relaxation time T_{2G}, then the observed 2τ dependence of spin-echo intensity $m(2\tau)$ may be expressed,

$$\frac{m(2\tau)}{m(0)} = \exp\{-\frac{2\tau}{T_{1L}} - \frac{1}{2}\left(\frac{2\tau}{T_{2G}}\right)^2\}\frac{W(2\tau)}{W(0)}. \tag{4}$$

In the present study data for spin-echo decay curves has been analyzed based on this equation.

4. Results and discussion

Figure 3 shows spin-echo decay data for $H = 7.2$ T $\parallel, \perp$ c-axis at 100 K. Since the local site symmetry for Si is tetragonal $4mm$, no anisotropy is observed for applied field in the (001) plane. At this temperature Yb magnetic moments are considered to have strongly localized character [6, 7, 8], since the spin–lattice relaxation time T_1 is almost independent of T. Figure 4 shows spin-echo decay with different second–pulse widths for $H = 7.2$ T $\parallel$ c-axis at 100 K. As the second pulse width decreases, the spin-echo decay becomes slower, indicating that the spin-echo

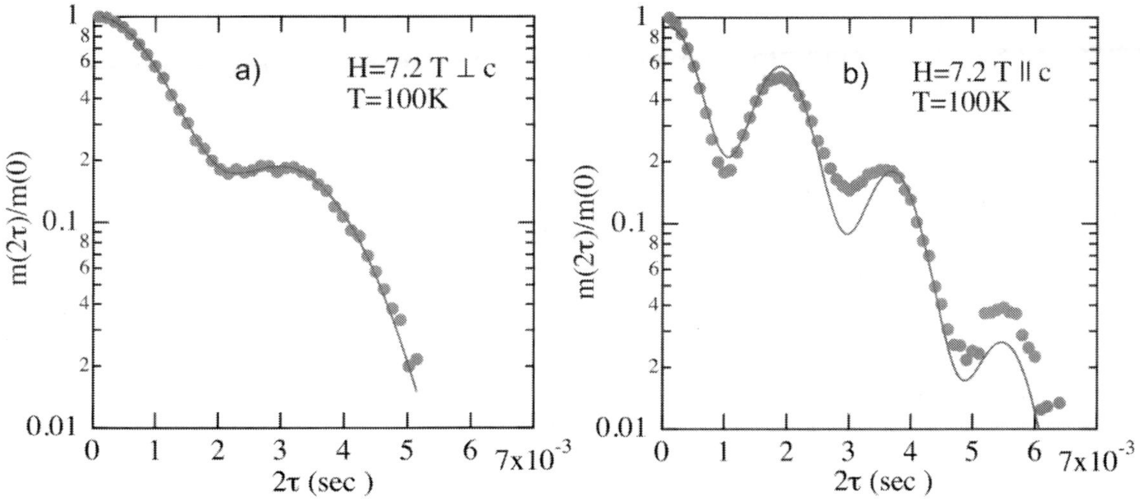

Figure 3. Spin-echo decay curves for a $\pi/2 - \pi$ pulse sequence at $H = 7.2\mathrm{T}\ \|, \perp$ the c-axis at 100 K. Solid lines were obtained using least square fits based on Eqs. 2 and 4.

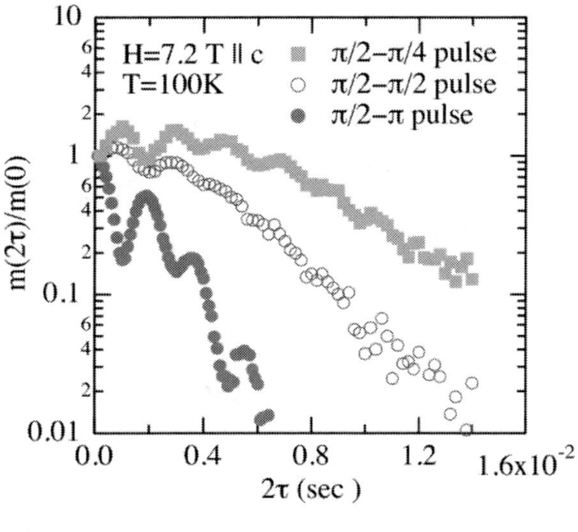

Figure 4. Spin-echo decay curves with different pulse sequences for $H = 7.2$ T $\|, \perp$ c-axis at 100 K. As the second pulse width decreases, the spin-echo decay becomes substantially slower, indicating that the spin-echo decay is static, *i.e.* driven by the π pulse, aside from the T_{1L} process.

decay process is dominated by pulse modulation of static couplings, except for the (weak) T_{1L} relaxation process. In this situation, the echo is relaxed by the reorientation of neighbor spins by the refocusing second pulse. As is clearly seen in Fig. 3, the oscillation frequency is larger for $H \parallel$ c than $H \perp$ c. The echo-decay curves are fitted using Eq. 4. Here the previously determined $1/T_1$ values [6, 7, 8] have been adopted as $1/T_{1L}$. Parameters J, A_r and $1/T_{2G}$ have been optimized with least–squares fits. Values of $1/T_{2G}$ and J so obtained are presented in

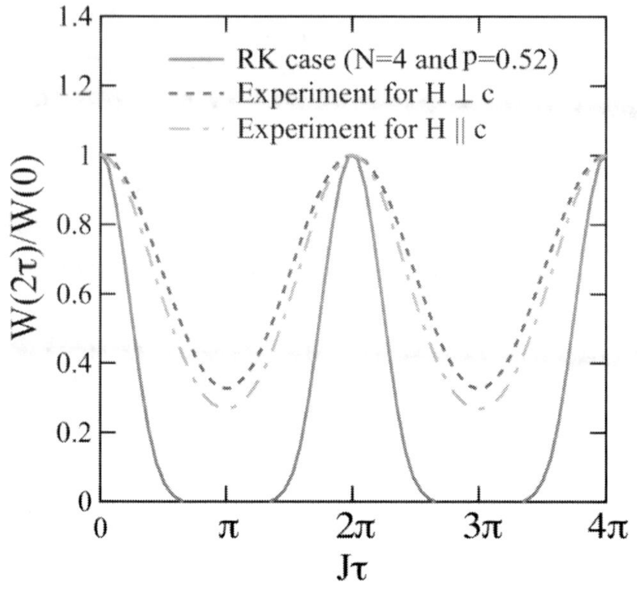

Figure 5. Normalized oscillation factor for spin-echo decay $W(2\tau)/W(0)$. Solid line corresponds to the ideal RK case expressed by Eq. 2 with $N = 4$ and $p = 0.52$. Dashed and broken lines are obtained by fitting to the results presented in Fig. 3 using Eqs. 2 and 4. Deviations from the solid line indicate either that the RK interaction is anisotropic in tetragonal symmetry or that the PD interaction is not negligible.

Table. 1. Since the value deduced for A_r is somewhat different from A_{rbn}, the value obtained for $W(2\tau)$ is slightly different from the ideal RK case as shown in Fig. 5. These deviations indicate that either RK coupling is anisotropic or PD contributions are appreciable. Such effects have been observed in previous studies [12].

In contrast to the scalar RK interaction (Eq. 1), the PD interaction is a tensor quantity [12],

$$B_{ij}(\mathbf{I}_i \cdot \mathbf{I}_j - 3I_j^z I_k^z), \tag{5}$$

$$B_{ij} = \frac{1}{2} b_{ij}(3\cos^2\theta_{ij} - 1), \tag{6}$$

where r_{ij} is the distance between the ith and the jth nuclei, θ_{ij} is the angle between r_{ij} and the applied magnetic field, and b_{ij} is the effective coefficient for dipolar-type interaction B_{ij} between the ith and jth nuclei. Thus, the PD interaction depends on the direction of applied field. If the PD interaction is included, the observed τ-dependence of $W(2\tau)$ can be accounted for better. However, the observed oscillation frequency may not correspond to J exactly, and the real J value may be somewhat larger than the one given here [12].

In the present study, the anisotropic response of J against applied field is suggested. However, in the present RK and PD models (Eqs. 1 and 6), the crystal *i.e.* tetragonal anisotropy is not included explicitly. It is necessary to consider a relevant model for tetragonal system such as $YbRh_2Si_2$ in order to estimate J and B_{ij} more precisely. Principally, the anisotropy of J reflects the anisotropy of the Fermi-surface shape. Further analysis based on an improved model will be presented elsewhere.

Acknowledgments

We are grateful for stimulating discussions with S. Watanabe, K. Izawa, K. Ishida, H. Ikeda, J.-P. Brison,Y.-F. Yang, Z. Fisk and K. Kubo. A part of this work was supported by a Grant-in-Aid for Scientific Research on Innovative Areas Heavy Electrons (No. 20102006 and No. 20102007) from the MEXT, and the REIMEI Research Program of JAEA.

[1] Gegenwart P, Custers J, Geibel C, Neumaier K, Tayama T, Tenya K, Trovarelli O and Steglich F 2002 *Phys. Rev. Lett.* **89** 056402.
[2] Millis A J 1993 *Phys. Rev. B* **48** 7183.
[3] Si Q, Rabello S, Ingersent K and Smith J L 2001 *Nature* (London) **413** 804.
[4] Trovarelli O, Geibel C, Mederle S, Langhammer C, Groshe F M, Gegenwart P, Lang M, Sparn G and Steglich F 2000 *Phys. Rev. Lett.* **85** 626.
[5] Custers J, Gegenwart P, Wilhelm H, Neumaier K, Tokiwa Y, Trovarelli O, Geibel C, Steglich F, Pépin C and Coleman P 2003 *Nature* (London) **424** 524.
[6] Kambe S, Sakai H, Tokunaga Y, Lapertot G, Matsuda T D, Knebel G, Flouquet J and Walstedt R.E. 2014 *Nature Physics* **10** 840.
[7] Kambe S, Sakai H, Tokunaga Y, Lapertot G, Matsuda T D, Knebel G, Flouquet J and Walstedt R.E. 2015 *Phys. Rev. B* **91** 161110(R).
[8] Kambe S, Sakai H, Tokunaga Y, Lapertot G, Matsuda T D, Knebel G, Flouquet J and Walstedt R.E. 2015 *J. Phys. Conf. Series* **592** 012085.
[9] Froidevaux C. and Weger M. 1964 *Phys. Rev. Lett.* **12** 123.
[10] Bloembergen N. and Rowland T. J. 1955 *Phys. Rev.* **97** 1679.
[11] Ruderman M. A. and Kittel C. 1954 *Phys. Rev.* **96** 99.
[12] Alloul H. and Froidevaux C. 1967 *Phys. Rev.* **163** 324.

Quantum Criticality Beneath the Superconducting Dome in β-$YbAlB_4$

T Tomita[1], K Kuga[2], Y Uwatoko[1], and S Nakatsuji[1]

[1]Institute for Solid State Physics, The University of Tokyo, Kashiwanoha, Kashiwa, Chiba 277-8581, Japan

[2]RIKEN SPring-8 Center, Sayo, Hyogo 679-5148, Japan

E-mail: ttomita@issp.u-tokyo.ac.jp

Abstract. Yb-based heavy fermion superconductor β-$YbAlB_4$ at 0 K and 0 T at ambient pressure is located near the quantum critical point with strong mixed valiancy. In this type of Yb electron system, we expect that the magnetic order connected to the quantum critical point derives from the applied pressure. We built a pressure–temperature phase diagram for β-$YbAlB_4$ by measuring the electrical resistivity of high quality single crystal at temperatures down to 40 mK under an applied pressure. A strange metal region appeared, showing non-Fermi liquid $\rho_{ab} \propto T^{1.5}$ behavior, which is stable with applied pressure up to 0.4 GPa, even when below the superconducting dome excluded by a magnetic field of 0.1 T. By increasing of pressure above 2.5 GPa, a magnetic order is first generated. Such ambient quantum criticality/superconductivity is unconventional and is detached from the magnetic order.

1. Introduction

Materials near the quantum critical point (QCP) have attracted much attention because they exhibit unconventional superconductivity and exotic strange metal behavior. In particular, high-temperature superconductivity in Cu-based and Fe-based materials near the QCP and unconventional metal behavior in heavy fermion (HF) materials near the QCP are well known phenomena. In $4f$-electron HF systems, the QCP appearing at 0 K between distinct ordered and disordered quantum phases, is intrinsically unstable and exhibits unconventional metal behavior, becoming a non-Fermi liquid dominated by quantum fluctuations rather than thermal fluctuations. The antiferromagnetic quantum fluctuations near the QCP also are believed to produce unconventional superconductivity, which is indeed connected to the presence of a magnetically ordered phase and is mediated by magnetism. [1, 2, 3] Conversely, these superconducting states emerge to stabilize the destabilized region near the QCP. The non-Fermi liquid state—also known the strange metal near the QCP, and which sometimes appears with unconventional superconductivity—shows characteristic power-law temperature-dependent physical properties such as electrical resistivity $\rho(T)$, specific heat $C(T)/T$, and magnetization $m(T)$ depend anomalously on temperature. Moreover, the quantum phase transitions across the QCP can be accessed by varying a physical parameter—such as magnetic field, external pressure, or chemical composition. Thus, by slightly tuning a physical parameter in an HF material at its QCP, the material easily transitions to a stabilized phase, such as a magnetically ordered phase or Fermi liquid phase. This quantum phase transition and its criticality have been studied and is well described by the conventional Doniach phase, mostly in the Ce-based HF systems.

Journal of Physics: Conference Series **683** (2016) 012007 doi:10.1088/1742-6596/683/1/012007

Recently, however, exotic QCP behaviors have been reported in Yb-based HF system, which is interesting because of the different electron-hole analogy of $4f^1$-Ce^{3+} and $4f^{13}$-Yb^{3+}. Yb-based HF systems are model materials with strong mixed valency for studying quantum criticality, which is not found in Ce-based HF systems.

The prototypical Yb-based HF system $YbRh_2Si_2$ is a weak antiferromagnet below $T_N = 0.07$ K with the intermediate valence state of Yb ($Yb^{\sim 2.9+}$) and can transform into an unconventional non-Fermi liquid when doped with 5% Ge in its Si site.[4, 5] To investigate the behavior of $YbRh_2Si_2$ near the QCP, they substituted Rh atoms with A atoms in $Yb(Rh_{1-x}A_x)_2Si_2$ as a form of chemical pressure, substituting Co and Ir for A as positive and negative chemical pressure, respectively. At the QCP, $YbRh_2Si_2$ shows an antiferromagnetic transition temperature with Kondo breakdown. Conversely, Co-doped $YbRh_2Si_2$ shows the signature of Kondo breakdown within the magnetically ordered phase. For Ir-doped $YbRh_2Si_2$, the Fermi liquid state is not connected with the antiferromagnetic order at a single QCP. These phase diagrams cannot be described by a Doniach phase, so the Yb-based HF system promises to display new behavior near the QCP. [6]

In contrast, β-$YbAlB_4$ shows a QCP at 0 K without tuning any parameter such as field or composition. [7] This material also exhibits a strong intermediate valence state ($Yb^{\sim 2.73+}$ at 20 K). [8] Moreover, its material is the first superconductor $T_c = 85$ mK) near a QCP observed in the Yb-based HF family. The existence of T_c in HF compounds is also an indicator of how far the current state is from the QCP. The fact that a superconducting state exists in ambient β-$YbAlB_4$ indicates that it is very near to quantum criticality. Indeed, its anomalous paramagnetic properties suggest that NFL exists in transport ($\rho_{ab} \propto T^{3/2}$), magnetization ($\chi_c \propto T^{-1/2}$), and specific heat ($C_M/T \propto -\ln T$) at low temperatures. [9, 7] The T_c is also very sensitive to reductions in the residual resistivity ratio (RRR): T_c decreases with increasing RRR and also disappears near 1% chemical Fe doping in the Al site. The low degree of impurity concentration or lattice deformation likely destroys the superconducting state in the pure Kondo lattice system and presumably affects the quantum critical behavior at very low temperatures. Thus, in studying the QCP it is important to tune its state with maintaining material quality, which is only possible in exactly the same high-quality single crystal.

Applying pressure is a non-thermal way to tune the QCP, and applying physical pressure is better than applying chemical pressure from changing the composition because chemical doping induces local lattice disorders or impurity effects. Here, to tune the ground state without causing lattice disorder, we pressurized an extremely high-quality single crystal of β-$YbAlB_4$ and investigated its QCP beneath the superconductivity dome.

2. Experimental details

Single crystals of β-$YbAlB_4$ were grown by the aluminum self-flux method. [10] To test their crystalline quality, we measured their residual resistivity ratio RRR=$\rho_{ab}(300\text{ K})/\rho_{ab}(0\text{ K})$. By testing 20 to 30 crystals, we obtained a few single crystals with an RRR greater than 250. We spot-welded electrical contacts to the surface of the crystals using 20 $\mu m\phi$ Au wires. The in-plane AC electrical resistivity ρ_{ab} was measured down to 40 mK by using a $^3He/^4He$ dilution refrigerator equipped with low-temperature transformers, which provide amplification of a factor of 30. Pressure was applied using a hybrid CuBe/NiCrAl piston-cylinder-type cell with a transmission medium of Daphne 7373. [11] No hysteresis in resistivity appeared between warming and cooling runs. The thermal gradient across the pressure cell, monitored by two thermometers (RuO_2 tips) at the top and bottom of the cell, was less than 5 mK. The superconducting transition temperatures, Al ($T_c = 1.19$ K) and Pd ($T_c = 3.9$ K), were used as accurate measures of pressure at low temperature. [12, 13]

3. Results and Discussion

To investigate the quantum criticality of Yb-based heavy fermion materials, we probed a ultrapure single crystal (RRR=300) of β-YbAlB$_4$ and performed highly sensitive transport measurements at low temperature. We measured these transport properties under pressures as high as 3 GPa and at temperatures of 0.04–1.5 K. Figure 1(a) shows the temperature-dependent electrical resistivity $\rho_{ab}(T)$ under zero field parallel to the ab-plane. Over a wide temperature range, $\rho_{ab}(T) \sim T^{1.5}$ appeared under a pressure of 0.25 GPa. Moreover, the applied pressure shifted $\rho_{ab}(T)$ by a constant resistivity, but the trends in resistivity remained the same. [14, 15] This result shows that the pronounced NFL behavior of resistivity did not change under an applied pressure of 0.25 GPa, although the pressure did decrease T_c.

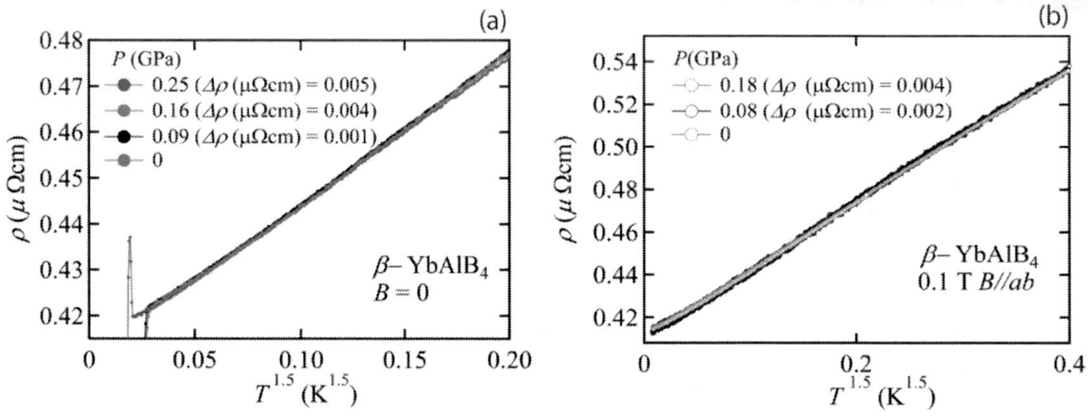

Figure 1. Zero-field resistivity data under various pressures, nearly overlapping but shifted by a constant resistivity indicated in the parentheses. (b) Resistivity data taken under 0.1 T, nearly overlapping but shifted by a constant resistivity indicated in the parentheses.

β-YbAlB$_4$ is reported to be the anisotropic type-II superconductor. This critical field B_{c2} shows an anisotropic upper critical field of $B_{c2} \sim 300$ G along $B//c$, which is lower than $B_{c2} \sim 1500$ G along the ab-plane. A magnetic field of 0.1 T restores the anomalous normal state, reducing T_c to 40 mK. [16] In contrast, applying a magnetic field parallel to ab-plane shifts the material into the Fermi liquid region. However, the resistivity was not sensitive to the field along the ab-plane, as shown in our previous study. In an applied field along ab-plane, the non-Fermi liquid state hidden below T_c emerges. Experimentally, the power law α determined from $\rho_{ab}(T)$ under a field of 0.1 T shows no pressure dependence, as shown in Figure 1(b). [14]

To evaluate the quantum criticality near 0 K in this material, T_c is an unwelcome complication. Figure 2 shows a contour plot giving the resistivity power-law exponent versus pressure. At a low pressure of $P \leq 0.72$ GPa, the resistivity data under 0.1 T were used to exclude T_c. The original data under an ambient field was measured under high pressures of $0.72 \leq P \leq 2.7$ GPa. The temperature dependence of the power-law exponent is determined by $\alpha = \partial \ln(\rho_{ab}(T) - \rho_0)/\partial T$. [14, 15] The Fermi-liquid temperature T_{FL} was determined as $\alpha = 2 \pm 0.1$. From the phase diagram, we obtained the following results: (1) At high pressure, a strange metal region appeared, the so-called "quantum-critical like phase," beneath superconducting dome, although this material is known to exhibit intrinsic zero-field quantum criticality at ambient pressure. [7] (2) A wide range of quantum criticality hidden by the superconducting phase is isolated from the border of magnetism. (3) Neither the quantum critical-like phase nor the isolated superconducting phase are predicted by the standard theory based on quantum spin fluctuations. A similar quantum-critical-like phase separated from the

Journal of Physics: Conference Series **683** (2016) 012007 doi:10.1088/1742-6596/683/1/012007

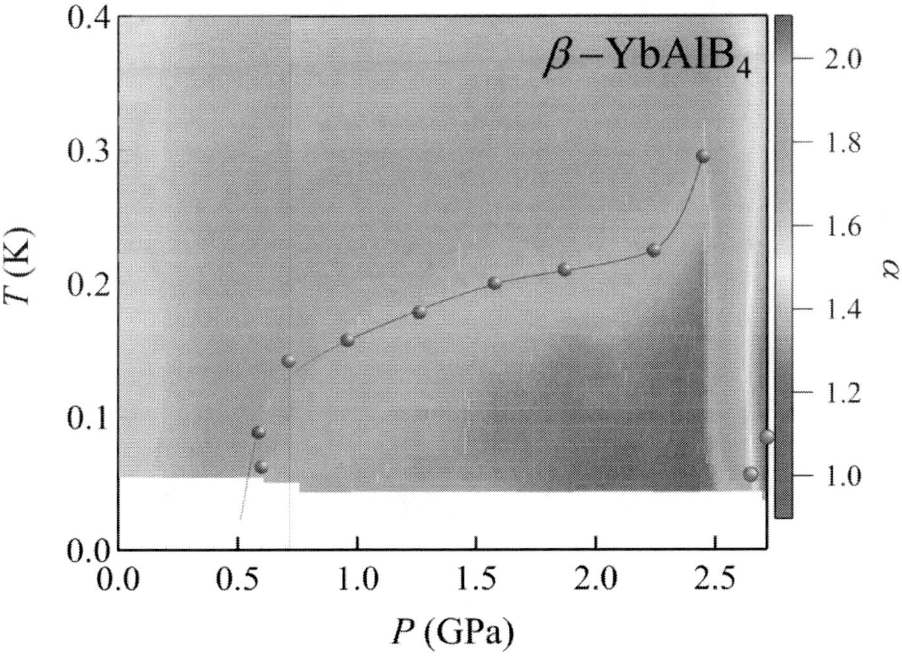

Figure 2. Contour plot of the resistivity power-law exponent α as a function of temperature and pressure. At the low pressure of $P \leq 0.72$ GPa, the data taken under 0.1 T ($B//ab$-plane) were used to extrapolate the analysis down to the lowest temperature by suppressing the superconductivity. T_{FL} is defined as the temperature where the exponent reaches 2. The values of T_{FL} under zero field and 0.1 T are given by blue and purple circles, respectively. The solid blue line is a guide for the eye. The vertical solid black line indicates $P = 0.72$ GPa.

conventional antiferromagnetic QCP appears in Ir-doped $YbRh_2Si_2$ in an applied magnetic field. [6] At 6% Ir, the phase reveals a strong divergence of C/T, presumably showing a new type of metallic spin-liquid ground state. [17] The same type of mechanism presumably arise, for instance, by frustrating the Yb honeycomb lattice in β-$YbAlB_4$. The extended Doniach phase diagram—the QK phase diagram of the Kondo screening (K) and spin zero-point fluctuations (Q)—suggested by recent theoretical approaches would likely explain a number of experimental systems, including YbAgGe and doped $YbRh_2Si_2$, and even β-$YbAlB_4$. [4] Another possible reason for the stabilized phase may be the valence fluctuations of Yb ions. [8] In β-$YbAlB_4$ we confirmed the possibility of a quantum-critical-like phase by studying a very pure sample with a pure Kondo lattice under zero field.

4. Summary

This study of β-$YbAlB_4$ suggests that a new unstable *region* exists, a non-Fermi liquid state, between the stabilized phases, that is not similar to a single critical point in conventional critical phenomena. This finding appears to be the first evidence of a quantum phase hidden by the superconducting dome in HF materials. The Fermi liquid behavior appeared over a wide pressure range ($0.4 \leq P \leq 2.5$ GPa) between the non-Fermi liquid region and the magnetically stable region, but under ambient pressure the material exhibited a real QCP, separated from 2nd QCP near 2.5 GPa. Here, the T_c was as low as 85 mK near the quantum critical point. Thus, the new region below the superconducting dome could appear in a small field. This type of study would be impossible for high-T_c superconductors, such as Cu-based and Fe-based high-

T_c superconductors, because an extremely high magnetic field is needed to exclude the high temperature superconducting state and will destroy the ground state.

Acknowledgments

The authors wish to thank Y. Matsumoto, K. Matsubayashi, and Y. Shimura for support in experiments and K. Miyake and S. Watanabe for fruitful discussions. The authors are also grateful to N. Horie for experimental assistance. This work was supported by Grants-in-Aids (Grant Nos. 25707030 and 24740243) and Scientific Research on Innovative Areas "Heavy Electrons" from MEXT of Japan (Grant No. 21102507). This work is partially supported by PRESTO, Japan Science and Technology Agency, Grants-in-Aid for Scientific Research (No. 25707030), by Grants-in-Aids for Scientific Research on Innovative Areas (15H05882, 15H05883) and Program for Advancing Strategic International Networks to Accelerate the Circulation of Talented Researchers (No. R2604) from the Japanese Society for the Promotion of Science.

References

[1] Yuan H Q, Grosche F M, Deppe M, Geibel C, Sparn G and Steglich F 2003 *Science* **302** 2104–2107
[2] Bianchi A, Movshovich R, Vekhter I, Pagliuso P G and Sarrao J L 2003 *Phys. Rev. Lett.* **91** 257001
[3] Mathur N D, Grosche F M, Julian S R, Walker I R, Freye D M, Haselwimmer R K W and Lonzarich G G 1998 *Nature* **394** 39–43
[4] Custers J, Gegenwart P, Geibel C, Steglich F, Coleman P and Paschen S 2010 *Phys. Rev. Lett.* **104** 186402
[5] Nakai H, Ebihara T, Tsutsui S, Mizumaki M, Kawamura N, Michimura S, Inami T, Nakamura T, Kondo A, Kindo K, and Matsuda Y H 2013 *J. Phys. Soc. Jpn.* **82** 124712
[6] Friedemann S, Westerkamp T, Brando M, Oeschler N, Wirth S, Gegenwart P, Krellner C, Geibel C and Steglich F 2009 *Nature Phys.* **5** 465–469
[7] Matsumoto Y, Nakatsuji S, Kuga K, Karaki Y, Horie N, Shimura Y, Sakakibara T, Nevidomskyy A H and Coleman P 2011 *Science* **331** 316–319
[8] Okawa M, Matsunami M, Ishizaka K, Eguchi R, Taguchi M, Chainani A, Takata Y, Yabashi M, Tamasaku K, Nishino Y, Ishikawa T, Kuga K, Horie N, Nakatsuji S and Shin S 2010 *Phys. Rev. Lett.* **104** 247201
[9] Nakatsuji S, Kuga K, Machida Y, Tayama T, Sakakibara T, Karaki Y, Ishimoto H, Yonezawa S, Maeno Y, Pearson E, Lonzarich G, Balicas L, Lee H and Fisk Z 2008 *Nature Phys.* **4** 603–607
[10] Macaluso R T, Nakatsuji S, Kuga K, Thomas E L, Machida Y, Maeno Y, Fisk Z and Chan J Y 2007 *Chem. Mater* **19** 1918–1922
[11] Mori N, Takahashi H and Takeshita N 2004 *High Press. Res.* **24** 225–232
[12] Garfinkel M and Mapother D E 1961 *Phys. Rev.* **122** 459–468
[13] Levy M and Olsen J L 1964 *Solid State Communications* **2** 137–139
[14] Tomita T, Kuga K, Uwatoko Y, Coleman P and Nakatsuji S 2015 *Science* **349** 506–509
[15] Tomita T, Kuga K, Uwatoko Y and Nakatsuji S 2015 *Journal of Physics: Conference Series* **592** 012019
[16] Kuga K, Karaki Y, Matsumoto Y, Machida Y and Nakatsuji S 2008 *Phys. Rev. Lett.* **101** 137004
[17] Senthil T, Vojta M and Sachdev S 2004 *Phys. Rev. B* **69** 035111

Journal of Physics: Conference Series **683** (2016) 012008 doi:10.1088/1742-6596/683/1/012008

Site-selective ^{11}B NMR studies on $YbAlB_4$

S Takano[1], M S Grbic[1], K Kimura[1], M Yoshida[1], M Takigawa[1], E C T O'Farrell[1], K Kuga[1], S Nakatsuji[1] and H Harima[2]

[1] Institute for Solid State Physics, University of Tokyo, Kashiwa, 277-8581, Japan

[2] Graduate School of Science, Kobe University, Kobe 657-8501, Japan

E-mail: shun.takano@issp.u-tokyo.ac.jp

Abstract. β-$YbAlB_4$ is a distinctive heavy fermion superconductor that exhibits unconventional quantum criticality without tuning in a strongly intermediate valence state. In this paper, we report the result of ^{11}B NMR measurements on the single crystals of β-$YbAlB_4$ and α-$YbAlB_4$, the locally isostructural polymorph of β-$YbAlB_4$. All ^{11}B NMR lines for both samples were successfully assigned to inequivalent crystallographic sites by comparing the experimental results and the ab-initio calculation of the electric field gradient. In both compounds, the anisotropy of the Knight shift exhibits a characteristic radial pattern, indicating approximate axial symmetry of the hybridization between the Yb-4f electrons and the conduction electrons.

1. Introduction

Heavy fermion systems have been studied over decades with great interest in unconventional superconductivity and non-Fermi liquid behavior in the vicinity of quantum critical points (QCP). β-$YbAlB_4$ is the first Yb-based heavy fermion superconductor (T_c = 80 mK) [1] and exhibits pronounced non-Fermi liquid behavior which can be seen in the temperature dependences of the resistivity ($\rho \sim T^{1.5}$ up to 0.1 K), the susceptibility ($\chi \sim T^{-0.5}$ up to 2 K), and the magnetic part of the specific heat ($C_M/T \sim \ln T$ up to 3 K) in zero magnetic field and ambient pressure [2]. Moreover, the hard -ray photoemission spectroscopy (H PES) measurement has revealed an intermediate valence of $Yb^{2.75+}$, providing the first example of quantum criticality in mixed valence systems [3]. On the other hand, α-$YbAlB_4$, the locally isostructural polymorph of β-$YbAlB_4$, shows conventional Fermi liquid behavior in zero magnetic field though α-$YbAlB_4$ also has an intermediate valence ($Yb^{2.73+}$ [3]) and similar crystal structure with β-$YbAlB_4$ [4] (figure 1).

Figure 1 shows the crystal structure of α-$YbAlB_4$ and β-$YbAlB_4$ [4]. Both polymorphs of $YbAlB_4$ have orthorhombic structure but with different space group, *Pbam* for α-$YbAlB_4$ and *Cmmm* for β-$YbAlB_4$. In both cases, the Yb and Al atoms reside within the same *ab*-plane which is sandwiched between two B layers. Yb atoms are centered between two heptagonal B-rings, and Al atoms are centered between pentagonal B-rings. The local structure of the two polymorphs are quite similar but distinguished by the packing pattern of the Yb hexagonal rings in the *ab*-plane. In this paper, we report the results of ^{11}B NMR measurements on the single crystals of both polymorphs.

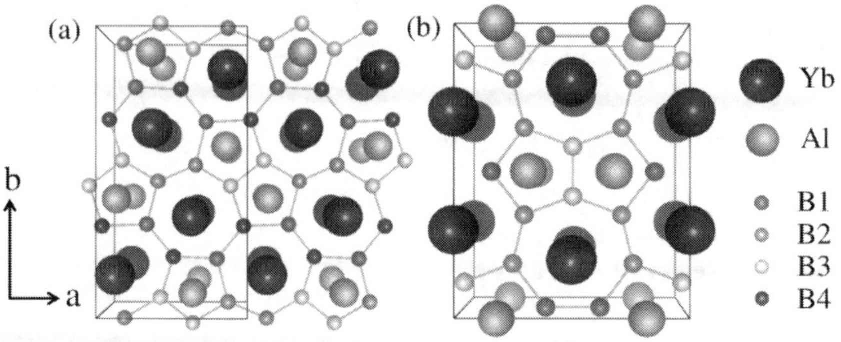

Figure 1. The crystal structure of (a)α-$YbAlB_4$ and (b)β-$YbAlB_4$. Solid lines represent a unit cell.

2. Experiment

Single crystals of α-$YbAlB_4$ and β-$YbAlB_4$ were prepared by Al self flux method reported in [4]. We can distinguish these polymorphs by the typical sizes of the crystal, $1 \times 1 \times 5$ mm^3 for α-$YbAlB_4$ and $1 \times 1 \times 0.1$ mm^3 for β-$YbAlB_4$ though both materials are synthesized together. The ^{11}B NMR spectra were obtained by Fourier transforming the spin echo signals at discrete frequencies and summing them up. The measurements for α-$YbAlB_4$ (β-$YbAlB_4$) were performed in a magnetic field of 6.0 T (5.0 T) and at $T = 20$ K (4.2 K).

3. Results

α-$YbAlB_4$ (β-$YbAlB_4$) has four (three) crystallographically inequivalent B sites, B1-B4 (B1-B3) as shown in figure 1. The Wyckoff numbers and site symmetries of these sites are summarized in table 1. Figure 2(a) (2(b)) shows ^{11}B NMR spectra for $\alpha(\beta)$-$YbAlB_4$ observed at 3.0 K (4.2 K) under the magnetic field applied along the c-direction. Since ^{11}B nuclei have spin $I = 3/2$, the electric field gradient (EFG) splits the NMR spectrum into equally spaced three lines due to the first order effect of the nuclear quadrupole interaction (quadrupole splitting). Then the spectrum can be decomposed of 4 (3) sets of resonance lines denoted A-D (E-G). In order to assign these four (three) sets of resonance lines to four (three) B sites, we measured the spectra under the field rotated in the ab-plane.

Figures 3(a) and 3(b) show the angle dependences of ^{11}B NMR frequencies under the magnetic field in the ab-plane. In general, resonance frequency $\nu_m(\theta)$ should obey the following equation.

$$\nu_m = (1 + K(\theta))\gamma H_0 + (m - 1/2)\nu_Q(\theta) \quad (1)$$

Table 1. The list of the Wycoff number, the site symmetry and the direction(s) of the principal axis (axes) for each B site.

	α-$YbAlB_4$				β-$YbAlB_4$		
Site	B1	B2	B3	B4	B1	B2	B3
Wyckoff number	$4h$	$4h$	$4h$	$4h$	$4h$	$8q$	$4j$
Site symmetry	$..m$	$..m$	$..m$	$..m$	$2mm$	$..m$	$m2m$
Principal axes	c	c	c	c	a, b and c	c	a, b and c

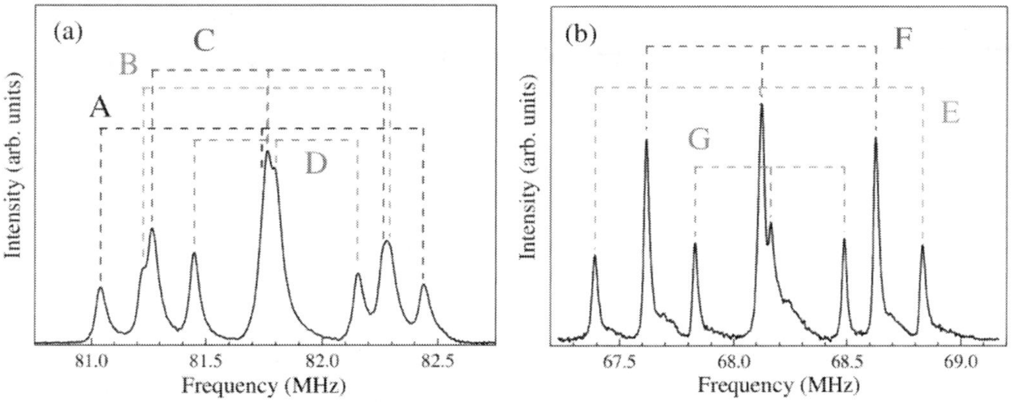

Figure 2. ^{11}B NMR spectra for (a)α-$YbAlB_4$ ($H\|c, H = 6.0$ T, $T = 3.0$ K), and (b)β-$YbAlB_4$ ($H\|c, H = 5.0$ T, $T = 4.2$ K).

Here, γ is the nuclear gyromagnetic ratio ($^{11}\gamma = 13.66012$ MHz/T), θ is the angle between the magnetic field and the a-axis, $K(\theta)$ is the Knight shift, $\nu_Q(\theta)$ is the quadrupole splitting, and m specifies the nuclear transitions $I_z = m \leftrightarrow m-1$. Only the frequencies of quadrupole satellite lines ($m = 3/2$ or $m = -1/2$) are plotted in figures 3(a) and 3(b) since the center lines ($m = 1/2$) from different sites overlap and cannot be well resolved. We also note that, except for B1 and B3 sites for β-$YbAlB_4$, the resonance lines from individual sites split into two sets of lines under the magnetic field in the ab-plane (see the discussion below). $K(\theta)$ and $\nu_Q(\theta)$ are determined by the following equations.

$$K(\theta) = (\nu_{3/2} + \nu_{-1/2} - 2\gamma H_0)/2\gamma H_0 \quad (2)$$
$$\nu_Q(\theta) = (\nu_{3/2} - \nu_{-1/2})/2 \quad (3)$$

The angle dependences of $K(\theta)$ and $\nu_Q(\theta)$ are shown in figures 3(c)-(f), which are well fitted by the following equations.

$$K(\theta) = K_{xx} + (K_{yy} - K_{xx})\cos^2(\theta - \theta_K) \quad (4)$$
$$\nu_Q(\theta) = \nu_{xx} + (\nu_{\alpha\alpha} - \nu_{xx})\cos^2(\theta - \theta_Q) \qquad (\alpha = y \text{ or } z) \quad (5)$$

The directions of the principal axes (PA), x, y and z, are defined to satisfy the relations $|K_{xx}| \le |K_{yy}| \le |K_{zz}|$ and $|\nu_{xx}| \le |\nu_{yy}| \le |\nu_{zz}|$. Here, θ_K and θ_Q indicate the y-direction of the Knight shift tensor and α-direction for EFG tensor respectively. α represents the y-axis of the EFG tensor of all B sites except for D and G. For these sites, the y-axis of the EFG tensor is the c-axis and α becomes the z-axis. We summarize the principal values, the directions of the PA, and anisotropic parameters $\eta = (\nu_{yy} - \nu_{xx})/\nu_{zz}$ obtained by the fit in table 2.

In case of β-$YbAlB_4$, a- and b-axes are the PA of both the Knight shift and the EFG tensors for B1 and B3 sites whereas they are not for the B2 site. Therefore, $K(\theta)$ and $\nu_Q(\theta)$ for B2 site show neither a maximum nor a minimum along the a- and b-axes. Since the crystal structure has the mirror symmetry with respect to the a- and b-planes, the angle dependence of the whole spectrum must be symmetric about a- and b-axes. Therefore, a second set of resonance lines originating from the B2 site must exist to satisfy this requirement, as is indeed observed experimentally (solid and dashed lines in figure 3). Then, the set of lines F is immediately assigned to B2 but the other B sites (E and G) cannot be assigned by considering only the site symmetry. NMR spectra for all B sites (B1-B4) of α-$YbAlB_4$, on the other hand, exhibit qualitatively the same angle dependences with B2 site of β-$YbAlB_4$ since their site symmetries are the same and the crystal structure has the glide symmetries with respect to the a- and b-planes.

It has been recognized that the state-of-art technique of ab-initio calculation based on the density functional theory provides reliable estimate of the EFG [5, 6]. Table 3 shows the results of ab-initio calculation of the quadrupole splitting parameters $^{11}\nu^{cal}$, $|\theta_Q^{cal}|$ and η^{cal}. Comparison of the experimental and the calculated results of the EFG parameters immediately allows us to assign all the B sites as shown in the last line of table 2.

Table 2. The list of the principal values and the angle of the y-axis from a-axis of the Knight shift tensor (upper part). The principal values, the angles of the y-axis (z-axis for D and G) and the anisotropic parameters η of the EFG tensor are listed in the middle part. The last line shows crystallographic sites assigned with A-G.

		α-$YbAlB_4$				β-$YbAlB_4$		
Site		A	B	C	D	E	F	G
$^{11}K_{xx}$	[ ]	0.006	−0.022	−0.026	−0.025	0.014	−0.013	−0.018
$^{11}K_{yy}$	[ ]	0.083	0.103	0.097	0.093	0.103	0.106	0.104
$^{11}K_{cc}$	[ ]	−0.248	−0.234	−0.228	−0.198	−0.313	−0.369	−0.384
$\|\theta_K\|$	[deg.]	37.3	71.0	34.3	36.4	90.0	20.5	90.0
$^{11}\nu_{xx}$	[MHz]	−0.296	−0.067	−0.052	−0.032	−0.322	−0.020	−0.018
$^{11}\nu_{yy}$	[MHz]	−0.396	−0.464	−0.440	−0.378 ($y \parallel c$)	−0.397	−0.487	−0.329 ($y \parallel c$)
$^{11}\nu_{zz}$	[MHz]	0.701 ($z \parallel c$)	0.532 ($z \parallel c$)	0.500 ($z \parallel c$)	0.379	0.720 ($z \parallel c$)	0.506 ($z \parallel c$)	0.343
$\|\theta_Q\|$	[deg.]	24.6	4.5	75.4	54.4	90.0	53.9	0.0
η		0.143	0.747	0.775	0.913	0.105	0.922	0.907
Site		B3	B2	B4	B1	B3	B2	B1

Table 3. The results of ab-initio calculation for the quadrupole splitting parameters. $|\theta_Q^{cal}|$ indicates angles of y-axis from a-axis.

		α-$YbAlB_4$				β-$YbAlB_4$		
Site		B1	B2	B3	B4	B1	B2	B3
$^{11}\nu_{xx}^{cal}$	[MHz]	−0.034	−0.129	−0.352	−0.134	−0.086	−0.092	−0.378
$^{11}\nu_{yy}^{cal}$	[MHz]	−0.446	−0.542	−0.484	−0.488	−0.361	−0.563	−0.507
$^{11}\nu_{cc}^{cal}$	[MHz]	0.480	0.670	0.836	0.621	0.447	0.655	0.885
$\|\theta_Q^{cal}\|$	[deg.]	52.8	1.5	0.2	73.5	0.0	33.9	90.0
η^{cal}		0.858	0.616	0.159	0.570	0.615	0.719	0.146

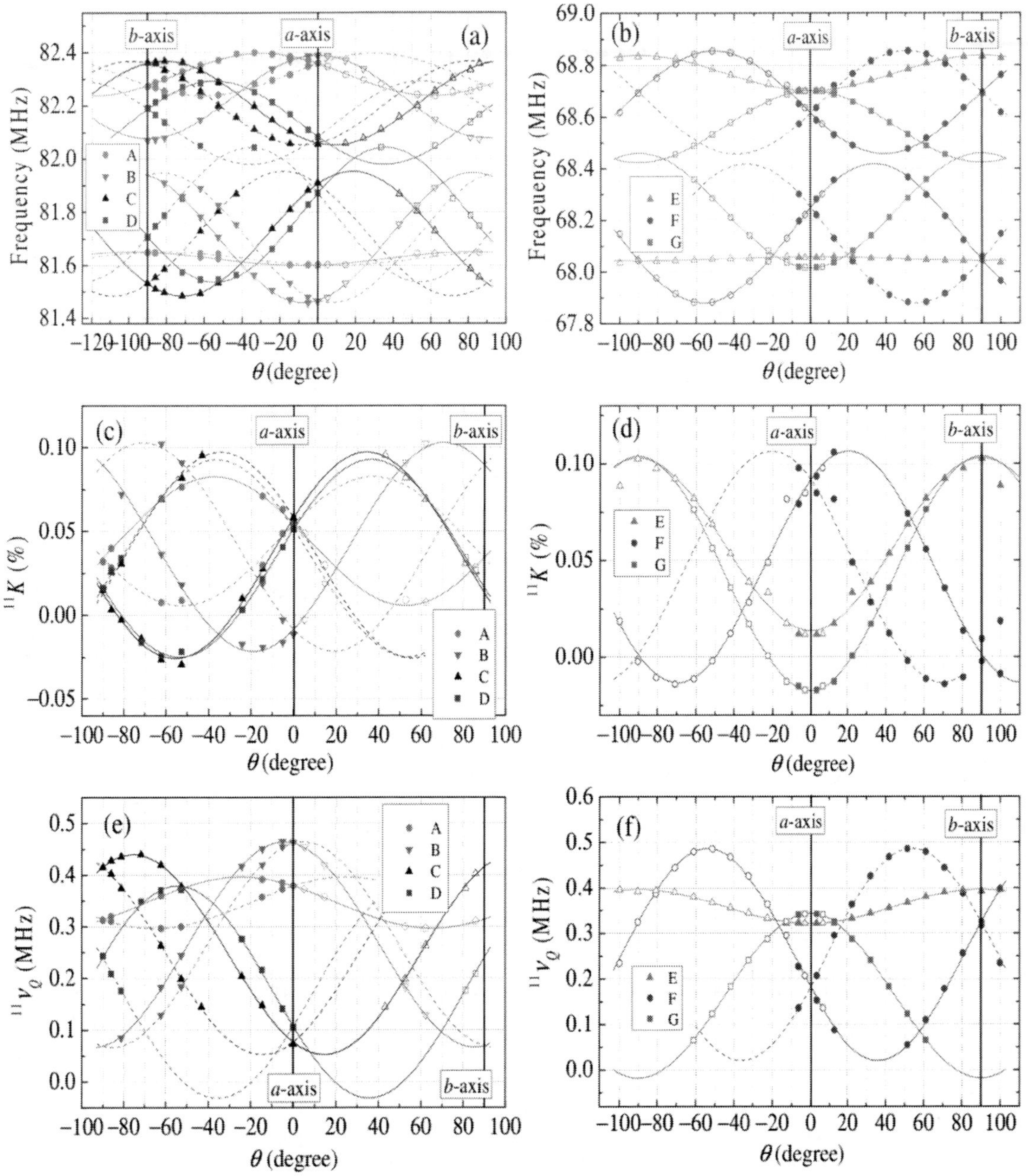

Figure 3. The angle dependences of NMR frequency ((a), (b)), the Knight shift ((c), (d)) and the quadrupole splitting ((e), (f)). (a), (c) and (e) are data for α-$YbAlB_4$, (b), (d) and (f) for β-$YbAlB_4$. Observed data are shown by the solid symbols while open symbols are the mirror reflection around a-axis. NMR frequency, $K(\theta)$ and $\nu_Q(\theta)$ are fitted by equation (1), (4) and (5), respectively as shown by solid and dashed lines.

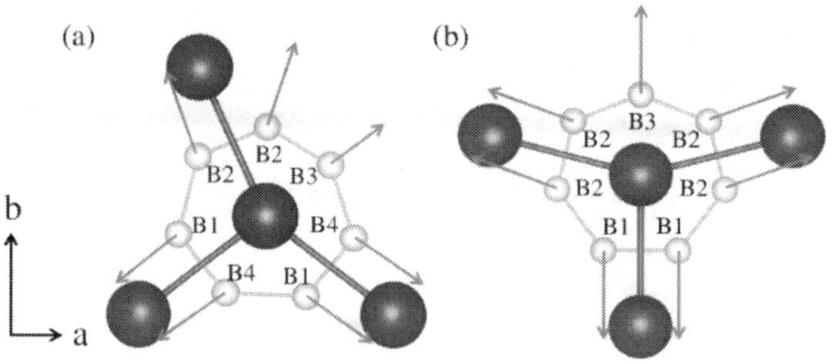

Figure 4. The principal values K_{yy}, and the directions of the y-axis for the B sites of (a)α-$YbAlB_4$, (b)β-$YbAlB_4$. The length of the red arrows represent the magnitude of K_{yy}. The direction towards outside of heptagonal B ring is defined as positive.

4. Discussions

The principal values of the Knight shift satisfy the relation $|K_{cc}| \gg |K_{yy}| \gg |K_{xx}|$ for all B sites. The value of K_{cc} does not depend much on the sites, $K_{cc} = -0.2 \sim -0.25$ for α-$YbAlB_4$ and $K_{cc} = -0.31 \sim -0.38$ for β-$YbAlB_4$. This is also the case for K_{yy} ($0.08 \sim 0.10$ for α-$YbAlB_4$ and $0.09 \sim 0.10$ for β-$YbAlB_4$). However, the direction of the y-axis (θ_K) varies significantly with the sites. In order to visualize the anisotropy of the Knight shift, we show the magnitude of K_{yy} and the direction of the y-axis by the red arrows in figure 4. We can see that y-axis directs almost radially outward with respect to the heptagonal B ring for both polymorphs. This suggests that hybridization between the Yb-4f states and conduction B states has approximate axial symmetry. Yb-4f orbital has axial symmetry around c-axis and directs towards B-ring which surrounds Yb atom and the orbital may hybridize with s or p orbitals of conduction B atoms. This result is expected from the local crystal structure around Yb atoms and considered in the theoretical models [7, 8]. This also suggests that we can expect overall similar electronic structure for both α-$YbAlB_4$ and β-$YbAlB_4$, even though physical properties at low temperatures and low magnetic fields are quite different.

Acknowledgments

This work was supported by Grant-in Aids for the Japan Society for the Promotion Science (JSPS) KAKENHI (B) (No. 25287083). Shun Takano was supported by the JSPS through the Program for Leading Graduate Schools (MERIT, Materials Education program for the future leaders in Research, Industry, and Technology).

References

[1] Nakatsuji S, Kuga K, Machida Y, Tayama T, Sakakibara T, Karaki Y, Ishimoto H, Yonezawa S, Maeno Y, Pearson E, Lonzarich G G, Balicas L, Lee H and Fisk Z 2008 *Nature Physics* **4** 603
[2] Matsumoto Y, Nakatsuji S, Kuga K, Karaki Y, Horie N, Shimura Y, Sakakibara T, Nevidomskyy A H and Coleman P 2011 *Science* **331** 316
[3] Okawa M, Matsunami M, Ishizaka K, Eguchi R, Taguchi M, Chainani A, Takata Y, Yabashi M, Tamasaku K, Nishino Y, Ishikawa T, Kuga K, Horie N, Nakatsuji S, and Shin S 2010 *Phys. Rev. Lett.* **104** 247201
[4] Macaluso R T, Nakatsuji S, Kuga K, Thomas E L, Machida Y, Maeno Y, Fisk Z and Chan J Y 2007 *Chem. Mater.* **19** 1918
[5] Blaha P, Schwarz K and Herzig P 1985 *Phys. Rev. Lett.* **54** 1192
[6] Harima H 2006 *Physica B* **378** 246
[7] Nevidomskyy A H and Coleman P 2009 *Phys. Rev. Lett.* **102** 077202
[8] Ramires A, Coleman P, Nevidomskyy A H and Tsvelik A M 2012 *Phys. Rev. Lett.* **109** 176404

Journal of Physics: Conference Series **683** (2016) 012009 doi:10.1088/1742-6596/683/1/012009

High Magnetic Transition Temperature and Semiconductor-like Transport Properties of Mn-doped α-$YbAlB_4$

S Suzuki[1], T Tomita[1], Y Shimura[1], K Kuga[2], Y Matsumoto[1] and S Nakatsuji[1]

[1]Institute for Solid State Physics, University of Tokyo, Kashiwa, Chiba 277-8581, Japan
[2]RIKEN SPring-8 Center, Sayo, Hyogo 679-5148, Japan

E-mail: s_suzuki@issp.u-tokyo.ac.jp

Abstract. We succeeded in growing single crystals of α-$YbAl_{1-x}Mn_xB_4$ by the substitution of Mn for the Al site in the intermediate-valence rare-earth heavy fermion system α-$YbAlB_4$. We performed specific heat and resistivity measurement on single crystals of α-$YbAl_{1-x}Mn_xB_4$ (x = 0.40). In the specific heat measurement on α-$YbAl_{0.60}Mn_{0.40}B_4$, a sharp peak was observed at magnetic transition temperature T_M = 9.7(6) K. The resistivity measurements on x = 0.40 also show a semiconductor-like behavior in the entire temperature range from 2 - 300 K. An inclination change was also observed at 10(1) K in resistivity measurement. These results indicate a magnetic order. The magnetic transition temperature is quite high compared to its temperatures observed in the other Yb-based heavy fermion compounds.

1. Introduction

The strongly correlated electron systems such as Cu- and Fe-based high-T_c superconductors and heavy fermion materials show a rich variety of interesting phenomena. Particularly, the various rare earth heavy fermion (HF) compounds with large effective mass have been reported to show exotic behaviors at low temperatures [1-3]. The ground states in the typical HF systems are essentially understood within the framework of the Doniach's phase diagram. When an energy scale characterized by Kondo effect (T_K) is larger than that by a RKKY interaction scale (J_{RKKY}), its ground state produces a non-magnetic Fermi liquid state. In contrast, when $T_K < J_{RKKY}$, the ground state becomes magnetic ordered state. As these two energy scales competes at absolute zero temperature, a second order phase transition emerges at a so-called quantum critical point (QCP). Near this QCP, a new type of metallic state not obeying Landau Fermi liquid theory and unconventional superconductivity emerge. Various characteristic energy scales (i.e., T_K, magnetic transition temperature $T_{M...}$) associated with their large effective mass in the HF rare-earth systems are relatively low compared with those in strongly correlated 3d-electron systems, namely the ground state is easily tuned by non-thermal control parameters. Therefore, the phenomena near the QCP in the HF systems have been investigated with applying pressure, magnetic field and chemical doping mostly in the Ce-based HF systems [1,2]. Some of these properties around a QCP are successfully described by Self-Consistent-Renormalization (SCR) theory in mostly Ce-based HF materials [4,5]. The Yb-based HF systems can be viewed as having one 4f-hole for Yb^{3+} in contrast to one 4f-electron for Ce^{3+}. So, the Yb-based HF systems are expected to have QCP interpreted as one hole system and had been less studied about the behaviors around QCP compared with the Ce-based system. A prototypical Yb-based HF system, $YbRh_2Si_2$, has a weak antiferromagnetism with T_N = 65 mK and has been intensively studied as a unconventional antiferromagnetic quantum criticality [6]. The 5% Ge-doped sample indeed exhibits a reduction of T_N and is found to show non-Fermi liquid behaviors [7].

β-$YbAlB_4$ is the first example showing HF superconductivity (T_C = 80 mK) in the Yb-based HF family [8,9]. This material also shows non-Fermi liquid behaviors without tuning any control parameter such as composition, pressure or magnetic field [10]. The temperature dependences of the electrical resistivity, magnetization, and specific heat exhibit nontrivial temperature-power law exponents that are not able to be explained by the conventional SCR theory [8-10]. In contrast, α-$YbAlB_4$, an isomorphism phase of β-$YbAlB_4$, shows the Kondo lattice and Fermi liquid behavior at low temperatures [11]. The specific heat at low temperatures becomes almost constant with Sommerfeld coefficient $\gamma \sim 130$ mJ/mol-K^2. The magnetic susceptibility also shows saturating tendency below 8 K. The temperature dependence of the resistivity, moreover, exhibits *T*-square below 100 mK. Notably, these two compounds both show valence fluctuation behavior (non-integer state) with $Yb^{2.73(2)+}$ for α-$YbAlB_4$ and $Yb^{2.75(2)+}$ for β-$YbAlB_4$ as revealed by the Hard X-ray Photoemission Spectroscopy measurements [12]. As usual, Kondo lattice and valence fluctuation behaviors do not coexist. An average occupation number of *f*-electron states is close to an integer in the typical HF system forming the Kondo lattices. The valence fluctuation behavior has never been seen in the Ce-based HF compounds.

In order to investigate the intriguing coexistence of Kondo lattice system and valence fluctuating system, the chemical doping effect of the Fe substitution on Al site in both α and β-$YbAlB_4$ has been studied [13]. The chemical substitution mainly serves as the combination of electron doping and the chemical pressure effect. In a previous study, α-$YbAl_{0.93}Fe_{0.07}B_4$ shows a kink at 7.5(5) K in the magnetic susceptibility measurements and a peak at 6.7(3) K in the heat capacity measurements, respectively [13]. These results suggest forming an antiferromagnetic order below ~7 K. The magnetism should not arise from the Fe ion because the magnetization in α-$LuAl_{0.79}Fe_{0.21}B_4$ does not detect any magnetic transition. X-ray diffraction measurement supports that the Fe doping can be regarded as a positive chemical pressure. This chemical pressure effect is reported on $YbCu_{5-x}Al_x$ as well [14]. If the Doniach's picture is applicable in α-$YbAl_{1-x}Fe_xB_4$, it is expected that the QCP emerges in the range $0 < x < 0.07$. This T_N is high compared with that of the other Yb-based HF systems [6,14-16].

On the periodic table, since Fe is not located in the vicinity of Al, it is difficult to judge whether the electron doping effect or the chemical pressure is important to understand the mechanism of high T_N antiferromagnetism. We need to study the effect in other substitution. Here, we synthesized with using Mn as a chemical doping instead of Fe in order to compare these two effects.

2. Experimental

We have grown single crystals of α-$YbAl_{1-x}Mn_xB_4$ ($0 \leq x \leq 0.57$) by excess Al flux method. The pure elements, Yb (Ames laboratory, ingot 99.99%), Al (Kyusyu-Mitsui Aluminum Co., Ltd., piece, 99.999%), B (Ceradyne. Inc., piece, 99.9999%), and Mn (Rare Metallic Co., Ltd., piece, 99.99%), were placed into an alumina crucible. The crucible and its contents were heated under an Ar atmosphere. After the crucible was removed from the furnace, excess aluminum flux was removed by etching in 1 mol/*l* NaOH solution. One of the single crystals we synthesized is shown in figure 1. The growth direction in 40% Mn-doped crystal is preferentially along *ab*-plane, which is different from the non-doped crystal grown along *c*-axis. Our synthesized samples α-$YbAl_{1-x}Mn_xB_4$ were confirmed to have same orthorhombic structure as α-$YbAlB_4$ from powder X-ray diffraction pattern. This measurement also detected the lattice parameters as a = 5.9006(2), b = 11.411(1), c = 3.4401(1) Å for α-$YbAl_{0.60}Mn_{0.40}B_4$. From the lattice parameters (a = 5.9220(2), b = 11.4730(3), c = 3.5060(5) Å) for α-$YbAlB_4$ [17], the volume in Mn doping affects positive chemical pressure as observed in Fe doping. The Mn concentration x was estimated by Inductively Coupled Plasma Atomic Emission Spectroscopy (ICP-AES) within 0.6% error. To evaluate the physical properties, temperature dependence of electric resistivity and specific heat were measured with a commercial physical property measurement system

Figure 1. Photograph of a single crystal of α-$YbAl_{0.60}Mn_{0.40}B_4$ grown in Al self-flux. Blue squares indicate 1 mm × 1 mm size.

(Quantum Design, PPMS) on α-$YbAl_{0.60}Mn_{0.40}B_4$ because single crystals with enough size were obtained to measure specific heat and resistivity in both along *ab*-plane and *c*-axis. The specific heat was obtained using the relaxation method. For the resistivity, standard four-probe method was used. Electrical contacts were made to the crystal by spot welding technique. The electrical current I of 0.5 mA was applied parallel to *ab*-plane and *c*-axis in the temperature range of 2-300 K.

3. Results and Discussion

We performed the specific heat (C) and electrical resistivity (ρ) measurements on α-$YbAl_{0.60}Mn_{0.40}B_4$ single crystals. Figure 2 shows the specific heat data divided by temperature for 2 K $\leq T \leq$ 40 K. Here, the magnetic contribution C_{mag}/T was calculated by subtracting C/T of α-$LuAlB_4$ from that of α-$YbAl_{0.60}Mn_{0.40}B_4$. α-$YbAlB_4$ data was taken from the reference [11]. As shown in Fig. 2, a peak was observed at T_M = 9.7(6) K across a transition. The same type of the peak is also observed in the heat capacity measurements for α-$YbAl_{1-x}Fe_xB_4$ and is confirmed to be an antiferromagnetic transition temperature from its magnetization results [13]. Assuming that this peak comes from antiferromagnetic order, the T_N for Mn doping is higher than the maximum of T_N (~7 K) for α-$YbAl_{0.93}Fe_{0.07}B_4$ [13]. Additionally, the T_N value is higher than that of the typical Yb-based HF systems, i.e., $YbRh_2Si_2$ (T_N = 65 mK) [6], $YbCu_3Al_2$ (T_N = 2 K) [14], $YbNi_2Ge_2$ (T_N = 2.1 K) [15], $YbAuCu_4$ (T_N = 0.6 K) [16], and $YbPbCu_4$ (T_N = 0.8 K) [16]. In high pressure transport study of β-$YbAlB_4$, it is reported that a magnetic phase transition is observed and reached 30 K under pressure of 8 GPa [18]. In the chemical substitution study of β-$YbAl_{1-x}Fe_xB_4$ (x = 0.06), the same type of the magnetic phase transitions are also observed at 9.5(5) K under ambient pressure and reached 20 K under pressure of 4 GPa [13,19]. The origin of such a high magnetic transition temperature with a 2-digits order in the β-phase may be related with that in pure α-phase and its Fe or Mn doped systems.

The temperature dependences of the electrical resistivity both along *ab*-plane (in-plane) and along *c*-axis of α-$YbAl_{0.60}Mn_{0.40}B_4$ are shown in Figure 3. α-$YbAlB_4$ data were taken from the reference [11]. The resistivity value at room temperature (300 K) is over 500 μΩ cm both in *ab*-plane and in *c*-axis of α-$YbAl_{0.60}Mn_{0.40}B_4$. These resistivity values are higher than those of pure α-$YbAlB_4$, where the Mn-doped resistivity values are 2 times larger than non-doped values for the *ab*-plane, 14 times larger in *c*-axis components, respectively. Moreover, the pure α-$YbAlB_4$ indicates $\rho_{ab} > \rho_c$ in contrast to $\rho_{ab} < \rho_c$

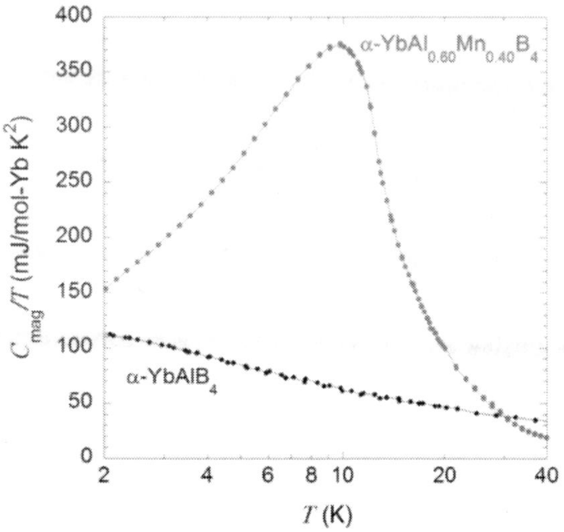

Figure 2. The zero-field specific heats divided by temperature C_{mag}/T vs. T are shown for α-$YbAl_{0.60}Mn_{0.40}B_4$ (red) and α-$YbAlB_4$ (black).

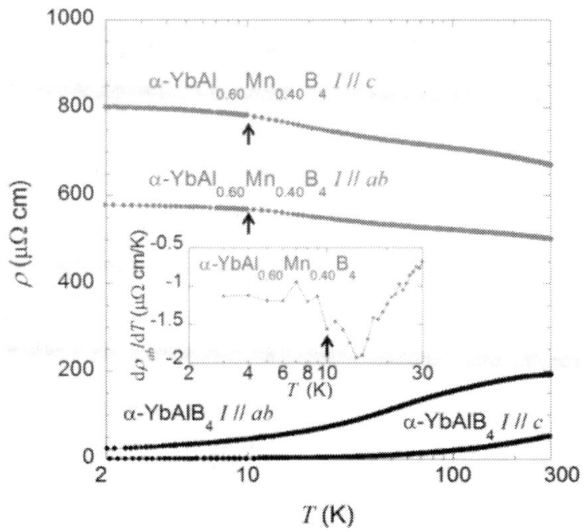

Figure 3. The temperature dependences of the resistivities of α-$YbAl_{0.60}Mn_{0.40}B_4$ (red) and α-$YbAlB_4$ (black) along *ab*-plane and *c*-axis at zero field. The inset shows temperature derivative of resistivity of α-$YbAl_{0.60}Mn_{0.40}B_4$ along *ab*-plane at zero field. Arrows at 10(1) K refer the point where inclination changes of resistivity which is defined by the temperature derivative of resistivity.

in the Mn-doped α-$YbAlB_4$. Interestingly, the resistivity value in α-$YbAl_{0.60}Mn_{0.40}B_4$ shows semiconducting-like behavior in contrast to the metallic-like one in α-$YbAlB_4$. As usual, applying pressure makes band touching, so this effect makes materials more metallic [20]. The inset of Figure 3 shows temperature derivative of resistivity of α-$YbAl_{0.60}Mn_{0.40}B_4$ along *ab*-plane. A decreasing of the slope in $\rho(T)$ both along the *ab*-plane and along the *c*-axis is observed at 10(1) K around T_M obtained from the peak of the specific heat measurement. The slope change of the $\rho(T)$ around T_M also indicates that the scattering of electron is suppressed with forming magnetic order.

4. Summary

We have obtained single crystals that are partially substituted with Mn for Al in α-$YbAlB_4$. The Mn doping provides a positive chemical pressure effect as in the Fe doping case. The heat capacity measurements on α-$YbAl_{0.60}Mn_{0.40}B_4$ founds a magnetic transition at T_M = 9.7(6) K. This high magnetic transition temperature is one of the characteristic properties of α,β-$YbAlB_4$. The electrical resistivities under chemical pressure behave as semiconductor-like as shown in α-$YbAl_{0.60}Mn_{0.40}B_4$, even α-$YbAlB_4$ remains metallic under external pressure [21]. At 10(1) K, there are incremental changes in $\rho(T)$ both along *ab*-plane and along *c*-axis. It comes from forming magnetic order which suppresses electronic scattering.

Here, there is one problem about origin of semiconductor-like transport. We need to think about two effects for Mn doping, namely doping induced lattice disorder and conduction-electron contributions. Specially, in the latter case, assuming that it comes from the hybridization between conduction electrons and localized *f*-electrons, α-$YbAl_{1-x}Mn_xB_4$ is one of Kondo insulators with orthorhombic structure. All Kondo insulators had been observed in only cubic structure, such as YbB_{12}

[22], SmB_6 [23] and so on, until Kondo insulator-like behaviors were observed in CeNiSn and CeT_2Al_{10} (T = Ru, Os) with orthorhombic structure [24-26]. The nodal hybridization theory pointed out that transport behavior in α,β-$YbAlB_4$ can be the results of a doped Kondo insulator [27]. Therefore, we need further experiments, especially on transport, in the series of the Mn-doped α-$YbAlB_4$ to clarify the contribution of these two effects and to get wide knowledge for stabilizing a Kondo insulator from QCP in β-$YbAlB_4$.

5. Acknowledgement

We thank K. Sone, K. Takubo, M. Suzuki, and A. Nevidomskyy for fruitful discussion. This work was partially supported by Japan Society for the Promotion of Science through Program for Leading Graduate Schools (MERIT). This work is partially supported by PRESTO, Japan Science and Technology Agency, Grants-in-Aid for Scientific Research (No. 25707030), by Grants-in-Aids for Scientific Research on Innovative Areas (15H05882, 15H05883) and Program for Advancing Strategic International Networks to Accelerate the Circulation of Talented Researchers (No. R2604) from the Japanese Society for the Promotion of Science.

References

[1] Yuan H Q, Grosche F M, Deppe M, Geibel C, Spern G and Steglich F 2003 *Science* **302** 2104
[2] Paglione J, Tanatar M A, Hawthorn D G, Boaknin. E, Hill R W, Ronning F, Sutherland M, Taillefer L, Petrovic C and Canfield P C 2003 *Phys. Rev. Lett.* **91** 246405
[3] Petrovic C, Pagliuso P G, Hundley M F, Movshovich R, Sarrao J L, Thompson J D, Fisk Z and Monthoux P 2001 *J. of Phys: Cond. Mat.* **13** L337
[4] Stockert O, Arndt J, Faulhaber E, Geibel C, Jeevan H S, Kirchner S, Loewenhaupt M, Schmalzl K, Schmidt W, Si Q and Steglich F 2011 *Nat. Phys.* **7** 119–123
[5] Regnault L P, Erkelens W A C, Rossat-Mignod J, Lejay P and Flouquet J 1988 *Phys. Rev.* B **38** 4481
[6] Trovarelli O, Gaibel C, Mederle S, Langhammer C, Grosche F M, Gegenwart P, Lang M, Sparn G and Steglich F 2000 *Phys. Rev. Lett.* **85** 626
[7] Custers J, Gegenwart P, Gaibel C, Steglich F, Coleman P and Paschen S 2010 *Phys. Rev. Lett.* **104** 186402
[8] Nakatsuji S, Kuga K, Machida Y, Tayama T, Sakakibara T, Karaki Y, Ishimoto H, Yonezawa S, Maeno Y, Pearson E, Lonzarich G, Balicas L, Lee H and Fisk Z 2008 *Nat. Phys.* **4** 603–607
[9] Kuga K, Karaki Y, Matsumoto Y, Machida Y and Nakatsuji S 2008 *Phys. Rev. Lett.* **101** 137004
[10] Matsumoto Y, Nakatsuji S, Kuga K, Karaki Y, Horie N, Shimura Y, Sakakibara T, Nevidomskyy A H and Coleman P 2011 *Science* **331** 316
[11] Matsumoto Y, Kuga K, Tomita T, Karaki Y and Nakatsuji S 2011 *Phys. Rev.* B **84** 125126
[12] Okawa M, Matsunami M, Ishizaka K, Taguchi M. Chainani A, Takata Y, Yabashi M, tamasaku K, Nishino Y, Ishikawa T, Kuga K, Horie N, Nakatsuji S and Shin S 2010 *Phys. Rev. Lett.* **104** 247201
[13] Kuga K, Morrison G,Treadwell L, Chan J Y, and Nakatsuji S 2012 *Phys. Rev.* B **86** 224413
[14] Bauer E, Hauser R, Keller L, Fischer P, Trovarelli O, Sareni J G, Rieger J J and Stewart G R 1997 *Phys. Rev B* **56** 711
[15] Bonville P, Hodges J A, Imbert P, Jéhanno G, Jaccard D and Sierro J 1990 *J of Mag. and Mag. Matt.* **97** 178-186
[16] Rossel C, Yang K N, Maple M B, Fisk Z, Zirngiebl E and Thompson J D 1987 *Phys. Rev B* **35** 1914
[17] Makaluso R T, Nakatsuji S, Kuga K, Thomas E L, Machida Y, Maeno Y, Fisk Z and Chan J Y 2007 *Chem. mater.* **19** 1918-1922
[18] Tomita T, Kuga K, Uwatoko Y and Nakatsuji S 2015 *J. of phys. Conf. Ser.* **592** 012019
[19] Tomita T, Kuga K, Uwatoko Y, Coleman P and S Nakatsuji 2015 *Science* **349** 506

[20] Matsumoto Y, Motoki S, Aoki K, Shimizu Y, Kimura N, Komatsubara T, Aoki H, Kimata M, Terashima T and Uji S 2011 *J of Phys Soc of Japan* **80** 074715
[21] Shimura Y *private communication*
[22] Kasaya M, Iga F, Negishi K, Nakai S and Kasuya T 1987 *J. of Mag. and Mag. Mat*. **31-34** 437-438
[23] Kasuya T, Takegahara K, Fujita T, Tanaka T and Bannai E 1979 *J. de Phys. Collo*. **40(C5)** C5-308-C5-313
[24] Takabatake T, Teshima F, Fujii H, Nishigori S, Suzuki T, Fujita T, Yamaguchi Y, Sakurai J and Jaccard D 1990 *Phys. Rev. B* **41** 9607
[25] Strydom A M 2009 *Physica B* **404** 2981
[26] Nishioka T, Kawamura Y, Takesaka T, Kobayashi R, Kato H, M Matsumura, K Kodama, Matsubayashi K and Uwatoko Y 2009 *J. Phys. Soc. Jpn.* **78** 123705
[27] Ramires A, Coleman P, Nevidomskyy A H and Tsvelik A M 2012 *Phys. Rev. Lett.* **109** 176404

Journal of Physics: Conference Series **683** (2016) 012010 doi:10.1088/1742-6596/683/1/012010

Quantum tricritical fluctuations driving mass enhancement and reentrant superconductivity in URhGe

Y. Tokunaga[1], D. Aoki[2,3], H. Mayaffre[4], S. Krämer[4], M.-H. Julien[4], C. Berthier[4], M. Horvatić[4], H. Sakai[1], S. Kambe[1], T. Hattori[1] and S. Araki[2,5]

[1]ASRC, Japan Atomic Energy Agency Tokai, Ibaraki 319-1195, Japan
[2]INAC/SPSMS, CEA-Grenoble/UJF, 38054 Grenoble, France
[3]IMR, Tohoku University, Ibaraki 311-1313, Japan
[4]LNCMI, UPR 3228, CNRS-UJF-UPS-INSA, 38042 Grenoble, France
[5]Department of Physics, Okayama University, Okayama 700-8530, Japan

E-mail: tokunaga.yo@jaea.go.jp

Abstract.
The field-induced reentrant superconductivity (RSC) discovered near a quantum critical point (QCP) in a ferromagnetic superconductor URhGe highlights the close interplay between superconductivity and magnetism. While the origin of the RSC is broadly thought to be associated with quantum critical fluctuations, their exact nature had not been well identified. Here we review our recent ^{59}Co NMR study in a single crystal of $URh_{0.9}Co_{0.1}Ge$. Our measurements of the NMR spin-spin relaxation reveal a divergence of electronic spin fluctuations in the vicinity of the field-induced QCP at $H_R \approx 13$ T. The fluctuations observed are characteristic of a tricritical point, followed by a phase bifurcation toward quantum wing-critical points. We show that these tricritical fluctuations enhance the effective mass of the conduction electrons and, further, drive the RSC near the H_R.

1. Introduction

A quantum phase transition emerges between two competing electronic ground states driven by fluctuations of quantum mechanical rather than thermal. A textbook example of magnetic-field-induced quantum phase transitions occurs in the one-dimensional Ising chain [1, 2]. At zero temperature (T), the ferromagnetic (FM) order switches to a paramagnetic state at a critical value H_{cr} of a transverse field applied perpendicular to the Ising axis. The value of H_{cr} defines the quantum critical point (QCP).

An analogous quantum phase transition induced by a transverse field occurs in the heavy-fermion compound URhGe [3]. Although the uranium (U) $5f$ electrons exhibit itinerant character, their spin moments still possess a strong Ising anisotropy along the c axis, with zigzag chains of nearest-neighbour U ions along the a axis. At zero magnetic field, U moments of 0.4 μ_B are aligned ferromagnetically along the c axis below the Curie temperature $T_{Curie} = 9.5$ K [4]. A transverse field H_b applied along the b axis gradually decreases the temperature of the FM transition, and eventually aligns the U moments along b at a critical field $H_R \sim 12$ T, resulting in a sudden jump of magnetization in a narrow field range (i.e., metamagnetism)[3, 5, 6]. Reentrant superconductivity (RSC) emerges in this critical region [3, 7]. The RSC transition temperature T_{sc}=0.42 K is higher than that observed at zero field (which disappears for $H > 2$ T, Fig. 1), and is broadly thought to be associated with quantum critical fluctuations possibly emerging near H_R[3, 5, 7, 8, 9, 10, 11].

In this paper, we review our recent ^{59}Co NMR study in a single crystal of $URh_{0.9}Co_{0.1}Ge$ [12] and discuss the interplay between the RSC, electron-mass enhancement and quantum critical fluctuations, all induced near H_R by applied magnetic fields along the b-axis. Since both Rh and Ge nuclei have poor NMR sensitivity, we conducted ^{59}Co NMR in URhGe doped with 10 cobalt ($URh_{0.9}Co_{0.1}Ge$). The substitution of Co for Rh is isostructural and isoelectronic, although it precludes superconductivity, it only minimally affects the magnetic properties [13, 14]: H_R(13.4 T) and T_{Curie}(11.8 K) are only slightly enhanced by the Co substitution (Fig. 1), suggesting a gradual increase of magnetic correlation [13, 14], and the spin system retains its strong Ising character along the c axis. The resistivity data are also very similar to those previously reported for $x = 0$ (see for example the inset of Fig. 1) [10, 15]. Therefore, $URh_{0.9}Co_{0.1}Ge$ appears to be a suitable proxy, allowing us indirectly to probe by NMR the microscopic nature of magnetic fluctuations in URhGe.

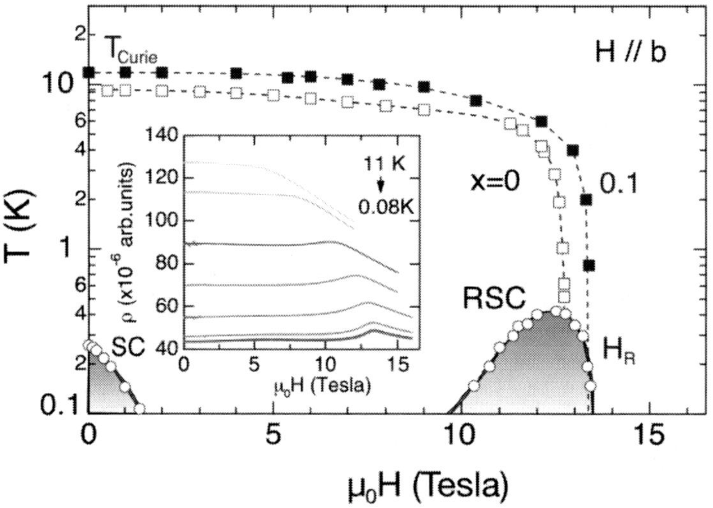

Figure 1. The temperature-field phase diagrams of $URh_{1-x}Co_xGe$ constructed from resistivity measurement in single crystals of $x = 0$ and 0.1 [12]. The inset shows the field ($\|b$) dependence of resistivity $\rho(T, H)$ at different temperatures for $x = 0.1$.

2. Experimental Methods

We prepared a high quality single crystal of $URh_{0.9}Co_{0.1}Ge$ by the Czochralski pulling method [4, 10] and measured the field strength (H) and angle (θ) dependence of the spin-spin relaxation time (T_2) by using a sample rotator inside a superconducting magnet, where the θ is the angle from the b axis in the (bc) crystal plane, and the magnetic field components are thus $(H_b, H_c) = (H\cos\theta, H\sin\theta)$. ^{59}Co nuclei possess a high nuclear spin value, $I = 7/2$, and a relatively large nuclear gyromagnetic ratio, $^{59}\gamma_N = 10.03$ MHz/T, providing high efficiency of the NMR excitation pulses as well as high sensitivity of NMR detection. These advantages enabled the use of very short NMR pulses, typically around 1 μsec, which was necessary to trace the spin-echo decay with extremely short T_2. All the NMR measurements reported here have been performed at $T = 1.6$ K, well below the $T_{\rm Curie}$ of 11.8 K of our crystal. The T_2 values were determined by fitting the τ dependence of the spin-echo intensity, measured on a center peak of NMR spectrum, to an exponential function, $M(2\tau) \propto \exp(-2\tau/T_2)$.

3. Results and discussion

Figure 1 show the T-H_b phase diagrams obtained for single crystals of $x = 0$ and 0.1. The transition temperature to the FM ordered state, $T_{\rm Curie}$, is gradually suppressed by the magnetic field applied along the b-axis, H_b, and reaches zero at the critical field, $H_{\rm R}$, where the U moments are aligned along the field direction. The inset of the figure shows the field dependence of the resistivity $\rho(T, H)$ for $x = 0.1$ at different temperatures. In general, magnetic transitions at high magnetic fields are not well defined in ferromagnets, because the application of the field transforms the phase transition into a broad crossover between the paramagnetic state and the FM state. However, when the field is perfectly aligned with the hard-magnetization axis in an Ising system, the magnetic transition remains clearly defined even at high fields [3, 10]. Indeed, as seen in the inset of Fig.1, the field dependence of the resistivity at low temperatures exhibits a distinct peak, defining $H_{\rm R}$ of 13.4 T in our crystal.

In URhGe ($x = 0$), the SC phase appears first below ~2 T, and then reappears at higher fields between ~8 and ~13.5 T with a higher $T_{\rm sc}$=0.42 K. On the other hand, neither the low field SC nor the RSC have been observed in substituted samples, including our crystal of $\rm URh_{0.9}Co_{0.1}Ge$. This, however, does not necessarily mean a lack of electronic pairing interactions in these systems. In URhGe, in contrast to the robust FM order, SC is extremely sensitive to sample quality; it has be observed only in high quality sample with a very small ρ_0 (typically less than 30 $\mu\Omega$ cm[10]). This implies that the SC is unconventional, with the phase of the SC order parameter changing as a function of orientation; thus, it is easily destroyed when the electronic mean free path is reduced to a value comparable to the superconductor's coherence length [4]. We can thus naturally expect that a strong reduction of electronic coherence brings a strong suppression of $T_{\rm sc}$ in the substituted systems. The SC transition would thus likely appear even in doped crystals if the electronic mean free path could be made long enough.

Figure 2 (a) shows the field dependence of $1/T_2$ for $\theta = 0°$ and 5°. Here the $1/T_2$ reflects the magnitude of longitudinal component (parallel to the applied field) of slow spin fluctuations near zero frequency ($\omega \sim 0$)[12]. As seen in the figure, $1/T_2$ for $\theta = 0°(H\|b)$ shows a strong field dependence associated with the divergence of magnetic fluctuations near the $H_{\rm R}$. The magnetic fluctuations become so strong near $H_{\rm R}$ that we lose the NMR spin-echo signal between 11.4 and 14 T, since for T_2 values smaller than 3 μs the NMR signal falls within the dead time of the spectrometer. On the other hand, this divergence of $1/T_2$ near $H_{\rm R}$ is rapidly suppressed by rotating the field (θ) from the b-axis. The field dependence of $1/T_2$ then changes to a crossover-like behavior for θ >5°, where a broad maximum appears at 15.5 T, higher than that for $H_{\rm R}(\theta = 0)$=13.4 T. The behavior is connected with a phase bifurcation above the $H_{\rm R}$, as see later.

In URhGe, the effective mass of the conduction electrons m^* has been reported to be enhanced in a relatively wide field range near $H_{\rm R}$, which was naively thought to be induced by critical fluctuations [10, 15]. This is now confirmed directly from our NMR. In Fig.2(b), we plot the field strength and angle dependences of the A coefficient of the resistivity, which were extracted by fitting low temperature resistivity data to $\rho(H, T) = A(H)T^2 + \rho_0(H)$ in URhGe [15]. According to the Kadowaki-Woods ratio, the A coefficient has a relation with the effective mass m^*, namely $A \propto \gamma^2 \propto m^{*2}$. At $\theta = 0°(H\|b)$, the A coefficient starts to increase above ~8 T and shows a clear maximum just before $H_{\rm R}$, and then suppressed with further increasing field. By tilting the field away form the b-axis, the peak structure is rapidly suppressed and broaden, and shifts to higher fields above $H_{\rm R}$. These characteristic behaviors of the A coefficient, and thus of the m^*,

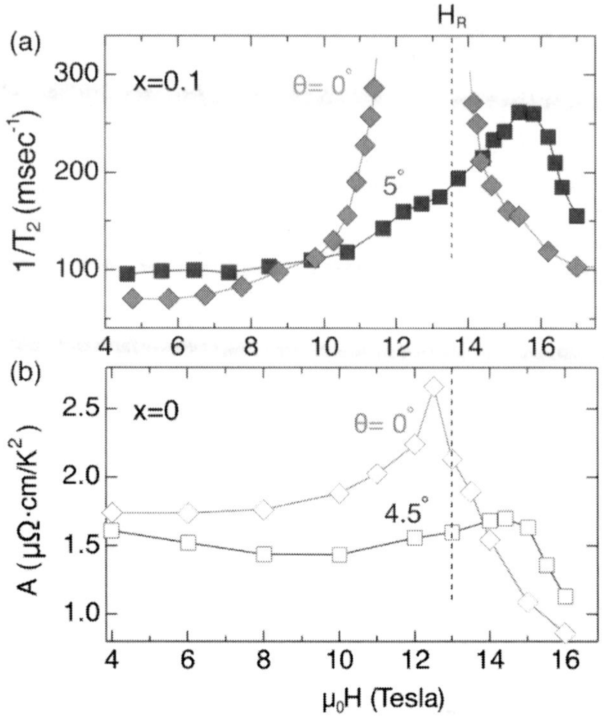

Figure 2. The field strength and angle dependences of (a) $1/T_2$ and (b) the A coefficient in $URh_{1-x}Co_xGe$. The A coefficient was extracted by fitting low temperature resistivity data to $\rho(H,T) = A(H)T^2 + \rho_0(H)$ [15]. The field axis of the figures is scaled at the H_R of each sample.

are well understood in terms of the field dependent FM fluctuations, for both the strength and angle dependences, as seen in Fig. 2(a) and (b). This definitely supports the mass enhancement mechanism mediated by the critical fluctuations.

Next we see how the critical fluctuations are developed in the proximity of H_R. The contour plot of the $1/T_2$ in Fig. 3(a) represents the map of strength distribution of longitudinal FM fluctuations in the H_b-H_c fields plane. The fluctuations are enhanced most strongly around H_R, where the NMR signal becomes undetectable due to the extremely short T_2 (hatched grey area). We can also see the development of another weak singularity around $(H_b, H_c) \simeq (15.4, \pm1.3)$[Tesla], providing a clear signature of the phase bifurcation above H_R. These characteristics of quantum fluctuations are all connected with the tricritical nature of the phase transition, as discussed below.

In the transverse Ising system considered here, the order parameter is the magnitude of FM moment along the Ising axis, $\langle M_c \rangle$. The H_b is a tuning parameter of the quantum fluctuations driving the phase transition itself, while the H_c adds a perturbation conjugate to the order parameter. Then two types of magnetic phase diagram can be expected in the H_b-H_c plane, depending on the criticality. If the long-range FM order is continuously connected to the field-induced paramagnetic state, the H_b-H_c phase diagram simply involves an ordinary QCP at $T = 0$. This ordinary QCP induces solely the diverging susceptibility of the order parameter, i.e. the divergence of the transverse fluctuations along the c axis. On the other hand, if the system possesses a tricritical point (TCP), which prevents the continuous connection of the long-range order to the disordered state, one can generally expect the emergence of the phase bifurcation above the TCP [16], where the first order planar surface at $H_c = 0$ into two wings that continue away from the H_b-axis and end up at quantum wing-critical points (QWCPs) at zero temperature for a finite H_c. The $1/T_2$ behavior clearly supports this latter case. Another characteristic features of a TCP, with regard to magnetic excitations, is that it induces a diverging susceptibility not only for the order parameter but also for the physical

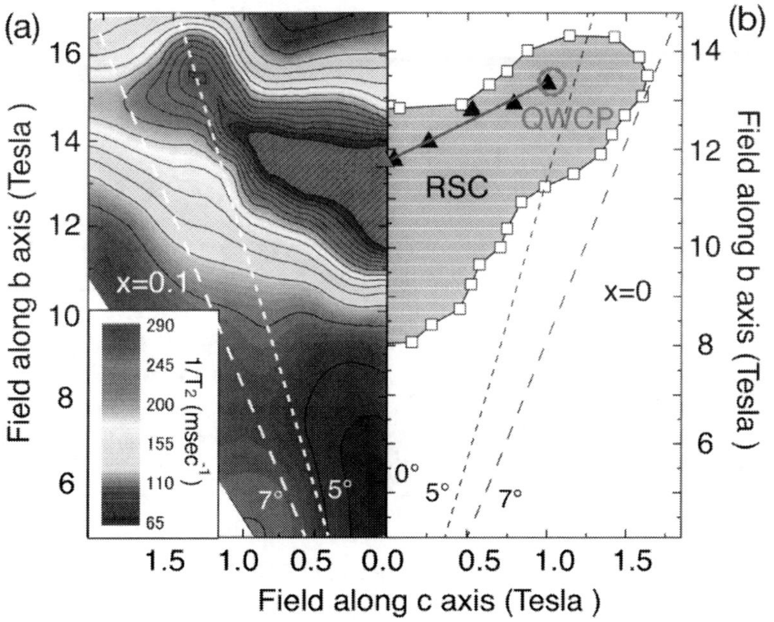

Figure 3. (a) The map of the strength distribution of magnetic fluctuations in the H_b-H_c fields plane around H_R. This contour plot of the $1/T_2$ at 1.6 K is constructed from the 104 data points in total measured with different combinations of H and θ, between H=4.7 and 17 T and between $\theta = 0°$ and 11°, respectively [12]. (b) The phase diagram constructed from resistivity measurements by Lévy *et al.* in a single crystal of URhGe[9]. The open squares indicate the region where the RSC appears at $T \sim 40$ mK and the filled triangles indicates positions where first-order-like metamagnetic transitions occurs at $T \sim 500$ mK). Since their crystal had a different H_R value (12 T) from that of our crystal (13.4 T), the field axes are scaled to the H_R.

quantity conjugate to the tuning parameter driving the phase transition [17, 18]. Here the tuning parameter is H_b, and thus we can expect the divergence of the longitudinal ($\|b$) component of magnetic fluctuations, as observed experimentaly.

Figure 3 further reveals the close interplay between the FM fluctuations and the RSC. In Fig. 3(b), we plot the H_b-H_c phase diagram constructed from resistivity measurements by Lévy *et al.* in a single crystal of URhGe[9]. The open squares indicate the region where the RSC has been observed at $T \sim 40$ mK and the filled triangles indicate the positions where first-order-like magnetic transitions have been observed at $T \sim 500$ mK. We can see that the RSC appears almost in the same limited region as that where we have observed the critical fluctuations on the H_b-H_c plane. Only about θ=7° misorientation from the b-axis suppresses the RSC[9, 19], and as seen in Fig. 3 (a), such suppression is now definitely connected to the rapid decline of the critical fluctuations for $\theta >7°$. The RSC also exhibits the highest T_{sc} and the largest diamagnetic signal around H_R (not around the QWCPs), where we have detected the diverging fluctuations [3, 6]. These close relationships between the FM fluctuations and the RSC strongly suggests these quantum fluctuations as the pairing glue responsible for the RSC.

Finally, we note that, although we revealed a divergence of the longitudinal component ($H\|b$) of slow fluctuations ($\omega \sim 0$) near the TCP, this does not necessarily mean the absence of a similar divergence for the transverse component ($H\|c$) at $\omega \sim 0$. The NMR relaxation measurements, both $1/T_1$ and $1/T_2$, can not probe such fluctuations (which can probe the transverse fluctuations at the NMR frequency of ~0.1 GHz), and we should recall that FM critical fluctuations may have strong enhance of fluctuations near zero frequency [12]. A divergence of the fluctuations is expected, in principle, for both the longitudinal and transverse components near the TCP, and we suggest that such tricritical nature of the fluctuations should be fully involved in understanding the mechanism of the RSC [20].

4. Summary

The experiments reported here indicates that URhGe possesses quantum tricriticality, associated with its low temperature TCP [7, 8], and hence develops strong longitudinal fluctuations perpendicular to the Ising axis (parallel to the applied field) at low temperatures. Such tricritical nature of the phase transition has already been proposed for URhGe; however, it was mostly relying on the observation of first-order-like behaviors near H_R [7, 8, 21, 22]. The present NMR go further, since they provide evidence by probing directly the nature of the associated fluctuations. We show that these tricritical fluctuations enhance the effective mass of the conduction electrons and, further, drive the RSC, most probably by reinforcing the SC pairing [11, 20, 23, 24], and partly through the mass enhancement [10, 15].

Acknowledgments

We are grateful for stimulating discussions with R. E. Walstedt, J. Flouquet, N. Tateiwa, K. Hattori, H. Ikeda, K. Aoyama, and V. P. Mineev. A part of this work was supported by JSPS KAKENHI Grant Numbers 26400375 and 15H05745, ERC starting grant (NewHeavyFermion), ICC-IMR, and the REIMEI Research Program of JAEA.

References

[1] see, for example, Sachdev S. Quantum Phase Transitions 1999 Cambridge Univ. Press
[2] Kinross A W *et al.* 2014 *Physical Review X* **4**, 031008
[3] Lévy F, Sheikin I, Grenier B, and Huxley A 2005 *Science* **309** 1343-1346
[4] Aoki D, *et al.* 2001 *Nature* **413** 613-616
[5] Hardy F, *et al.*, 2011 *Phys. Rev. B* **83** 195107
[6] Aoki D and Flouquet J 2014 *J. Phys. Soc. Jpn.* **83** 061011
[7] Lévy F, Sheikin I, and Huxley A, *Nature Phys.* **3** 460-463
[8] Huxley A *et al.* 2007 *J. Phys. Soc. Jpn.* **76** 051011
[9] Lévy F *et al.*, 2009 *J. Phys. Condens. Matter* **21** 164211
[10] Miyake A, Aoki D, and Fluquet J 2008 *J. Phys. Soc. Jpn.* **77** 094709
[11] Hattori K and Tsunetsugu H, 2013 *Phys. Rev. B* **87** 064501
[12] Tokunaga Y *et al.* 2015 *Phys. Rev. Lett.* **114** 216401
[13] Sakarya S *et al.* 2008 *J. Alloys and Comp.* **457** 51-56
[14] Huy N T, and de Visser A 2009 *Solid State Comm.* **149** 703-706
[15] Aoki D *et al.* 2011 *J. Phys. Soc. Jpn.* **80** SA008
[16] Kirkpartrick T R and Belitz D 2015 *Phys. Rev. Lett.* **115** 020402
[17] Misawa T, Yamaji Y and Imada M 2006 *J. Phys. Soc. Jpn.* **75** 064705
[18] Misawa T, Yamaji Y and Imada M 2008 *J. Phys. Soc. Jpn.* **77** 093712
[19] Yelland E A, *et al.* 2011 *Nat. Phys.* **7** 890-894
[20] Mineev V P 2015 *Phys. Rev. B* **91** 014506
[21] Aoki D, Knebel G, and Flouquet J, 2014 *J. Phys. Soc. Jpn.* **83** 094719
[22] Kotegawa H *et al.* 2015 *J. Phys. Soc. Jpn.* **84** 054710
[23] Hattori T *et al.*, 2012 *Phys. Rev. Lett.* **108** 066403
[24] Tada Y *et al.* 2013 *J. Phys.: Conf. Ser.* **449** 012029

TMU International Symposium 2015 IOP Publishing
Journal of Physics: Conference Series **683** (2016) 012011 doi:10.1088/1742-6596/683/1/012011

Effect of Ga Substitution on the Γ_3 Doublet Ground State in $PrIr_2Zn_{20}$

Takahiro Onimaru[1], Kazuto Uenishi[1], Yu Yamane[1], Kazuhei Wakiya[1], Keisuke T. Matsumoto[1], Kazunori Umeo[2] and Toshiro Takabatake[1,3]

[1]Department of Quantum Matter, Graduate School of Advanced Sciences of Matter, Hiroshima University, Higashi-Hiroshima 739-8530, Japan

[2]Cryogenics and Instrumental Analysis Division, N-BARD, Hiroshima University, Higashi-Hiroshima 739-8526, Japan

[3]Institute for Advanced Materials Research, Hiroshima University, Higashi-Hiroshima 739-8530, Japan

E-mail: onimaru@hiroshima-u.ac.jp

Abstract. Electrical resistivity ρ, specific heat C, and magnetic susceptibility χ measurements are reported on $PrIr_2Zn_{20-x}Ga_x$ for x=0.5 and 1.0. On cooling, $\rho(T)$'s monotonically decrease to the residual values which are two orders of magnitudes larger than that of $PrIr_2Zn_{20}$. A broad hump in $C_m(T)$'s at around 10 K can be reproduced by the two-level model with the Γ_3 doublet ground state and the Γ_4 triplet excited state. The magnetic entropy release shifts to higher temperatures with increasing x. The increase in $\chi(T)$ at $T < 5$ K contrasts with the saturation of $\chi(T)$ for x=0. These observations indicate the splitting of the ground state doublet due to the lowering of the local symmetry at the Pr site by the Ga substitution.

1. Introduction

Praseodymium-based cubic compounds with $4f^2$ configuration attract much attention because they show various phenomena arising from (quasi-)degenerated ground states. In a Pr-filled skutterudite $PrOs_4Sb_{12}$, the feasibility of superconductivity mediated by quadrupolar fluctuations has been pointed out by the measurements of neutron scattering and nuclear quadrupole resonance [1, 2, 3]. A well-known example of a $4f^2$ cubic system is $PrPb_3$ with the $AuCu_3$-type structure, where the cubic crystalline electric field (CEF) splits the ninefold multiplet of $J = 4$ into four multiplets. The CEF ground state of the Γ_3 doublet carries no magnetic dipole moment but electric quadrupoles. Thereby, the low temperature property is governed by the interaction of the quadrupoles and the conduction electrons. In fact, $PrPb_3$ undergoes an antiferroquadrupole (AFQ) transition at $T_Q = 0.4$ K [4]. The combined analysis of magnetization and neutron diffraction experiments identified the AFQ order parameter as the quadrupole O_2^0 [5, 6]. The quadrupole moments are aligned with an incommensurate sinusoidally modulated structure even in the ground state. Thereby, the indirect RKKY-type interaction between the quadrupoles plays the essential role. On the other hand, substituting La for Pr in $PrPb_3$ destroys the AFQ order at a small amount of $x = 0.03$ in $Pr_{1-x}La_xPb_3$ [7]. For x $\geq$ 0.95, on the other hand, there appears non-Fermi liquid (NFL) behavior. Because the electric quadrupole conserves the time-reversal symmetry, it has two orbital channels. If the localized quadrupole is overcompensated by the conduction electrons, the following NFL behaviors would

Journal of Physics: Conference Series **683** (2016) 012011 doi:10.1088/1742-6596/683/1/012011

appear; the magnetic susceptibility $\chi \sim \ln T$, specific heat $C/T \sim \ln T$ and electrical resistivity $\rho - \rho_0 \propto 1 + A\sqrt{T}$, where ρ_0 is residual resistivity and A is a coefficient [8]. More important is the residual entropy of $(1/2)R\ln 2$ at zero temperature. The absence of AFQ order in $PrAg_2In$ [9] and $PrMg_3$ [10, 11] with the Γ_3 doublets ground state has been discussed by taking the chemical disorder effect into consideration.

Recently, much attentions are drawn to a family of a cubic system PrT_2X_{20} (X=Al and n) which show a variety of phenomena such as quadrupole order, heavy-fermion superconductivity, and structural transitions [12, 13, 14]. The CEF ground state of most compounds is the non-Kramers Γ_3 doublet possessing no magnetic dipoles but electric quadrupoles [15, 16]. Among them, $PrIr_2$ n_{20} undergoes an AFQ order at T_Q=0.11 K and a superconducting transition at T_c= 0.05 K, suggesting a possible contribution of the quadrupole fluctuations to the superconducting Cooper-pair formation [12, 13]. The coexistence of the AFQ order and the superconducting state has been also observed in the isostructural $PrRh_2$ n_{20}, $PrTi_2Al_{20}$, and PrV_2Al_{20} [17, 18, 19]. In $PrTi_2Al_{20}$, T_c=0.2 K at ambient pressure is enhanced up to 1 K under high pressure, for which a quantum critical phenomenon due to the quadrupolar degrees of freedom has been proposed [20]. On the other hand, in $PrIr_2$ n_{20}, in the temperature range 0.2 $<T<$ 0.8 K, above T_Q, the magnetic specific heat C_m and $\rho(T)$ exhibit NFL behavior; C_m/T and ρ are proportional to $-\ln T$ and $\sqrt{T}$, respectively [21]. All data set of C_m and $\rho(T)$ in magnetic fields $B \leq 6$ T are well scaled with characteristic temperature T_0. Because the constituent atoms are ordered periodically in the high quality crystal with low residual resistivity of 0.2 $\mu\Omega$ cm, these temperature variations of C_m and ρ strongly suggest the formation of a quadrupole Kondo lattice, in which quadrupole moments are periodically placed and overcompensated by the conduction electrons, leading to the NFL state with an energy scale of $k_B T_0$ [21, 22].

In the present work, the effects of the Ga substitution on the ground state of $PrIr_2$ $n_{20-x}Ga_x$ for x=0.5 and 1.0 were studied by the $\rho(T)$, $C(T)$, and $\chi(T)$ measurements. Since the electronic number of Ga is one more than that of n, it is expected that hybridization between the $4f$ and conduction electrons could be enhanced by the increase of $4p$ electron density due to the Ga substitution for n, and the NFL behavior would be modified.

2. Experimental

Single crystalline samples of $PrIr_2$ $n_{20-x}Ga_x$ were synthesized by the melt-growth method from the high-purity elements Pr (4N), Ir (3N), n (6N), and Ga (6N). The details of sample preparation were reported in the previous paper [12].

The samples were characterized by the -ray powder diffraction technique and electron-probe microanalysis (EPMA). The -ray diffraction patterns with using the Cu K_α radiation for x=0, 0.5 and 1.0 are shown in Fig. 1. Almost all of the peaks can be indexed by the cubic $CeCr_2Al_{20}$-type structure [23], although some impurity peaks are found in the pattern for x=1.0 as depicted with the arrows in Fig. 1. The inset of Fig. 1 shows the backscattered electron image for x=0.5. There are small amounts of secondary phases in addition to the main phases. The atomic compositions were determined by the EPMA using an electron beam of 20 keV on a JEOL J A-8200 analyzer. The crystal compositions of the main phases were determined as $PrIr_{1.9(1)}$ $n_{19.4(2)}Ga_{0.49(2)}$ and $PrIr_{1.9(1)}$ $n_{18.9(5)}Ga_{1.0(1)}$ for x=0.5 and 1.0, respectively, by averaging over 10 different regions for each crystal, where it was assumed that the Pr sites are fully occupied. The crystal compositions of Ga equal the starting ones. The results confirmed the chemical homogeneity of the samples used for the measurements of physical properties. The solid solubility limit was found to be x= 1.3. The secondary phase for the x= 0.5 sample is $PrIr_{4.6}$ $n_{15.4}Ga_{2.7}$. On the other hand, that for x= 1.0 is a nonmagnetic Ir(n,Ga)$_{5.9}$. The lattice parameters were determined by the Rietveld refinement of the -ray diffraction patterns using the program RIETAN-FP [24]. The lattice parameter for x=1.0 decreases by only 0.1 from that for x=0.

Journal of Physics: Conference Series **683** (2016) 012011 doi:10.1088/1742-6596/683/1/012011

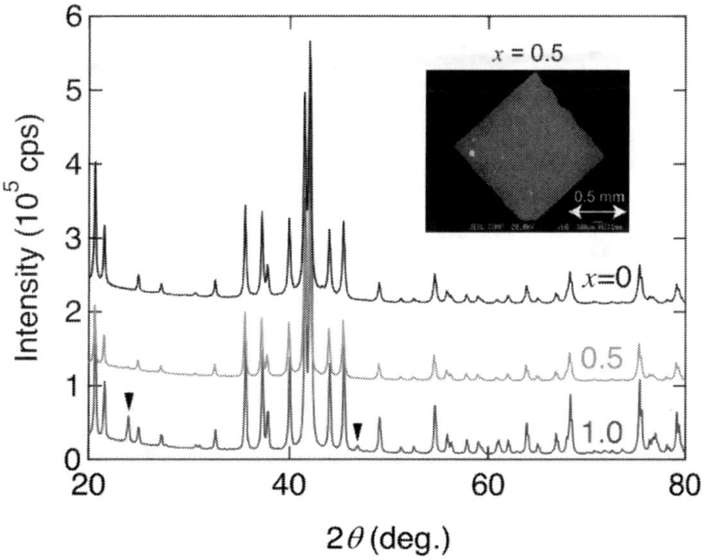

Figure 1. Powder -ray diffraction patterns of PrIr$_2$ n$_{20-x}$Ga$_x$ for x=0, 0.5 and 1.0. The data are offset for clarity. The arrows indicate the peaks due to an impurity phase. The inset shows the backscattering electron image of PrIr$_2$ n$_{20-x}$Ga$_x$ for x=0.5.

The electrical resistivity ρ was measured by a standard four-probe AC method using a laboratory built system. A Gifford-McMahon type refrigerator was used in the temperature range for $3 \leq T \leq 300$ K. The specific heat C between 0.4 and 20 K was measured by a thermal relaxation method using a Quantum Design physical property measurement system. Magnetic susceptibility χ was measured using a commercial SQUID magnetometer (Quantum

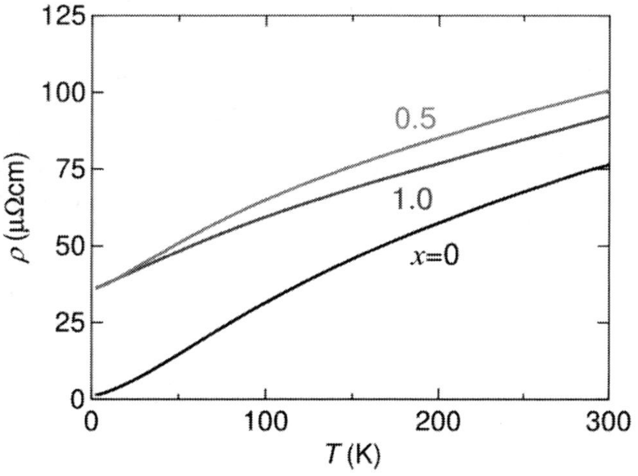

Figure 2. Temperature dependence of the electrical resistivity of PrIr$_2$ n$_{20-x}$Ga$_x$ for x=0, 0.5, and 1.0.

Journal of Physics: Conference Series **683** (2016) 012011 doi:10.1088/1742-6596/683/1/012011

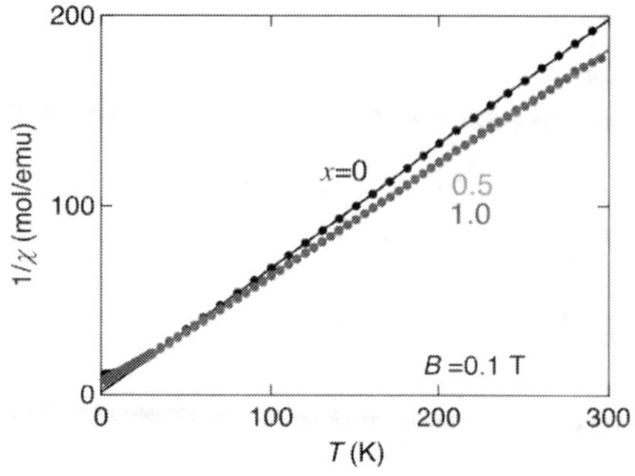

Figure 3. Temperature dependence of the magnetic susceptibility of $PrIr_2$ $n_{20-x}Ga_x$ for x=0 [12], 0.5, and 1.0 measured in the magnetic field of B=0.1 T.

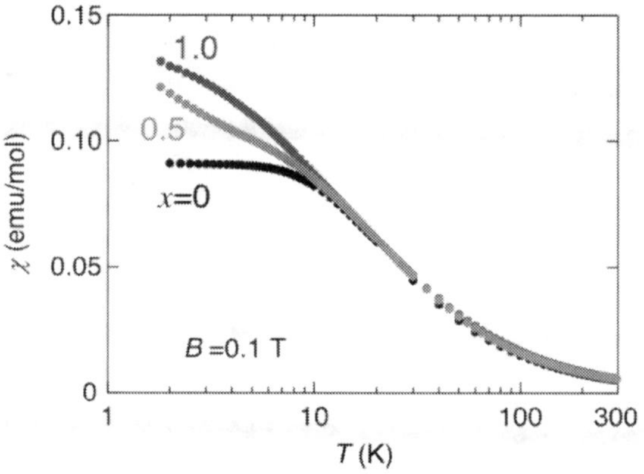

Figure 4. Temperature dependence of the magnetic susceptibility of $PrIr_2$ $n_{20-x}Ga_x$ for x=0 [12], 0.5, and 1.0 measured in the magnetic field of B=0.1 T.

Design MPMS) between 1.8 and 300 K.

3. Results and discussion

The temperature dependence of the electrical resistivity $\rho(T)$ of $PrIr_2$ $n_{20-x}Ga_x$ for x=0, 0.5 and 1.0 are shown in Fig. 2. On cooling, $\rho(T)$ monotonically decreases with an upward curvature. The values of residual resistivity for x=0.5 and 1.0 are 36 $\mu\Omega$ cm, whose values are two orders of magnitudes larger than that for x=0. The values of the residual resistivity ratio (RRR) defined as ρ(300K)/ρ(2K) for x= 0.5 and 1.0 are 2.8 and 2.6, respectively. The decrease of RRR by the Ga substitution probably results from the atomic disorder due to the Ga substitution, which prevents us from detecting a change of the $\sqrt{T}$ dependence of $\rho(T)$ observed for x=0.

Figure 3 shows the temperature dependence of the inverse magnetic susceptibility $\chi^{-1}(T)$ for x=0 [12], 0.5 and 1.0 measured in a field of B= 0.1 T. Above 20 K, $\chi^{-1}(T)$'s for x=0.5 and 1.0 are linear in T as is expected from the Curie-Weiss law. The effective magnetic moments of $\mu_{\rm eff}$ =3.66 $\mu_{\rm B}$/Pr-ion for x=0.5 and 1.0 are close to 3.49 $\mu_{\rm B}$/Pr-ion for x=0. The paramagnetic Curie temperatures θ_p's for x=0.5 and 1.0 are −5.2 and −5.6 K, respectively, indicating that the intersite magnetic interactions between the magnetic moments of CEF excited states of the trivalent Pr ions are weakly antiferromagnetic. Figure 4 shows $\chi(T)$ of $PrIr_2$ $n_{20-x}Ga_x$ measured in a magnetic field of B=0.1 T. As T goes down to 2 K, $\chi(T)$'s for x=0.5 and 1.0 continue to increase, although that for x=0 is saturated below 5 K [12]. The origin of the increment of $\chi(T)$ at T< 5 K for x=0.5 and 1.0 will be discussed later.

We turn our attention to the change in the specific heat by the Ga substitution. Figure 5 shows the temperature dependence of the magnetic specific heat divided by temperature $C_{\rm m}/T$ for $PrIr_2$ $n_{20-x}Ga_x$ with x=0 [13], 0.5, and 1.0. The values of $C_{\rm m}$ for x=0.5 and 1.0 were estimated by subtracting the specific heat of $LaIr_2$ $n_{20-x}Ga_x$ for x=0.5 and 1.0, respectively, as the lattice and conduction electron contributions from those of $PrIr_2$ $n_{20-x}Ga_x$. For x= 0, the broad peak appears at around 10 K as was previously reported, which was reproduced by the Γ_3 doublet - Γ_4 triplet two-level model with an energy split of 30 K, as shown with the solid curve [13]. For x= 0.5, the peak is much broadened than that for x= 0. The magnitude at T > 7 K is comparable to the calculation for the Γ_3 doublet ground state rather than that for the the Γ_1 singlet ground state shown with the dashed curve in Fig. 5. For x= 1.0, $C_{\rm m}/T$ is monotonically increases on cooling. The magnitude of $C_{\rm m}/T$ for T > 7 K is more comparable

Journal of Physics: Conference Series **683** (2016) 012011 doi:10.1088/1742-6596/683/1/012011

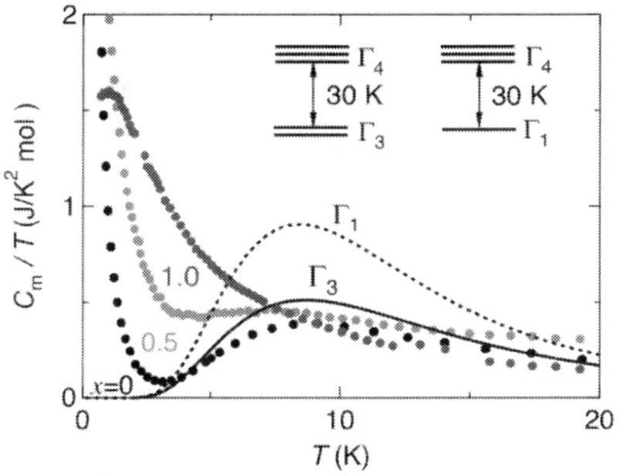

Figure 5. Temperature dependence of the specific heat of PrIr$_2$ n$_{20-x}$Ga$_x$ for x=0 [13], 0.5, and 1.0. The solid and dashed curves are calculations using two-level models with the Γ_3 doublet and the Γ_1 singlet ground states, respectively, as shown in the inset.

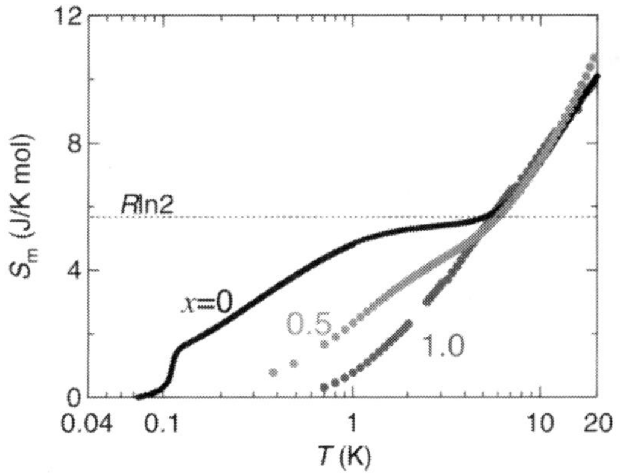

Figure 6. Temperature variation of the magnetic entropy $S_{\rm m}$ of PrIr$_2$ n$_{20-x}$Ga$_x$ for x=0 [13], 0.5, and 1.0, which are estimated by integrating $C_{\rm m}/T$ with respect to the temperature. It was assumed that $S_{\rm m}$ reaches Rln2 at 5 K.

to the calculation of the two-level model with the Γ_3 doublet ground state than that with the Γ_1 singlet ground state. Considering the moderate agreement between the experimental data and the calculation for $T > 7$ K, we expect that the CEF ground state of the Pr^{3+} ion in the Ga substitution system would remain in the Γ_3 doublet as in PrIr$_2$ n$_{20}$. Note that $C_{\rm m}/T$'s for x=0.5 and 1.0 at $T < 7$ K diverge upward from the two-level model calculation with the Γ_3 doublet ground state. This divergence suggests that the doubly degenerated state is split by the Ga substitution. With respect to the NFL behavior for the specific heat, the $-\ln T$ dependence of $C_{\rm m}/T$ observed in PrIr$_2$ n$_{20}$ varnishes for x=1.0, and instead, $C_{\rm m}/T$ shows a broad peak at around 1 K, indicating collapse of the NFL state.

The magnetic entropy $S_{\rm m}$ for x=0.5 and 1.0 estimated by integrating $C_{\rm m}/T$ with respect to the temperature is shown in Fig. 6. Here, it was assumed that $S_{\rm m}$'s commonly reach Rln2 at 5 K. The curve of $S_{\rm m}(T)$ shifts to higher temperatures with increasing x, indicating that the entropy of Rln2 due to the two-fold degeneracy of the Γ_3 doublet is removed at higher temperature. It is reasonable because $C_{\rm m}/T$'s for x=0.5 and 1.0 diverge from the two-level model calculation at $T < 7$ K as mentioned above.

Let us discuss possible mechanisms for the enhancement of $\chi(T)$ at $T < 5$ K and the release of the magnetic entropy at the higher temperatures in the Ga substituted system. One is the enhancement of the hybridization between the 4f and conduction electrons due to the increase of 4p electron density. Recently, emergence of a Fermi liquid state accompanied with a free magnetic spin in the vicinity of the non-Fermi liquid state has been theoretically pointed out for Pr-based compounds with the Γ_3 doublet ground state, where competition between the magnetic and quadrupole Kondo effects was considered on the basis of an extended two-channel Kondo model.[25] If this is the case, $\chi(T)$ could be enhanced compared with that for x=0. In addition, if the NFL state would collapse by the Ga substitution, then the characteristic temperature shifts to higher temperatures and the magnetic entropy is released at the higher temperatures.

However, it is elusive whether the hybridization could be strengthen by the small amount of the substitution for $x \leq 1.0$. More plausible mechanism is the lowering of local symmetry at the Pr site by the atomic disorder. In this case, the Γ_3 doublet is split into two singlets, whose energy is statistically distributed. Thereby, the magnetic entropy is released at higher temperatures.

Furthermore, $\chi(T)$ is increased on cooling until the thermal population of the upper singlet becomes negligible. If the splitting energy is less than 1 K, then $\chi(T)$ would continuously increase down to 2 K. The splitting of the doublet ground state probably makes the two-channel orbitals anisotropic. By the channel anisotropy, emergence of a crossover from NFL to FL state was theoretically proposed.[26] In any events, the NFL state in $PrIr_2$ n_{20} arising from the formation of the quadrupole Kondo lattice becomes unstable by the Ga substitution. To reveal the ground state of the Ga substituted system, further measurements at lower temperatures below 0.4 K are on going.

4. Conclusion

The transport and magnetic properties of Ga substituted system $PrIr_2$ $n_{20-x}Ga_x$ for x=0.5 and 1.0 were studied by the measurements of the $\rho(T)$, $C_m(T)$, and $\chi(T)$. $\rho(T)$ monotonically decreases on cooling. The high residual resistivity is attributed to the atomic disorder. The hump in $C_m(T)/T$ at around 10 K can be moderately reproduced by the two-level model with the Γ_3 doublet ground state. The release of the magnetic entropy estimated from C_m shifts to higher temperatures. $\chi(T)$ increases on cooling at $T < 5$ K, which is in contrast to the saturation behavior for x=0. These observations indicate the splitting of the ground state doublet due to lowering of the local symmetry at the Pr site by the Ga substitution.

Acknowledgments

The authors would like to thank H. Kusunose, K. Izawa, T. Yoshida, A. Tsuruta and K. Miyake for helpful discussions. We also thank Y. Shibata for the electron-probe microanalysis performed at N-BARD, Hiroshima University. The magnetization measurements with MPMS and specific heat measurements with PPMS were carried out at N-BARD, Hiroshima University. This work was financially supported by JSPS KAKENHI Grant Numbers 21102516, 23102718, 26707017, and 15H05886 (J-Physics), and by The Mazda Foundation Research Grant, Japan and Hiroshima University Fujii Research Fund.

References

[1] Bauer E D, Frederick N A, Ho P C, Zapf V S and Maple M B 2002 *Phys. Rev. B* **65** 100506
[2] Kuwahara K, Iwasa K, Kohgi M, Kaneko K, Metoki N, Raymond S, Méasson M -A, Flouquet J, Sugawara H, Aoki Y and Sato H 2005 *Phys. Rev. Lett.* **95** 107003
[3] Yogi M, Nagai T, Imamura Y, Mukuda H, Kitaoka Y, Kikuchi D, Sugawara H, Aoki Y, Sato H and Harima H 2006 *J. Phys. Soc. Jpn.* **75** 124702
[4] Tayama T, Sakakibara T, Kitami K, Yokoyama M, Tenya K, Amitsuka H, Aoki D, Ōnuki Y and Kletowski Z 2001 *J. Phys. Soc. Jpn.* **70** 248
[5] Onimaru T, Sakakibara T, Harita A, Tayama T, Aoki D and Ōnuki Y 2004 *J. Phys. Soc. Jpn.* **73** 2377
[6] Onimaru T, Sakakibara T, Aso N, Yoshizawa H, Suzuki H S and Takeuchi T 2005 *Phys. Rev. Lett.* **94** 197201
[7] Kawae T, Kinoshita K, Nakaie Y, Tateiwa N, Takeda K, Suzuki H S and Kitai T 2006 *Phys. Rev. Lett.* **96** 027210
[8] Cox D L and Zawadowski Z 1998 *Adv. Phys.* **47** 599
[9] Yatskar A, Beyermann W P, Movshovich R and Canfield P C 1996 *Phys. Rev. Lett.* **77** 3637
[10] Tanida H, Suzuki H S, Takagi S, Onodera H and Tanigaki K 2006 *J. Phys. Soc. Jpn.* **75** 073705
[11] Morie T, Sakakibara T, Suzuki H S, Tanida H and Takagi S 2009 *J. Phys. Soc. Jpn.* **78** 033705
[12] Onimaru T, Matsumoto K T, Inoue Y F, Umeo K, Saiga Y, Matsushita Y, Tamura R, Nishimoto K, Ishii I, Suzuki T and Takabatake T 2010 *J. Phys. Soc. Jpn.* **79** 033704
[13] Onimaru T, Matsumoto K T, Inoue Y F, Umeo K, Sakakibara T, Karaki Y, Kubota M and Takabatake T 2011 *Phys. Rev. Lett.* **106** 177001
[14] Sakai A and Nakatsuji S 2011 *J. Phys. Soc. Jpn.* **80** 063701
[15] Iwasa K, Kobayashi H, Onimaru T, Matsumoto K T, Nagasawa N, Takabatake T, Ohira-Kawamura S, Kikuchi T, Inamura Y and Nakajima K 2013 *J. Phys. Soc. Jpn.* **82** 043707
[16] Sato T J, Ibuka S, Nambu Y, Yamazaki T, Hong T, Sakai A and Nakatsuji S 2012 *Phys. Rev. B* **86** 184419
[17] Onimaru T, Nagasawa N, Matsumoto K T, Wakiya K, Umeo K, Kittaka S, Sakakibara T, Matsushita Y and Takabatake T 2012 *Phys. Rev. B* **86** 184426

[18] Sakai A, Kuga K and Nakatsuji S 2012 *J. Phys. Soc. Jpn.* **81** 083702
[19] Tsujimoto M, Matsumoto Y, Tomita T, Sakai A and Nakatsuji S 2014 *Phys. Rev. Lett.* **113** 267001
[20] Matsubayashi K, Tanaka T, Sakai A, Nakatsuji S, Kubo Y and Uwatoko Y 2012 *Phys. Rev. Lett.* **109** 187004
[21] Onimaru T *et al.* unpublished
[22] Tsuruta A and Miyake K 2015 *J. Phys. Soc. Jpn.* **84** 114714
[23] Nasch T, Jeitschko W and Rodewald U C 1997 *Z. Naturforsch. B* **52** 1023
[24] Izumi F and Momma K 2007 *Solid State Phenom.* **130** 15
[25] Kusunose H and Onimaru T 2015 *J. Phys.: Conf. Ser.* **592** 012099
[26] Yotsuhashi S and Maebashi H 2002 *J. Phys. Soc. Jpn.* **71** 1705

Journal of Physics: Conference Series **683** (2016) 012012 doi:10.1088/1742-6596/683/1/012012

Very Low Temperature Magnetoresistance in the Quadrupole Ordered System PrV_2Al_{20}

Y Shimura[1], M Tsujimoto[1], B Zeng[2], Q Zhang[2], L Balicas[2], A Sakai[1,3], and S Nakatsuji[1]

[1]Institute for Solid State Physics, University of Tokyo, Kashiwa 277-8581, Japan
[2]National High Magnetic Field Laboratory, Florida State University, Tallahassee, Florida 32310, USA
[3]I. Physikalisches Institut, Georg-August-Universität Göttingen, 37077 Göttingen, Germany

E-mail: simu@issp.u-tokyo.ac.jp

Abstract. We measured magnetoresistance of the quadrupole ordered system PrV_2Al_{20} with a Γ_3 doublet ground state under high-DC magnetic field up to 30 T. The field dependence of the magnetoresistance is strongly different between $H \parallel [111]$ and $H \parallel [110]$ in spite of the cubic crystal structure. For $H \parallel [110]$, we observed a jump with distinct hysteresis between 13 and 16 T and a shoulder structure in the field of 8 T at 23 mK.

1. Introduction

Cubic Pr-based compounds with Γ_3 non-magnetic doublet ground state recently attracts much interest. The Γ_3 doublet does not carry magnetic moments but only multipole moments, two electric quadrupole moments O_2^2, O_2^0 and magnetic octapole moments T_{xyz}. Therefore, we can study pure multipole contribution at low temperature if the Γ_3 doublet ground state is sufficiently separated from the crystal-electric-field excited state. In usual case, the degeneracy of the Γ_3 doublet is lifted by the quadrupole ordering at low temperature. For example, $PrPb_3$, $PrIr_2$ n_{20} and $PrRh_2$ n_{20} are found to exhibit antiferro-quadrupolar ordering at 0.4 K, 0.11 K, and 0.06 K, respectively [1, 2, 3]. These magnetic phase diagrams are anisotropic among [100], [110] and [111] in spite of the cubic symmetry. The anisotropy of the magnetic phase diagram gives us important clue to determine the quadrupolar order parameter.

PrV_2Al_{20} with the Γ_3 doublet ground state also exhibits an antiferro-quadrupole ordering at 0.65 - 0.75 K [4, 5]. In the quadrupole ordered state, superconducting transition with heavy quasi-particle mass was recently found at 0.05 K [5]. The phase diagram of the quadrupole ordered state is little anisotropic for three principal field directions [100], [110] and [111] below 9 T [4]. However, above 9 T, the magnetic phase diagram is strongly anisotropic. For $H \parallel [111]$, the quadrupole phase diagram is closed at the critical field of 11 T [6]. Near the critical field, the field dependence of magnetoresistance $\rho(H)$ exhibits a peak, suggesting the enhancement of the residual resistivity. In addition, the field dependence of A coefficient in $\rho(T) = \rho_0 + AT^2$, where ρ_0 is residual resistivity, divergently increases by approaching the critical field from 30 T. These results suggest the emergence of the field-induced quantum critical point at the critical field of 11 T. By contrast, for $H \parallel [100]$, another high-field ordered phase was found above 11 T [7]. This high-field phase transition probably is due to the switching of the quadrupolar ordered

Journal of Physics: Conference Series **683** (2016) 012012 doi:10.1088/1742-6596/683/1/012012

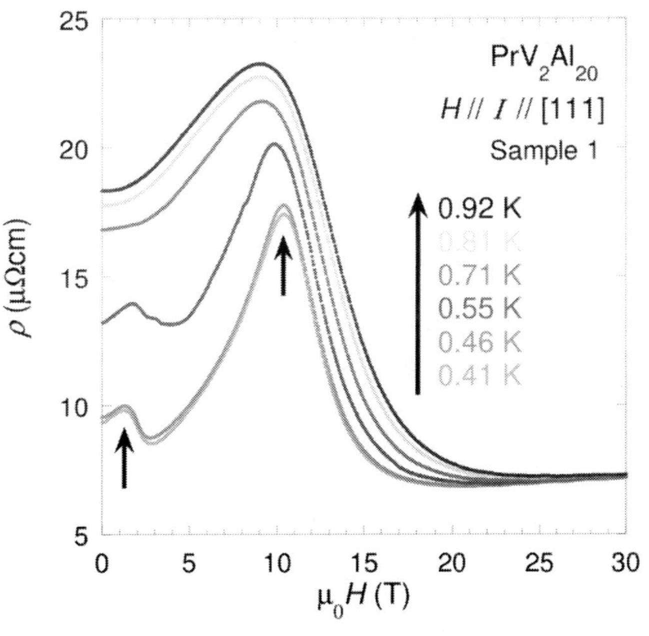

Figure 1. The field dependence of magnetoresistance $\rho(H)$ of PrV_2Al_{20} (sample 1), measured at the various temperature points in the field aligned for $I \parallel H \parallel [111]$.

parameter between O_2^2 and O_2^0. Thus, the phase diagram is strongly anisotropic under high magnetic field. The results about magnetic phase diagram above 9 T for $H \parallel [110]$ has not been reported in magnetoresistance as well as we know.

2. Experimental

Single crystal samples of PrV_2Al_{20} were grown by Al-self-flux method [4, 5, 6]. In this paper, we show results of the magnetoresistance for two samples. For one sample (sample 1), residual resistive ratio (RRR) and the current direction are RRR $\sim$ 10 and $I \parallel [111]$, respectively. Those for the other sample are RRR $\sim$ 12 and $I \parallel [110]$, respectively. For both samples, emergence of the multipole ordered transition at $\sim$ 0.6 K was confirmed.

3. Results and Discussion

Figure 1 shows the field dependence of magnetoresistance $\rho(H)$ in PrV_2Al_{20} (sample 1) at various temperatures between 0.41 K and 0.92 K for the magnetic field and current parallel to the [111] direction. Below $\sim$ 10 T, the value of the resistivity dramatically increases from 0.46 K to 0.71 K. This change is due to the quadrupole ordered transition observed at $\sim$ 0.6 K. In the field sweep, two characteristic anomalies were observed. One is a small peak in the field of $\sim$ 1 T, and the other is the peak observed at $\sim$ 11 T. While the former peak is observed in the temperature only below $T_Q = 0.6$ K, the latter peak remains even above the order temperature. The emergence of the peak even above T_Q is evidence of the development of the critical quantum fluctuation at $\sim$ 11 T. Above 11 T, no-anomalies due to the transition were detected in the high-magnetic field up to 30 T. This result suggests that para-quadrupole state remains above 11 T. The detailed temperature dependence above 11 T and the analysis have been already reported [6].

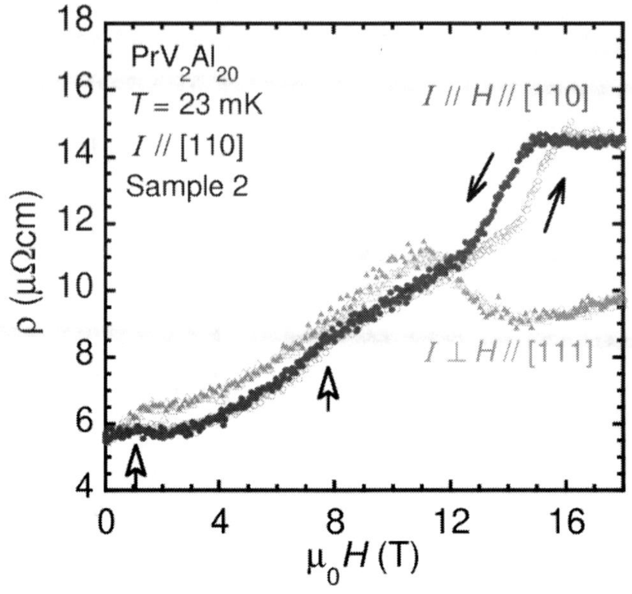

Figure 2. $\rho(H)$ in PrV_2Al_{20} (sample 2) for the $I \parallel [110] \perp H \parallel [111]$ (triangle) and $I \parallel H \parallel [110]$ (circle) at 23 mK. Open (close) symbols indicate the field-increasing (decreasing) process. Two open arrows show a small anomaly at ~ 1 T and a shoulder structure at ~ 8 T for $I \parallel H \parallel [110]$.

Figure 2 exhibits $\rho(H)$ in PrV_2Al_{20} (sample 2) for the current parallel to [110] at the very low temperature of 23 mK, deep in the quadrupolar ordered phase. We measured $\rho(H)$ for two field directions of $H \parallel [111]$ (triangle) and $H \parallel [110]$ (circle). For $H \parallel [111]$, two peaks at ~ 1 T and ~ 11 T were also observed, consistent with the results as shown in Fig. 1. For these peaks, distinct hysteresis was not detected. Under the high magnetic field for $H \parallel [111]$, the magnetoresistance in Fig. 2 is enhanced, compared with that in Fig. 1. Difference of these two data is mainly arising from the relation between current direction and magnetic field direction (H parallel or perpendicular to the current direction) For $H \parallel [110]$, $\rho(H)$ is dramatically different from that for $H \parallel [111]$. $\rho(H)$ for $H \parallel [110]$ shows a jump accompanied with distinct hysteresis in the field between 13 - 16 T, in addition to the small anomaly at ~ 1 T and a shoulder structure at ~ 8 T. These two anomalies at ~ 1 T and ~ 8 T remain only inside the quadrupole ordered state (not shown). The field where jump is observed for $H \parallel [110]$ is higher than the critical field of 11 T for $H \parallel [111]$. This result suggests that the critical field of the ordered phase for $H \parallel [110]$ is higher than that for $H \parallel [111]$. The anisotropy of the magnetization indicates that Γ_5 magnetic triplet is the first-excited state lying above the Γ_3 doublet ground state [8]. In that case, the gap in the Γ_3 doublet for $H \parallel [111]$ is larger than that for $H \parallel [110]$ under the magnetic field. The difference of the gap size under magnetic field is probably associated with the anisotropic critical field. The detailed field/temperature dependence and the phase diagram for $H \parallel [110]$ will be published elsewhere.

Anisotropic phase diagram has been studied in detail in the cubic $PrPb_3$ with a Γ_3 doublet ground state, indicating an antiferro-quadrupole ordering due to the O_2^0 quadrupole moments at 0.4 K [1, 9]. For $H \parallel [110]$, above 7 T, another high-field phase was recently found by specific heat measurements [10]. This ordered phase remains, at least, up to 13 T [11]. By contrast, for $H \parallel [111]$, the quadrupole ordered phase closes at 6 T, and the high-field ordered phase as observed for $H \parallel [110]$ was not observed [12].

In the case of PrV_2Al_{20}, as shown in Fig. 2, a shoulder structure at 8 T probably due to the transition was observed for $H \parallel [110]$. This result suggests emergence of the high-field ordered state in the field between 8 T and 15 T. Such an anomaly was not found for $H \parallel [111]$. Noted that both of PrV_2Al_{20} and $PrPb_3$ exhibit high-field ordered phases for $H \parallel [100]$ above $\sim$ 11 T and $\sim$ 6 T, respectively [13, 7]. Thus, for both systems, the high-field ordered state is observed only for $H \parallel [110]$ and $H \parallel [100]$. In $PrPb_3$, for $H \parallel [110]$, neutron scattering measurements revealed that the order parameter of the high-field phase above 7 T is O_2^2 quadrupole moments, which is different from that of O_2^0 moments at zero field [9, 14]. From the analogy to the phase diagram in $PrPb_3$, a shoulder structure at 8 T in PrV_2Al_{20} for $H \parallel [110]$ may be due to switching of the ordered parameter between Γ_3-type quadrupole moments O_2^0 and O_2^2.

4. Conclusion

We reported the magnetoresistance in the quadrupole ordered system PrV_2Al_{20} with Γ_3 doublet ground state at very low temperature of 23 mK under DC-high magnetic field. The field dependence between $H \parallel [111]$ and $H \parallel [110]$ is strongly anisotropic. The critical field of ordered phase for $H \parallel [110]$ is higher than that for $H \parallel [111]$.

Acknowledgments

This work is partially supported by PRESTO, Japan Science and Technology Agency, Grants-in-Aid for Scientific Research (No. 25707030, 15J08663 and 25887015), by Grants-in-Aids for Scientific Research on Innovative Areas (15H05882, 15H05883) and Program for Advancing Strategic International Networks to Accelerate the Circulation of Talented Researchers (No. R2604) from the Japanese Society for the Promotion of Science. Y.S. is partially supported by the Institute of Complex Adaptive Matter (ICAM). The NHMFL is supported by NSF through NSF-DMR-0084173 and the State of Florida. L.B. is supported by DOE-BES through award DE-SC0002613. This work was supported in part by NSF Grant No. PHYS-1066293 and the hospitality of the Aspen Center for Physics.

References

[1] Bucher E, Andres K, Gossard A and Maita J 1974 *J. Low Temp. Phys.* **2** 322
[2] Onimaru T, Matsumoto K T, Inoue Y F, Umeo K, Sakakibara T, Karaki Y, Kubota M and Takabatake T 2011 *Phys. Rev. Lett.* **106**
[3] Onimaru T, Nagasawa N, Matsumoto K T, Wakiya K, Umeo K, Kittaka S, Sakakibara T, Matsushita Y and Takabatake T 2012 *Phys. Rev. B* **86** 184426
[4] Sakai A and Nakatsuji S 2011 *J. Phys. Soc. Jpn.* **80** 063701
[5] Tsujimoto M, Matsumoto Y, Tomita T, Sakai A and Nakatsuji S 2014 *Phys. Rev. Lett.* **113** 267001
[6] Shimura Y, Tsujimoto M, Zeng B, Balicas L, Sakai A and Nakatsuji S 2015 *Phys. Rev. B* **91** 241102
[7] Shimura Y, Ohta Y, Sakakibara T, Sakai A and Nakatsuji S 2013 *J. Phys. Soc. Jpn.* **82** 043705
[8] Araki K, Shimura Y, Kase N, Sakakibara T, Sakai A and Nakatsuji S 2014 *JPS Conf. Proc.* **3** 011093
[9] Onimaru T, Sakakibara T, Aso N, Yoshizawa H, Suzuki H S and Takeuchi T 2005 *Phys. Rev. Lett.* **94** 197201
[10] Sato Y, Makiyama S, Morodomi H, Inagaki Y, Kawae T, Suzuki H S and Onimaru T 2012 *J. Phys.: Conf. Ser.* **391** 012060
[11] Onimaru T 2005 *Ph.D. Thesis, The University of Tokyo*
[12] Tayama T, Sakakibara T, Kitami K, Yokoyama M, Tenya K, Amitsuka H, Aoki D, Ōnuki Y and Kletowski Z 2001 *J. Phys. Soc. Jpn.* **70** 248
[13] Sato Y, Morodomi H, Ienaga K, Inagaki Y, Kawae T, Suzuki H S and Onimaru T 2010 *J. Phys. Soc. Jpn* **79** 093708
[14] Onimaru T, Aso N, Prokes K, Suzuki H, Sato T and Sakakibara T 2007 *J. Phys. Chem. Soli.* **68** 2091

Journal of Physics: Conference Series **683** (2016) 012013 doi:10.1088/1742-6596/683/1/012013

Heavy Fermion Superconductivity in Non-magnetic Cage Compound PrV_2Al_{20}

Yosuke Matsumoto, Masaki Tsujimoto, Takahiro Tomita, Akito Sakai and Satoru Nakatsuji

Institute for Solid State Physics, University of Tokyo, Kashiwa 277-8581, Japan

E-mail: `matsumoto@issp.u-tokyo.ac.jp`

Abstract. PrT_2Al_{20} (T = Ti, V) are ideal systems to study the quadrupole Kondo effect and quantum criticality arising from orbital degrees of freedom. Both systems have the nonmagnetic cubic Γ_3 crystal electric field ground doublet with the well separated excited state. In particular, PrV_2Al_{20} exhibits anomalous metallic behavior above and below the multipolar ordering temperatures, reflecting the even stronger hybridization between f and conduction electrons possibly due to a proximity to a orbital quantum critical point. Here, we discuss the heavy fermion superconductivity (SC) of PrV_2Al_{20} in detail. The SC appears at $T_c = 0.05$ K with the highly enhanced effective mass ($m^*/m_0 \sim 140$) estimated using the temperature dependence of the upper critical field. In addition, large electronic specific heat coefficient of $\gamma \sim 0.9$ J/mol K^2 above T_c and the large specific heat jump at T_c of $\Delta C/T_c \sim 0.3$ J/mol K^2 provide direct evidences of the heavy fermion SC. This observation indicates the first realization of the novel SC arising from the orbital fluctuation of the f electrons at ambient pressure.

1. Introduction

In the f electron intermetallics, a lot of interesting quantum states, such as exotic superconductivity (SC) with highly enhanced effective mass and non-Fermi liquid, have been reported in the vicinity of magnetic quantum critical points (QCPs) and have been extensively studied for the past few decades[1, 2, 3]. In this case, a competition between Kondo effect and RKKY interaction results in a quantum phase transition between Fermi liquid and a magnetically ordered state of f electrons. On the other hand, it is an interesting open question what happens in the case of orbital QCPs which emerges on the border of f electrons' orbital ordered states. This is highly non-trivial problem because hybridization effect between f electrons' orbital degrees of freedom and conduction electrons is already exotic. The hybridization effect has been studied in the impurity model known as 2-channel (or quadrupolar) Kondo model in the nonmagnetic cubic Γ_3 crystal electric field (CEF) doublet of f^2 state[4], where non-Fermi liquid ground state and residual entropy were suggested. However, it is not established both theoretically and experimentally what happens in the lattice systems. Therefore, it is even more non-trivial problem what happens at the orbital QCP where the strong hybridization suppresses the orbital ordering.

In order to study this highly non-trivial problem, it is important to choose a material which has purely orbital degrees of freedom in the ground state. Such examples can be found in cubic non-magnetic Γ_3 doublet systems of a f^2 configuration. Intensive studies have revealed various interesting phenomena in the cubic Γ_3 doublet systems, in particular, in Pr-based intermetallic compounds[5, 6, 7, 8, 9]. However, Pr-based cubic Γ_3 doublet systems usually have well-localized f moments, and thus there has been no prototypical system where we can study the orbital instability by tuning the hybridization between f and conduction electrons.

On the other hand, recent studies have revealed that PrT_2Al$_{20}$ (T = Ti, V) are ideal systems to study the quadrupole Kondo effect and quantum criticality arising from orbital degrees of freedom, where we can tune the c-f hybridization by chemical substitution and by pressure[10, 11]. The both systems have the nonmagnetic cubic Γ_3 CEF doublet with the well separated excited state at $\Delta_{\rm CEF}$ ~60 K (Ti) and 40 K (V)[10, 12, 13]. In addition, strong hybridization is expected in these compounds, from a caged structure with 16 Al ions surrounding T ions. Indeed, the hybridization is strong as is evident in many physical properties[10, 14, 15, 16]. In particular, PrV$_2$Al$_{20}$ exhibits anomalous metallic behavior above and below the multipolar ordering temperatures, reflecting the even stronger hybridization due to a smaller size of the cage.

While a ferro-quadrupole (FQ) ordering (at $T = 2.0$ K) was observed in PrTi$_2$Al$_{20}$[10, 12, 17], double multipolar orderings at temperatures $T = 0.65$ and 0.75 K were observed in PrV$_2$Al$_{20}$[16, 18]. The observation of the multipolar orderings ensures that the degeneracy of Γ_3 doublet remains at least down to the ordering temperature. Note that these are non-Kramers systems and the degeneracy can be easily removed by disorders in the crystals.

Interestingly, both of these compounds exhibit heavy fermion SC inside the mutipole ordering phases[11, 16, 19]. In PrTi$_2$Al$_{20}$, the SC is highly enhanced under pressure of 8 GPa, where FQ ordering starts to be suppressed. The transition temperature (T_c) at ambient pressure (~0.2 K) increases up to ~ 1 K and the effective mass reaches $m^*/m_0 \sim 110$[11]. This strongly suggests a novel heavy fermion SC emerges in the vicinity of the orbital QCP. In this paper, we discuss the experimental observations made in the resistivity and specific heat measurements, which suggest strong orbital fluctuation in PrV$_2$Al$_{20}$. We also discuss the correspondence between the resistivity measurements and the specific heat measurements, where we observed the double transition. Then we discuss the heavy fermion SC of PrV$_2$Al$_{20}$ in detail. The SC appears at $T_c = 0.05$ K with the highly enhanced effective mass ($m^*/m_0 \sim 140$) even at ambient pressure[16]. In addition, large electronic specific heat coefficient of $\gamma \sim 0.9$ J/mol K^2 above T_c and the large specific heat jump at T_c of $\Delta C/T_c \sim 0.3$ J/mol K^2 provide direct evidences of the heavy fermion SC. This observation indicates the first realization of the novel SC arising from the orbital fluctuation of the f electrons at ambient pressure. Note that a part of the results presented in this paper has been already discussed in the previous works[10, 16, 20, 21].

2. Experimental

Single crystals of RT_2Al$_{20}$ (R = Pr, La, T = Ti, V) were grown by the Al self-flux method[10]. We have succeeded in growing high quality single crystals, whose residual resistivity ratio (RRR) reaches ~ 20, by tuning the starting ratio. Single and powder -ray diffraction and SEM-ED measurements indicate high quality and single phase of obtained crystals. The electrical resistivity $\rho(T)$ was measured by a low-frequency ac four-terminal method below ~ 1 K using ^{3}He-^{4}He dilution refrigerator. A dc four-terminal method was used above $T = 0.3$ K in conjunction with a commercial system (PPMS, Quantum Design). In this paper, residual resistivity ratio (RRR) is roughly defined as the ratio between the values at 300 K and lowest temperature above T_c of the SC transition. The specific heat was measured using a relaxation method using a specific heat cell installed in a ^{3}He-^{4}He dilution refrigerator. The ac- and dc-susceptibility measurements were performed by using a home-made magnetometer comprising a commercial SQUID sensor (Tristan Technologies) installed in a dilution refrigerator. For these specific heat and ac- and dc-susceptibility measurements, a pure single crystal with RRR ~ 20 (0.111 mg) was used. The details were already discussed in the previous work[16].

3. Results and Discussions

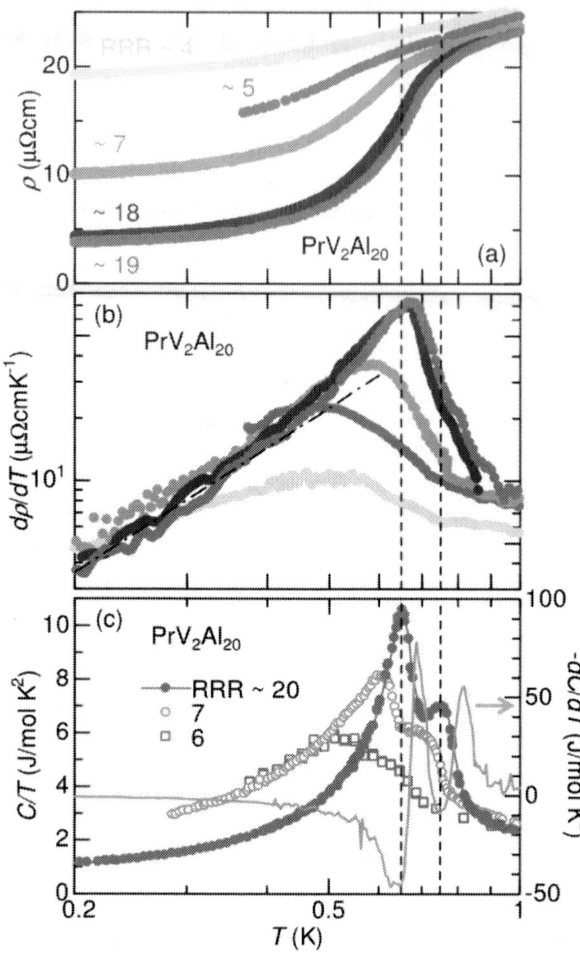

Figure 1. (a) Temperature dependence of the resistivity ρ for various single crystals of PrV_2Al_{20} with different sample qualities. (b) Full logarithmic plot of $d\rho/dT$ versus temperature for the data presented in (a). The dashed-dotted line indicates a power law temperature dependence of $d\rho/dT \propto T^2$. (c) Specific heat C divided by temperature T, C/T, for the samples with different RRRs (left axis) and $-dC/dT$ for the sample with RRR = 19 (right axis). Here, the lattice contribution is already subtracted by using C/T for LaV_2Al_{20}. The results for the sample with RRR = 6, 7 are from the previous works[10, 18]. The vertical broken lines in (a), (b), (c) indicate the transition temperatures T_Q = 0.75 K and T^* = 0.65 K for RRR = 20 defined as the peak temperatures in C/T in the previous works[16, 20].

First we present the temperature dependence of the resistivity ρ for various single crystals of PrV_2Al_{20} around multipolar ordering temperatures in Figure 1 (a). Depending on the sample quality, the resistivity drops below the ordering temperatures are different to each other. Correspondingly, the temperature derivative of the resistivity, $d\rho/dT$, indicates a systematic change depending on the sample quality as shown in Figure 1 (b). The samples with higher RRR indicate a sharper peak in $d\rho/dT$ at the multipolar ordering temperatures. Interestingly, $d\rho/dT$ for different quality samples overlap on top in the temperature range well below the multipolar ordering temperatures, especially for the sample with RRR $\sim$ 19, 18, 7. The observed temperature dependence is consistent with a power law behavior $d\rho/dT \propto T^2$ which is indicated by the dashed line in the figure and is consistent with the power law behavior $\Delta\rho \propto T^3$ reported in the previous work[21]. Here, $\Delta\rho$ is defined as $\Delta\rho \equiv \rho - \rho_0$, where ρ_0 is the residual resistivity estimated above T_c. The power law behavior suggests the strong orbital fluctuation even below the orbital ordering temperatures.

We also present the specific heat C divided by temperature T, C/T, for the samples with different RRRs in Figure 1 (c), in order to see the correspondence to the resistivity data. Here, the lattice contribution is already subtracted by using C/T for LaV_2Al_{20}. As already discussed in the previous works[16, 20], a broad single peak observed for RRR $\sim$ 6 separates into a sharp double peak structure for RRR $\sim$ 7 and 20. The peak structure which becomes sharper as RRR

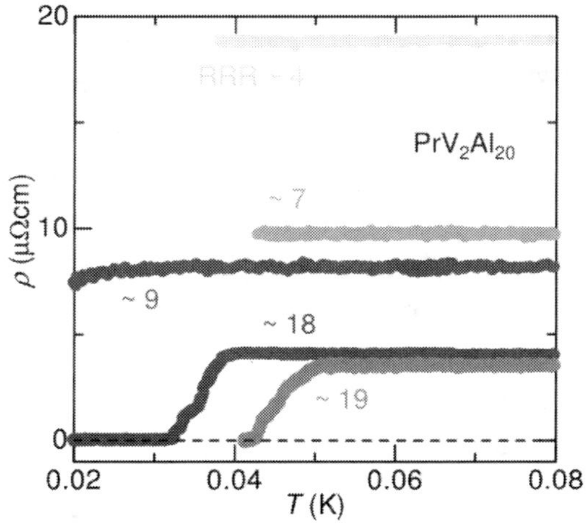

Figure 2. Temperature dependence of the resistivity ρ for various single crystals of PrV_2Al_{20} with different sample qualities at the lowest temperatures below $T \sim 0.08$ K.

increases strongly indicates that the double transition is intrinsic behavior in this compound. In the previous works[16, 20], the transition temperatures $T_Q = 0.75$ K and $T^* = 0.65$ K for RRR $= 19$ were defined by the peak temperatures in C/T as indicated by the vertical broken lines in the Figures 1 (a), (b), (c). Note that these temperatures are designated following the previous work[18] and we do not intend to refer to any specific order parameter in this paper. As seen in Figures 1 (a) and (b), there is no clear indication of the double transition in the resistivity data even for the samples with almost the same quality. If we dare to see the correspondence between ρ for the sample with RRR $= 19$ and C for the sample with RRR $= 20$, $T^* = 0.65$ may correspond to the peak in $d\rho/dT$ and $T_Q = 0.75$ may correspond to the temperature where $d\rho/dT$ becomes most steep (Fig. 1 (b)). We also present $-dC/dT$ for the sample with RRR $= 20$, which is indicated by the solid line in Figure 1 (c). Alternatively, it is also possible to associate the peak in $-dC/dT$ found at $T \sim 0.69$ K with the peak in $d\rho/dT$. Another peak in $-dC/dT$ found at $T \sim 0.82$ K, then, may correspond to the temperature where $d\rho/dT$ starts to rise on cooling. In order to see further correspondence or to obtain further insight on the order parameters of the multipolar orderings, more studies including those using macroscopic/microscopic probes are indispensable.

Next we discuss the SC found inside the multipolar ordered state of PrV_2Al_{20} under the strong orbital fluctuations. As clearly seen in Figure 2, the sample with RRR ~ 19 exhibits a resistivity drop due to the SC transition at $T_c = 0.05$ K. The SC transition was also found in another sample (RRR ~ 18) at a slightly lower $T_c = 0.039$ K. The sample with RRR ~ 9 exhibit a slight drop of the resistivity at $T \sim 0.02$ K. This is supposed to be the SC because it exhibits further drop reaching ~ 4 $\mu\Omega$cm in the lower temperatures where the thermometer is not calibrated and is suppressed under a weak field of the order of 10 mT applied along [110] direction (not shown). On the other hand, we did not find any sign of SC at least down to 38 and 43 mK for the sample with RRR ~ 4 and 7, respectively. The results suggest that T_c strongly depends on RRR, which may indicate an unconventional character of the SC. Apparently, further measurements are required to confirm the RRR dependence.

The bulk superconductivity was confirmed by dc- and ac-susceptibility measurements of a high quality crystal (RRR ~ 20). Figure 3 (a) shows the T dependence of the dc-susceptibility $\chi(T)$ of PrV_2Al_{20} under a field of $\mu_0 H = 0.1$ mT along the [110] direction. The unit of $\chi(T)$ is calibrated by using the perfect diamagnetic signal of the reference Al sample with the same dimension placed in the canceling coil. The volume fraction was estimated to be 82 (FC) and 47 (FC), indicating the bulk SC. Note that the susceptibility for the perfect diamagnetism of the sample is estimated to be -1.15 emu/cc, considering the diamagnetic correction factor

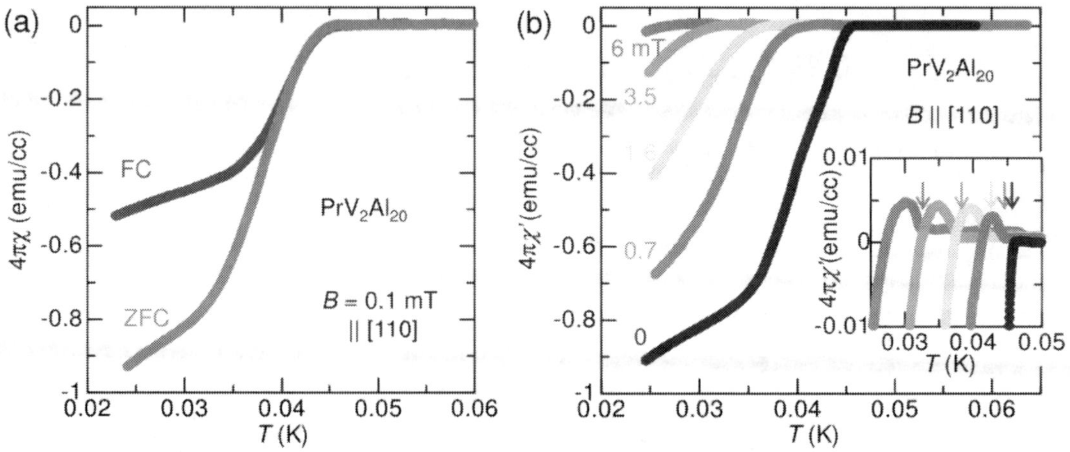

Figure 3. (a) Temperature dependence of the dc-susceptibility $\chi(T)$ for the single crystal of PrV_2Al_{20} (RRR $\sim$ 20) under the field of 0.1 mT for zero field cooled (FC) and field cooled (FC) sequences. (b) Real part of the ac-susceptibility $\chi'(T)$ for the same PrV_2Al_{20} sample. Inset shows an enlarged view around T_c. The arrows indicate T_c defined as the onset of the anomaly.

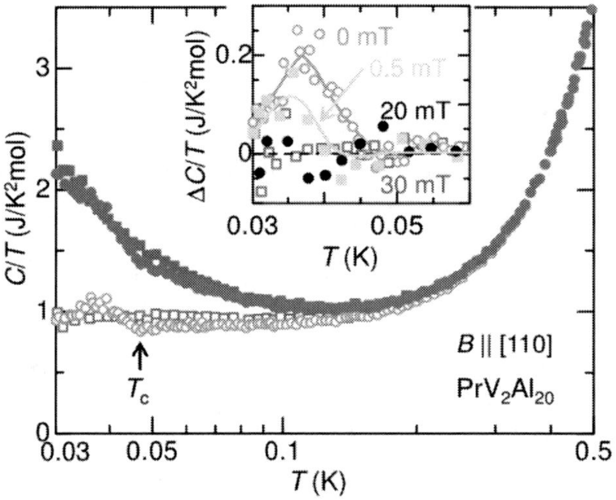

Figure 4. Temperature dependence of C/T (closed circles). Here, the lattice contribution is subtracted by using C/T for LaV_2Al_{20}. C_{4f}/T obtained after subtracting the nuclear contribution C_n/T is also shown (open circles). Inset shows $\Delta C/T$ obtained after subtracting γ from C_{4f}/T at the lowest temperatures around T_c. In addition to the zero-field data, those in the weak magnetic field along the [110] direction are also shown. See text for details.

$D \sim 0.13$. The large diamagnetic signal comparable to the Al shielding signal is also observed in the T dependence of the real part of the ac-susceptibility $\chi'(T)$ under zero dc-field and ac-field $\sim 0.1\ \mu$T as shown in Figure 3 (b). Here, T_c was determined by the onset of the anomaly of the ac-susceptibility as shown by the arrows in Figure 3 (b) inset. Compared to $PrTi_2Al_{20}$, the peak in the susceptibility around T_c due to differential paramagnetic effect is quite small in PrV_2Al_{20}[22], indicating strong pinning effect typical to the type II SC.

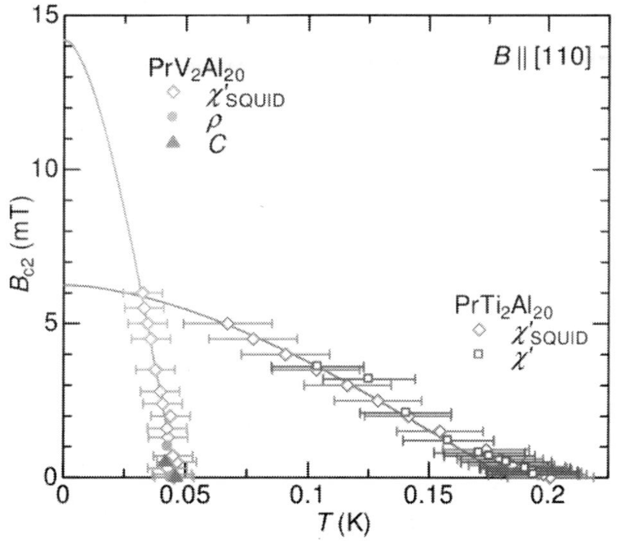

Figure 5. T dependence of the upper critical field B_{c2} for the single crystals of PrV_2Al_{20} (RRR ~ 20) and $PrTi_2Al_{20}$ (RRR ~ 150) under a field along [110]. The square/diamond/circle/triangle data points are determined by ac-susceptibility with a standard technique, ac-susceptibility by SQUID, resistivity, and specific heat measurements, respectively. Solid lines represent the fit based on the WHH model.

Furthermore, the bulk heavy fermion SC is confirmed by the specific heat data. We present C/T below 0.5 K in Figure. 4. In the lower temperatures, the data exhibit a slight increase on cooling below 0.12 K and finally indicate an anomaly due to SC transition at $T = 0.046$ K. The low T upturn in C/T becomes evident in the normal state stabilized under the magnetic field of 30 mT and is found to follow $C/T \sim T^2$ down to the lowest T of 30 mK. As already discussed in the previous work[16], this contribution is most likely arising from a hyperfine enhanced nuclear magnetism. The $4f$-electron contribution of the specific heat divided by T, C_{4f}/T is obtained after subtracting $C_n/T \sim T^{-2}$ coming from the nuclear magnetism. The zero-field SC anomaly is clearly seen in C_{4f}/T, which is shown in the figure by the open circles. This confirms the bulk character of the SC. The nearly constant C_{4f}/T in the normal state provides an estimate of the large electronic coefficient $\gamma \sim 0.9$ J/molK^2. The SC jump in C/T is evaluated using the temperature dependence of $\Delta C/T \equiv C_{4f}/T - \gamma$, as is shown in the inset of Figure 4. The jump reaches 0.3 J/molK^2, which provides direct evidence of the heavy-fermion SC. The ratio obtained from the jump, $\Delta C/\gamma T_c \sim 0.3$ is much smaller than the BCS value of 1.43. As already discussed in the previous work[16], the above estimated jump can be further increased if the sample quality is improved, which is the one of the important future issues. In the inset of Figure 4, $\Delta C/T$ in weak magnetic fields are also shown. The SC anomaly is slightly suppressed at $B = 0.5$ mT with $T_c = 0.042$ K and is not observed at $B = 20$ and 30 mT.

Figure 5 shows the T dependence of the upper critical field B_{c2} estimated using the ac-susceptibility, resistivity and specific heat data for PrV_2Al_{20}. For comparison, we show the results for $PrTi_2Al_{20}$ as well [19]. The solid line in Fig. 5 is the fit to our B_{c2} results based on the Werthamer-Helfand-Hohenberg (WHH) model [23, 24]. The best fitting was obtained using parameters of $T_c = 46.2$ mK and the slope of B_{c2} at T_c, $B'_{c2} \equiv dB_{c2}/dT = 0.41$ T/K. The model reproduces the experimental data well, indicating the orbital depairing effect is dominant. The resultant orbital critical field at $T = 0$, $B_{c2}^{orb}(0) = -0.727 B'_{c2} T_c$, and Ginzburg-Landau (GL) coherence length, $\xi = \sqrt{\Phi_0/2\pi B_{c2}^{orb}(0)}$, are $B_{c2}^{orb}(0) = 14.3$ mT and $\xi = 0.15$ μm, respectively.

Surprisingly, B'_{c2} of PrV_2Al_{20} is about 10 times larger than the Ti analog [19], indicating significantly heavier effective mass. Indeed, the effective mass is estimated to be $m^* = \hbar k_F/v_F \sim 140m_0$ by using the GL coherence length $\xi = 0.15$ μm, the Fermi velocity $v_F = \xi k_B T_c/(0.18\hbar) = 5.1$ km/s, $k_F = (3\pi^2 Z/\Omega) = 6.1 \times 10^9$ m^{-1}, where Z is the number of electrons per unit cell, and Ω is the unit-cell volume. The effective mass $m^*/m_0 \sim 140$ is much larger than $m^*/m_0 \sim 16$ estimated for $PrTi_2Al_{20}$ under ambient pressure [19], is comparable to $m^*/m_0 \sim 110$ under ~ 8 GPa in the vicinity of the quadrupolar quantum criticality [11]. Thus, the mass enhancement in PrV_2Al_{20} indicates not only the strong c-f hybridization, but its proximity to a quadrupolar QCP.

4. Conclusion

In this paper, we discussed the experimental observations made in the resistivity and specific heat measurements, which suggest strong orbital fluctuation in PrV_2Al_{20}. We also discussed the correspondence between the resistivity measurements and the specific heat measurements, where we observed the double transition. Then we discussed the properties of the heavy fermion SC in PrV_2Al_{20} in detail. The SC appears at $T_c = 0.05$ K for the sample with RRR = 19. The strong RRR dependence of T_c may indicate an unconventional character of the SC. The bulk superconductivity was confirmed by dc- and ac-susceptibility measurements and specific heat measurements made for the same high quality single crystal with RRR ~ 20. Large electronic specific heat coefficient of $\gamma \sim 0.9$ J/mol K^2 above T_c and the large specific heat jump at T_c of $\Delta C/T_c \sim 0.3$ J/mol K^2 provide direct evidences of the heavy fermion SC together with the highly enhanced effective mass ($m^*/m_0 \sim 140$) estimated using the temperature dependence of the upper critical field. This observation indicates the first realization of the novel SC arising from the orbital fluctuation of the f electrons at ambient pressure. In order to establish the role of orbital fluctuation in the heavy fermion SC found in PrV_2Al_{20}, comprehensive studies including those using macroscopic/microscopic probes in conjunction with high pressure and magnetic fields are quite important. Further improvement of the sample quality is also the important future issue.

Acknowledgments

We thank Y. Uwatoko, K. Matsubayashi, J. Suzuki, T. Sakakibara, Y. Shimura, K. Araki, Y.-B. Kim, K. Kuga, K. Hattori and K. Ueda for their support and the useful discussions. This Letter was partially supported by PRESTO of JST and Grants-in-Aid (Nos. 25707030 and 90422443) from the Japanese Society for the Promotion of Science (JSPS), by Grants-in-Aids for Scientific Research on Innovative Areas" Heavy Electrons " and Scientific Research on Innovative Areas (15H05882, 15H05883) of the Ministry of Education, Culture, Sports, Science and Technology, Japan. The use of the facilities of the Materials Design and Characterization Laboratory at the Institute for Solid State Physics, The University of Tokyo is gratefully acknowledged. One of the authors, (M. T.) was supported by the JSPS through Program for Leading Graduate Schools (MERIT).

References

[1] Stewart G R 2001 *Rev. Mod. Phys.* **73** 797
[2] Löhneysen H v, Rosch A, Vojta M and Wölfle P 2007 *Rev. Mod. Phys.* **79** 1015
[3] Monthoux P, Pines D and Lonzarich G G 2007 *Nature* **450** 1177
[4] Cox D L 1987 *Phys. Rev. Lett.* **59** 1240
[5] Morin P, Schmitt D and du Tremolet de Lacheisserie E 1982 *J. Magn. Magn. Mater.* **30** 257
[6] Suzuki H, Kasaya M, Miyazaki T, Nemoto Y and Goto T 1997 *J. Phys. Soc. Jpn.* **66** 2566
[7] Onimaru T, Sakakibara T, Aso N, Yoshizawa H, Suzuki H S and Takeuchi T 2005 *Phys. Rev. Lett.* **94** 197201

[8] Onimaru T, Matsumoto K T, Inoue Y F, Umeo K, Saiga Y, Matsushita Y, Tamura R, Nishimoto K, Ishii I, Suzuki T and Takabatake T 2010 *J. Phys. Soc. Jpn.* **79** 033704
[9] Onimaru T, Matsumoto K T, Inoue Y F, Umeo K, Sakakibara T, Karaki Y, Kubota M and Takabatake T 2011 *Phys. Rev. Lett.* **106** 177001
[10] Sakai A and Nakatsuji S 2011 *J. Phys. Soc. Jpn.* **80** 063701
[11] Matsubayashi K, Tanaka T, Sakai A, Nakatsuji S, Kubo Y and Uwatoko Y 2012 *Phys. Rev. Lett.* **109** 187004
[12] Sato T J, Ibuka S, Nambu Y, Yamazaki T, Hong T, Sakai A and Nakatsuji S 2012 *Phys. Rev. B* **86** 184419
[13] Koseki M, Nakanishi Y, Deto K, Koseki G, Kashiwazaki R, Shichinomiya F, Nakamura M, Yoshizawa M, Sakai A and Nakatsuji S 2011 *J. Phys. Soc. Jpn.* **80** SA049
[14] Matsunami M, Taguchi M, Chainani A, Eguchi R, Oura M, Sakai A, Nakatsuji S and Shin S 2011 *Phys. Rev. B* **84** 193101
[15] Tokunaga Y, Sakai H, Kambe S, Sakai A, Nakatsuji S and Harima H 2013 *Phys. Rev. B* **88** 085124
[16] Tsujimoto M, Matsumoto Y, Tomita T, Sakai A and Nakatsuji S 2014 *Phys. Rev. Lett.* **113** 267001
[17] Ito T U, Higemoto W, Ninomiya K, Luetkens H, Baines C, Sakai A and Nakatsuji S 2011 *J. Phys. Soc. Jpn* **80** 113703
[18] Araki K, Shimura Y, Kase N, Sakakibara T, Sakai A and Nakatsuji S 2014 *JPS Conf. Proc.* **3** 011093
[19] Sakai A, Kuga K and Nakatsuji S 2012 *J. Phys. Soc. Jpn* **81** 083702
[20] Tsujimoto M, Matsumoto Y and Nakatsuji S 2015 *J. Phys. Conf.: Ser.* **592** 01023
[21] Matsumoto Y, Tsujimoto M, Tomita T, Sakai A and Nakatsuji S *to appear in J. Magn. Magn. Mater.*
[22] Hein R A and Falge R L 1961 *Phys. Rev.* **123** 407
[23] Werthamer N R, Helfand E and Hohenberg P C 1966 *Phys. Rev.* **147** 295
[24] Helfand E and Werthamer N R 1966 *Phys. Rev.* **147** 288

Low-temperature thermal expansion measurements in PrV_2Al_{20}

A. Magata[1], **Y. Matsumoto**[1], **M. Tsujimoto**[1], **T. Tomita**[1], **R. Küchler**[2], **A. Sakai**[3] **and S. Nakatsuji**[1]

[1]Institute for Solid State Physics, The University of Tokyo, Kashiwa, Chiba, Japan
[2]Max Planck Institute for Chemical Physics of Solids, Nöthnitzer Strasse 40, 01187 Dresden, Germany
[3]Universtät Augsburg, Universitätsstrasse 2, 86159 Augsburg, Germany

E-mail: `magata@issp.u-tokyo.ac.jp`

Abstract. We have measured thermal expansion of PrV_2Al_{20} and LaV_2Al_{20} from room temperature down to 2 K, using a capacitance dilatometer. Linear thermal expansion $\Delta L/L$ along [111] direction decreases monotonically on cooling in both materials. The extracted $4f$ electrons contribution of the linear thermal expansion coefficient α clearly shows a broad peak at $\sim$ 30 K which may correspond to the crystal electric field excited state at 40 K suggested in the previous specific heat study.

1. Introduction

Recently, orbital degrees of freedom have attracted much attention in condensed matter physics and a lot of interesting phenomena have been studied extensively, such as spin-orbital disordered state in $Ba_3CuSb_2O_9$ [1] and superconductivity in Iron-based compounds [2]. While spin and orbital degrees of freedom are strongly coupled in these d electron systems, f electron intermetallics can provide an interesting case, such as non-magnetic Γ_3 crystal electric field (CEF) ground doublet of f^2 configuration in a cubic site symmetry. This doublet does not have magnetic dipolar degrees of freedom but consists of quadrupolar and octupolar degrees of freedom. Therefore, in this case, we can study phenomena arising from pure orbital degrees of freedom. In particular, quadrupole Kondo effect, originally suggested as a single ion effect in the nonmagnetic cubic Γ_3 doublet [3], has attracted a lot of attention because of its non-Fermi liquid ground state. Indeed, several cubic Pr-based compounds having cubic symmetry in crystal structure, such as $PrPb_3$ [4, 5, 6], $PrInAg_2$ [7] and $PrMg_3$ [8], have been studied extensively in this context. However, there has been no established example with strong hybridization between f electrons and conduction electrons without structural disorders. Note that structural disorders can easily lift the degeneracy of the non-Kramers doublet and hence high quality single crystal is quite important.

On the other hand, recent studies revealed that $PrTr_2Al_{20}$ (Tr: Ti, V) are ideal systems to study quadrupolar Kondo effect. $PrTr_2Al_{20}$ are cage compounds which have cubic $CeCr_2Al_{20}$-type structure with the space group Fd-3m [9]. As it is expected from the crystal structure where Pr^{3+} ion is surrounded by 16 Al atoms, the hybridization is strong as is evident in many physical properties [10, 11]. In our previous studies, we have already succeeded in synthesizing high-quality single crystal of PrTr_2Al_{20} by Al self-flux methods [10, 11]. The both have non-magnetic Γ_3 ground doublet with the well separated excited state at $\Delta_{CEF} = 60$ K (Ti) and 40 K (V) respectively, which have been confirmed by various experiments [10, 12, 13]. $PrTi_2Al_{20}$ exhibits ferroquadrupole ordering at $T_Q = 2$ K, and PrV_2Al_{20} exhibits double transitions at $T_Q = 0.75$ K and $T^* = 0.65$ K [10, 11, 14]. The origin of double transition in PrV_2Al_{20} is unknown still, but should come from quadrupole and/or octupole degree of freedom. In PrV_2Al_{20}, c-f hybridization is expected to be stronger than that of $PrTi_2Al_{20}$ due to smallness of Al cage surrounding Pr ions. Indeed, it exhibits anomalous metallic behavior above the multipolar transitions, such as anomalous temperature dependence of electrical resistivity ($\rho \sim T^{-1/2}$ and $\rho \sim \ln T$), magnetic susceptibility ($\chi \sim -T^{1/2}$), and specific heat($C/T \sim T^{-3/2}$) [10]. Furthermore, PrV_2Al_{20} exhibits heavy fermion superconductivity at $T_C = 50$ mK in the multipole ordered state even under the ambient pressure [11], suggesting a proximity to an orbital QCP, where an orbital ordering is suppressed due to a strong hybridization and orbital fluctuation.

Here, we have performed thermal expansion measurements in PrV_2Al_{20} and LaV_2Al_{20} from 300 K to 2 K. Thermal expansion is a quite useful quantity to detect a lattice strain which couples to quadrupolar degrees of freedom. For example, if there is a ferroquadrupole ordering, spontaneous strain of lattice with the same symmetry as that of the quadrupole ordering emerges due to co-operative Jahn-Teller effect [15, 16]. Thermal expansion measurements are useful also for studies of heavy fermion system. One example is the recent study on $EuNi_2P_2$, where it has been revealed that the heavy fermion state in this compound arises from Kondo effect for the first time in Eu-based compounds on the analogy of $CeRu_2Si_2$ [17, 18]. In our study, we observed obvious difference between PrV_2Al_{20} and LaV_2Al_{20} in linear thermal expansion coefficient α along [111] direction. The extracted $4f$ electrons contribution of the linear thermal expansion coefficient α clearly shows a broad peak at ~ 30 K, which may correspond to the crystal electric field excited state at 40 K, which was suggested from the previous specific heat measurements.

2. Experimental

Single crystals of PrV_2Al_{20} and LaV_2Al_{20} were prepared by Al self-flux method, using 4N (99.99)-Pr, 4N-La, 3N-V and 5N-Al, respectively. The sample length along [111] direction is 1.2 mm and 1.1 mm for PrV_2Al_{20} and LaV_2Al_{20}, respectively.

Thermal expansion measurements were performed at temperature range from 300 K down to 2 K, in a zero-magnetic field, by using a capacitance dilatometer installed in a commercial cryostat (PPMS, Quantum Design). Linear thermal expansion was measured with an accuracy of ~ 0.5 Å.

3. Results and discussion

The linear thermal expansion $\Delta L/L$ of PrV_2Al_{20} and LaV_2Al_{20} along [111] direction are shown in Figure 1. $\Delta L/L_{[111]}$ for both compounds exhibit monotonous decrease on cooling and no anomaly was observed in a temperature range from 300 K to 2K. The magnitude of the thermal expansion $\Delta L/L \sim$ 0.25-0.3 is close to the value for Cu $\Delta L/L \sim 0.33$ [19] and is relatively common value among metals. Thermal expansion consists of three contributions, namely, phonons, conduction electrons, and $4f$ electrons contributions. The length change ΔL can be written as $\Delta L = \Delta L_{\text{phonon}} + \Delta L_e + \Delta L_{4f}$. Because the former two contributions are supposed to be almost the same for both compounds, the difference between these two corresponds to the $4f$ electrons contribution.

Figure 2 shows thermal expansion coefficient α in PrV_2Al_{20} and LaV_2Al_{20} along [111] direction. Linear thermal expansion coefficient α is defined as the temperature derivative of normalized length change $\Delta L/L$, *i.e.*, $\alpha \equiv \frac{d}{dT}(\Delta L/L)$. α also decreases monotonically

Journal of Physics: Conference Series **683** (2016) 012014 doi:10.1088/1742-6596/683/1/012014

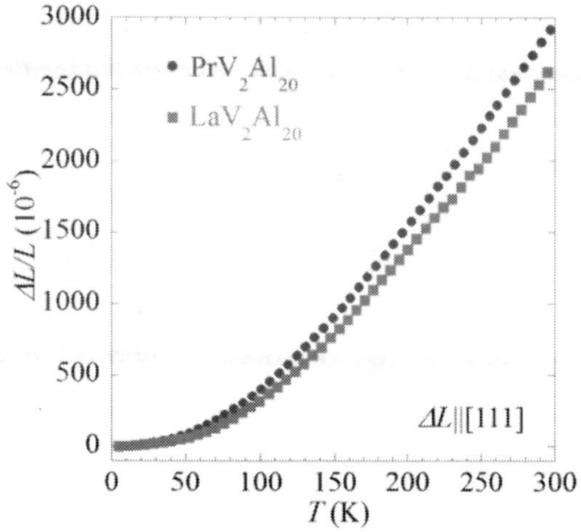

Figure 1. Linear thermal expansion along [111] direction for PrV_2Al_{20} and LaV_2Al_{20}.

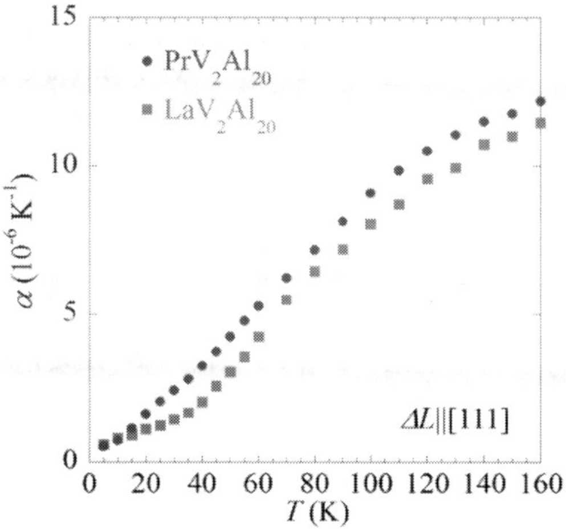

Figure 2. Linear thermal expansion coefficient α along [111] direction for PrV_2Al_{20} and LaV_2Al_{20}.

with decreasing temperature for both samples. In general, thermal expansion coefficient α is proportional to specific heat. Therefore, in the low enough temperatures, α for LaV_2Al_{20} will be expressed by a summation of T-linear and T^3 power law terms, which arise from conduction electrons and phonons, respectively. However, while the specific heat can be well expressed by a summation of T-linear and T^3 power law terms below 8.5 K [20], it is hard to see this behavior in the current thermal expansion data because of the accuracy and the temperature range of the measurements. More accurate measurements down to lower temperature are required. Note

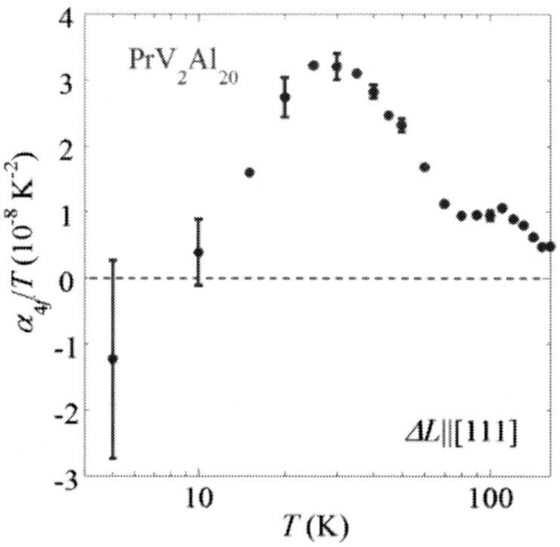

Figure 3. $4f$ electrons contribution of the linear thermal expansion coefficient α_{4f} along [111] direction divided by temperature T. For detail, see text.

that the Debye temperature θ_D of LaV_2Al_{20} has been already estimated as ~ 510 K from the specific heat measurements [20].

Figure 3 shows $4f$ electrons contribution of the linear thermal expansion coefficient α_{4f} divided by temperature. α_{4f} is obtained by subtracting the α for LaV_2Al_{20} from that for PrV_2Al_{20}, *i.e.*, $\alpha_{4f} \equiv \alpha_{PrV_2Al_{20}} - \alpha_{LaV_2Al_{20}}$. The extracted $4f$ electrons contribution of the linear thermal expansion coefficient α clearly shows a broad peak at ~ 30 K. Below ~ 30 K, it decreases rather steeply and reaches almost zero at the lowest temperature of $T \sim 5$ K. The broad peak found at ~ 30 K is supposed to arise from the CEF excited state at 40 K, which was suggested from the previous specific heat measurements. $4f$ electrons contribution of the specific heat C_{4f} exhibits broad peak at ~ 20 K [10]. In the cubic system with Γ_3 ground doublet, the excited CEF state should be either magnetic Γ_5 or Γ_4 triplet. In order to obtain further information on the excited CEF state from the thermal expansion measurements, measurements for other directions, such as [100] and [110] directions, and magnetic field angle resolved measurements will be useful, which are interesting future issues. In the lowest temperatures around 5 K, rather large experimental error makes it difficult to see if the observed negative value of α_{4f}/T is intrinsic or not. Negative thermal expansion is rather unusual behavior. Therefore, it is an interesting future issue to examine such a possibility in PrV_2Al_{20}. In f electron systems, there are a few examples of negative thermal expansion, such as those found in TmTe [21] and Yb-based heavy fermions [22]. In the case of TmTe, the negative thermal expansion is associated with a sign change of Grüneisen constant across the CEF level [21]. However this is unlikely in PrV_2Al_{20} because the possible negative thermal expansion occurs well below the CEF excited state at 40 K. On the other hand, in Yb-based heavy fermions, negative thermal expansion occurs due to the valence change from Yb^{3+} to Yb^{2+}, which is induced by c-f hybridization (Kondo effect) [22]. Likewise, in the case of PrV_2Al_{20}, one interesting possibility is a negative thermal expansion arising from a valence change due to a strong hybridization between f electrons orbital degrees of freedom and conduction electrons, *i.e.*, quadrupole Kondo effect. Apparently, further experiments and theoretical inputs are required to study this.

4. Conclusion

We have measured thermal expansion of PrV_2Al_{20} and LaV_2Al_{20} from room temperature down to 2 K, using a capacitance dilatometer. Linear thermal expansion $\Delta L/L$ along [111] direction decreases monotonically on cooling in both materials. The extracted $4f$ electrons contribution of the linear thermal expansion coefficient α clearly shows a broad peak at ~ 30 K, which may correspond to the CEF excited state at 40 K and is consistent with the previous specific heat measurements. Further measurements are required in order to investigate the possible negative thermal expansion in the lowest temperatures and the detail of the CEF excited state at 40 K.

5. Acknowledgments

We thank Y. Shimura for support and useful discussion. This work was partially supported by J-Physics : Physics of Conductive Multipole Systems.

References

[1] Nakatsuji S, Kuga K, Kimura K, Satake R, Katayama N, Nishibori E, Sawa H, Ishii R, Hagiwara M, Bridges F, Ito T U, Higemoto W, Karaki Y, Halim M, Nugroho A A, Rodriguez-Rivera J A, Green M A, Broholm C 2012 *Science* **336** 559
[2] Kamihara Y, Watanabe T, Hirano M and Hosono H 2008 *J. Am. Chem. Soc.* **130** 3296
[3] Cox D L 1987 *Phys. Rev. Lett.* **59** 1240
[4] Morin P, Schmitt D and E du Tremolet de Lacheisserie 1982 *J. Magn. Magn. Mater.* **30** 257
[5] Onimaru T, Sakakibara T, Aso N, Yoshizawa H, Suzuki H S and Takeuchi T 2005 *Phys. Rev. Lett.* **94** 197201

[6] Kawae T, Kinoshita K, Nakaie Y, Tateiwa N and Takeda K 2006 *Phys. Rev. Lett.* **96** 027210
[7] Yatskar A, Beyermann W P, Movshovich R and Canfield P C 1996 *Phys. Rev. Lett.* **77** 3637
[8] Tanida H, Suzuki H S, Takagi S, Onodera H and Tanigaki K 2006 *J. Phys. Soc. Jpn.* **75** 073705
[9] Kangas M J, Schmitt D C, Sakai A, Nakatsuji S and Chan J Y 2012 *J. Solid State Chem.* **196** 274
[10] Sakai A and Nakatsuji S 2011 *J. Phys. Soc. Jpn.* **80** 063701
[11] Tsujimoto M, Matsumoto Y, Tomita T, Sakai A and Nakatsuji S 2014 *Phys. Rev. Lett.* **113** 267001
[12] Sato T J, Ibuka S, Nambu Y, Yamazaki T, Sakai A and Nakatsuji S 2012 *Phys. Rev.* B **86** 184419
[13] Koseki M, Nakanishi Y, Deto K, Koseki G, Kashiwazaki R, Shichinomiya F, Nakamura M, Yoshizawa M, Sakai A and Nakatsuji S 2011 *J. Phys. Soc. Jpn.* **80** SA 049
[14] Tsujimoto M, Matsumoto Y and Nakatsuji S 2015 *J. Phys.: Conf. Ser.* **592** 012023
[15] Nemoto Y, Yamaguchi Y, Horino T, Akatsu M, Yanagisawa T and Goto T 2003 *Phys. Rev.* B **68** 184109
[16] Kazei Z A, Kolmakova N P, Krynetski I B, Sidorenko A A and Takunov L V 2000 *Physics of the Solid State* **42** 285
[17] Hiranaka Y, Nakamura A, Hedo M, Takeuchi T, Mori A, Hirose Y, Mitamura K, Sugiyama K, Hagiwara M, Nakama T and Onuki Y 2013 *J. Phys. Soc. Jpn.* **82** 083708
[18] Lacerda A, Visser A, Haen P, Lejay P and Flouquet J 1989 *Phys. Rev.* B **40** 8759
[19] Kroeger F R, Swenson C A 1977 *J. Appl. Phys.* **48** 853
[20] Sakai A *Unpublished*
[21] Ott H R, Lüthi B 1977 *Z. Physik* B **28** 141
[22] Takeuchi T, Yoda M, Matsushita M, Yoshiuchi S, Ohya M, Katayama K, Hirose Y, Yoshitani N, Honda F, Sugiyama K, Hagiwara M, Kindo K, Yamamoto E, Haga Y, Tanaka T, Kubo Y, Settai R and Onuki Y 2010 *J. Phys. Soc. Jpn.* **79** 064609

Journal of Physics: Conference Series **683** (2016) 012015 doi:10.1088/1742-6596/683/1/012015

Analysis of ^{27}Al- and ^{93}Nb-NMR spectra of $PrNb_2Al_{20}$ single crystal

Tetsuro Kubo[1,*], **Hisashi Kotegawa**[1], **Hideki Tou**[1,†]
Ryuji Higashinaka[2], **Akihiro Nakama**[2], **Yuji Aoki**[2], **and Hideyuki Sato**[2]

[1]Department of Physics, Graduate School of Science, Kobe University, 1-1 Rokkodai, Kobe, Hyogo 657-8501, Japan

[2]Department of Physics, Graduate School of Science and Engineering, Tokyo Metropolitan University, 1-1 Minamiohsawa, Hachioji, Tokyo 192-0397, Japan

E-mail: *tkubo@stu.kobe-u.ac.jp †tou@crystal.kobe-u.ac.jp

Abstract. We report the results of ^{27}Al- and ^{93}Nb- nuclear magnetic resonance (NMR) and nuclear quadrupole resonance (NQR) measurements in $PrNb_2Al_{20}$ single crystals. The field angle dependence of the observed ^{93}Nb-NMR lines for the single crystal at around 5 T and at 50 K is reproduced by a simulation using previously reported NQR parameters $\nu_{Q,Nb} \approx 1.82$ MHz and $\eta_{Nb} \approx 0$, while ^{27}Al-NMR lines were not reproduce by the previously reported NQR parameters $\nu_{Q,Al(3)} \approx 1.53$ MHz and $\eta_{Al(3)} \approx 0.17$. By reexamining and reconsidering the NMR and NQR lines for powder samples, we obtained a correct NQR parameters for Al(3) site to be $\nu_{Q,Al(3),new} \approx 1$ MHz and $\eta_{Al(3),new} \approx 0.4$.

1. Introduction

Pr-based cubic intermetallic compound PrT_2X_{20} (T = Ir, Rh, Ti, V, Nb, X = Zn, Al) shows various interesting behaviors such as quadrupole order, unconventional superconductivity and non-Fermi liquid (NFL) behavior [1, 2, 3, 4, 5, 6]. In this system, $4f$ electrons can have a nonmagnetic but multipole active crystalline electric field (CEF) ground state because of the cubic CEF effect with T_d point symmetry. Such a CEF ground state is characterized by Γ_3 doublet that has no magnetic dipole moment but has two electric quadrupole moment O_{20} and O_{22} and one magnetic octupole moment T_{xyz}. These multipole degrees of freedom are considered to be a key to understanding the intriguing phenomena at low temperatures in the PrT_2X_{20}.

In the PrT_2X_{20} series, it is shown that $PrNb_2Al_{20}$ does not show any phase transition down to 75 mK by bulk and nuclear quadrupole resonance (NQR) measurements [7, 8]. In addition, this compound shows NFL behaviors such as T-linear resistivity and logarithmic divergent specific heat divided by temperature at low temperatures below 2 K [3]. The relationship between these anomalous behaviors and the multipole degrees of freedom has been discussed in this compound [7, 8].

In order to investigate the low-energy excitation of the multipoles in $PrNb_2Al_{20}$, nuclear magnetic resonance (NMR) and NQR are desirable because these methods can obtain the information of the local electronic state from static and dynamic viewpoints. We have already reported possible NQR parameters at Al site from the ^{27}Al-NMR results measured at 30 K and around 2.8 T for a powder sample [9]. In the previous study, we assumed that the contributions of the hyperfine and classical dipolar interactions are negligible compared to the electric quadrupole interactions.

$PrNb_2Al_{20}$ has the cubic $CeCr_2Al_{20}$-type structure with the space group $Fd\bar{3}m$: Pr atoms at 8a site are encapsulated in highly symmetric Al_{16} cages, so called the Frank-Kasper cage formed by four Al(1) atoms and twelve Al(3) atoms. Note that Al atoms in 1-2-20 system have three crystallographically inequivalent Al sites; Al(1) is in the position of trigonal 16c site; Al(2) is in the orthorombic 48f site; Al(3) is in the monoclinic 96g site. Another structural aspect to be noted is that the Nb atoms at the trigonal 16d site form a pyrochlore lattice. Therefore, it is difficult to determine the electric field gradient (EFG) tensors at NMR active Al and Nb sites only by the powder NMR spectrum. Before discussing NMR results in detail, we need to gain a detailed EFG parameters. In this paper, we report the results of ^{93}Nb-NMR measurements using a single crystal and reanalysis of ^{27}Al-NMR results for the powder sample. In the present study, we have found that contributions from dipolar and hyperfine interactions are important to discuss the NMR line profile as well as field and temperature dependences of the NMR spectrum. Thus we have reexamined and reanalyzed the ^{27}Al-NMR powder spectrum together with ^{93}Nb-NMR measurements using a single crystal sample.

2. Experimental procedure

Single crystal samples of $PrNb_2Al_{20}$were prepared by Al-flux method. Details of sample preparation method are reported in the previous paper [3]. Field angle dependence of NMR lines are measured by using a single crystal sample that has the dimension of $\sim 2\times2\times2\,\mathrm{mm}^3$. To estimate NQR parameters, measurements were performed using powder samples, in which single crystal samples were crushed into powder form in order to gain signal intensity. NMR measurements were performed using 12 T and 17 T superconducting magnets and a phase-coherent pulsed spectrometer at magnetic fields around $\mu_0 H = 2.8$ and 5 T. ^{93}Nb- (nuclear spin $I = 9/2$, nuclear gyromagnetic ratio $^{93}\gamma_n/2\pi = 10.452\,\mathrm{MHz/T}$, and nuclear quadrupole moment $Q = -0.320\times10^{-28}\,\mathrm{m}^{-2}$) and ^{27}Al- ($I = 5/2$, $^{27}\gamma_n/2\pi = 11.103\,\mathrm{MHz/T}$, and $Q = -0.1466\times10^{-28}\,\mathrm{m}^{-2}$) NMR spectra were obtained by tracing spin-echo intensity as a function of the resonance field.

3. Results and discussion

First, we discuss the ^{93}Nb-NMR spectrum for a single crystal. Figure 1 shows a field swept NMR spectrum measured at a fixed frequency of $f = 48.751$ MHz and 50 K for the applied field nearly parallel to $\langle 110\rangle$. The linewidth of the obtained NMR lines is narrow about 30 G at around 5 T, and the each line splits well.

To reproduce the observed ^{93}Nb-NMR spectrum, the resonance fields were calculated by an exact diagonalization of the $(2I+1)\times(2I+1)$ nuclear spin hamiltonian [10],

$$\mathcal{H} = -\gamma_n\hbar\boldsymbol{I}\cdot\boldsymbol{H}_0 - \gamma_n\hbar\boldsymbol{I}\cdot\tilde{K}\cdot\boldsymbol{H}_0 - \gamma_n\hbar\boldsymbol{I}\cdot\boldsymbol{H}_d + \frac{1}{6}h\nu_Q\left\{3I_Z^2 - I(I+1) + \frac{\eta}{2}(I_+^2 + I_-^2)\right\} \quad (1)$$

where h is Planck constant and $\hbar \equiv h/2\pi$. The first term is the Zeeman term between the nuclear spin $\boldsymbol{I}$ and the applied field $\boldsymbol{H}_0$. The second term shows a hyperfine interaction between the nuclear spin and electrons. $\tilde{K}$ is a second rank Knight shift tensor, which includes an isotropic part K_{iso} and an anisotropic part K_i ($i = 1,2,3$). The anisotropic part is defined so as to fulfill the relations $\sum_i K_i = 0$ and $|K_3| \geq |K_2| \geq |K_1|$. The third term represents a classical dipole field contribution between the nuclear spin and the localized Pr-4f moment $\boldsymbol{\mu}_i = -g_J\mu_B\boldsymbol{J}_i$ at the ith Pr site, where g_J, μ_B, and $\boldsymbol{J}_i$ are Landé g-factor, Bohr magneton, and the 4f electron spin matrix, respectively. The dipole field at Nb, Al(1), Al(2), and Al(3) site is calculated by summing up the contributions of $11\times11\times11$ neighboring Pr atoms in the lattice as

$$\boldsymbol{H}_d = \sum_i\left[-\frac{\boldsymbol{\mu}_i}{r_i} + \frac{3(\boldsymbol{\mu}\cdot\boldsymbol{r}_i)\boldsymbol{r}_i}{r_i^5}\right], \quad (2)$$

where r_i is the distance between the site concerned and the ith Pr 4f moment. For instance, when we applied the magnetic field parallel to $\langle 001\rangle$ direction, $|\boldsymbol{H}_d|$ for each site is obtained to be $|\boldsymbol{H}_{d,\mathrm{Nb}}| \approx 8$ Oe, $|\boldsymbol{H}_{d,\mathrm{Al(2)}}| \approx 10$ Oe, and $|\boldsymbol{H}_{d,\mathrm{Al(3)}}| \approx 40$ Oe.

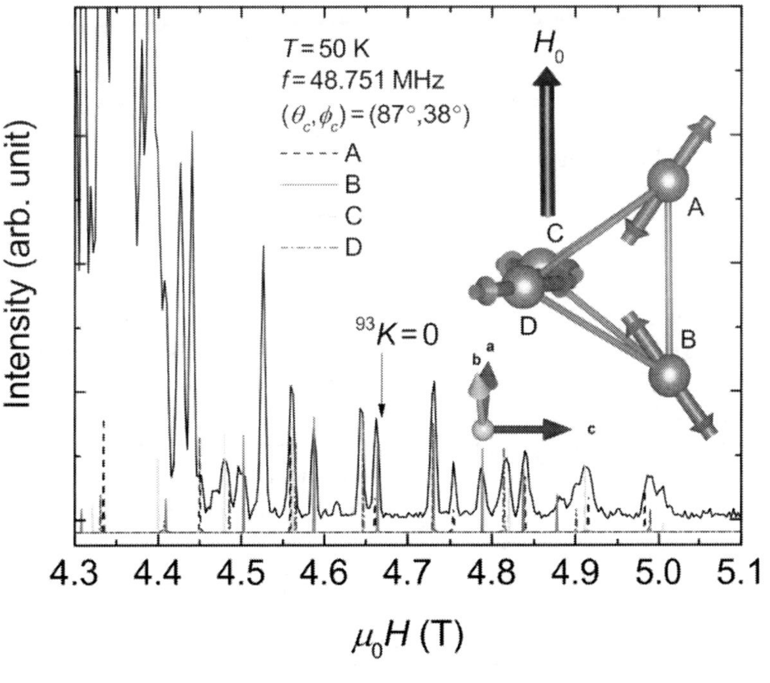

Figure 1. ^{93}Nb- and ^{27}Al-NMR lines obtained for a fixed frequency of 48.751 MHz and 50 K for a field nearly parallel to ⟨110⟩ direction of the crystal axes. The black dashed, red solid, green dotted, and blue dashed-dotted lines are the result of the simulation assuming that the magnetic field is applied for $(\theta_C, \phi_C) = (87°, 38°)$ direction. (θ_C, ϕ_C) are the polar angles between ⟨100⟩ direction and the applied field direction. The black arrow indicates the origin of the ^{93}Nb Knight shift. The inset shows four neighboring Nb sites. When the magnetic field is applied along ⟨110⟩ direction indicated by a black arrow, two inequivalent Nb sites (A, B) and (C, D) appear. Red arrows denotes V_{ZZ} directions that are parallel to one of ⟨111⟩ directions. Unassigned lines come from ^{27}Al nuclei.

The fourth term is the electric quadrupole interaction. Here $\nu_Q \equiv 3eQV_{ZZ}/2hI(2I-1)$ is a nuclear quadrupole frequency, where e is unit charge, $\eta \equiv (V_{XX} - V_{YY})/V_{ZZ}$ is an asymmetry parameter, and $V_{\alpha\alpha}$ ($\alpha = X, Y, Z$) are principal values of the electric field gradient (EFG) tensor defined to be $|V_{ZZ}| \geq |V_{YY}| \geq |V_{XX}|$, so as $0 \leq \eta \leq 1$. Note that the quantization axis is taken as the principal axes of the nuclear quadrupole interaction.

As mentioned above, the Nb site has axial symmetry and the maximum principal axes of the the Knight shift (K), classical dipole (D), and nuclear quadrupole interaction (Q) direct one of the ⟨111⟩ directions of the crystal coordinate (C) as shown in the inset of Fig. 1. In this case, we can take K, D, and Q coordinates to coincide.

In order to calculate NMR lines for an arbitrary field angle with respect to the crystal axis, we should perform a frame transformation from C, K, D, and Q coordinates to the Zeeman (Z) coordinates. We define the field angle by polar angles (θ_j, ϕ_j) of the field direction with respect to the C, K, D, and Q

coordinates. Then, nuclear spin matrices are transformed as:

$$I_x^Z = I_x^j \cos\theta_j \cos\phi_j + I_y^j \cos\theta_j \sin\phi_j - I_z^j \sin\theta_j, \tag{3}$$

$$I_y^Z = -I_x^j \sin\phi_j + I_y^j \cos\phi_j, \tag{4}$$

$$I_z^Z = I_x^j \sin\theta_j \cos\phi_j + I_y^j \sin\theta_j \sin\phi_j + I_z^j \cos\theta_j, \tag{5}$$

where j designates the one of the coordinates (C, K, D, Q).

When we applied the magnetic field parallel to the $\langle 110\rangle$ direction , $(\theta_C, \phi_C) = (90°, 45°)$, two inequivalent Nb sites should appear as shown in the inset of Fig. 1. And, at least eighteen ^{93}Nb-NMR lines are expected to be observed. In this field direction, the field angle is expressed as $(\theta_{K,Q,D}, \phi_{K,Q,D}) = (35°, 0°)$ for A and B sites in the inset of Fig. 1, and $(90°, 45°)$ for C and D sites in Fig. 1.

The black dashed, red solid, green dotted, and blue dashed-dotted lines is a result of the simulation using $(\theta_C, \phi_C) = (87°, 38°)$, $K_{\rm iso} = 0.3\%$, $K_1 = K_2 = -0.1\%$, and $K_3 = 0.2\%$, respectively. Here we used $\nu_{\rm Q,Nb} = 1.82$ MHz and of $\eta_{\rm Nb} = 0$ [9]. In this case, the classical dipole contribution is obtained to be $\approx$ 10 Oe for this direction and is negligibly small in this experimental accuracy. For this field direction, four Nb sites A, B, C, and D give distinct lines. Thus, thirty-six ^{93}Nb-NMR lines are expected to be observed. The agreement between experimental and calculated lines is sufficiently good. Thus, we determined the field direction by analyzing the obtained ^{93}Nb-NMR line. The misalignment from the $\langle 110\rangle$ direction is due to the setup of the sample rotator.

Then, we tried to assign ^{27}Al lines. As shown above, Al(3) site has a monoclinic local symmetry and there is one mirror plane that is parallel to one of $\langle 110\rangle$ directions. In such a case, one of the EFG principal axes is perpendicular to the mirror plane and the rest axes are in the mirror plane. However, we found that ^{27}Al-NMR lines cannot be reproduced by using the obtained field angle of $\theta_C = 87°$, $\phi_C = 38°$ and using the previous NQR parameters $\nu_{\rm Q,Al(3)} = 1.53$ MHz and $\eta_{\rm Al(3)} = 0.17$ for Al(3) site measured at $\approx$ 2.8 T and 30 K [9]. To reproduce the ^{27}Al-NMR line profile, the NQR parameters should depend drastically on temperature in the temperature range of 30 K and 50 K. However, ^{27}Al- and ^{93}Nb-NQR studies revealed that the NQR parameters hardly depend on temperature below 50 K [8, 9]. Thus, we should reconsider the simulation of the powder NMR lines by taking account of contributions from the hyperfine and nuclear dipolar interactions.

Figure 2 displays the frequency-swept ^{27}Al-NMR lines obtained at 2.8 T and 60 K. A red solid line is a result of the simulation for Al(2) sites by using the NQR parameters $\nu_{\rm Q,Al(2)} = 2.28$ MHz and $\eta = 0.17$ previously obtained in Ref. [9]. The experimental line is well reproduced by the simulation, confirming the validity of the estimated NQR parameters. On the other hand, as shown by a green dashed line in Fig. 2, there is a slight discrepancy between the experiment and simulation for Al(3) site. As mentioned previously, we neglected the contributions of the anisotropic Knight shift and classical magnetic dipole field in the previous analysis [9], where these contributions are affected by the Curie-Weiss like temperature dependence of magnetic susceptibility. The dipole contribution is small for Nb and Al(2) sites as mentioned before. On the other hand, the contribution is significant at Al(3) site as shown above. In addition, the effect of the anisotropic shift should be taken into consideration. In this context, it is reasonable to consider that the contributions from the hyperfine and dipolar interactions affect temperature and field dependences of NMR spectrum. In fact, we can fit the line profile better by using $\nu_{\rm Q,Al(3),new} \approx 1.0$ MHz and $\eta_{\rm Al(3),new} \approx 0.4$. A blue dotted line is the result of the simulation by using $\nu_{\rm Q,Al(3),new}$ and $\eta_{\rm Al(3),new}$. Thus, obtained $\nu_{\rm Q,Al(3),new}$ and $\eta_{\rm Al(3),new}$ are considered to be reliable.

Finally, we reconsider the previous NQR results. In the previous analysis [9], as shown by black solid lines in Fig. 3, we reported the NQR lines at around 1.5 MHz and 3.0 MHz were reproduced by using $\nu_{\rm Q,Al(3)}$ and $\eta_{\rm Q,Al(3)}$. However, powder samples used in the previous NQR study [9, 11] contain a small but NQR detectable amount of $NbAl_3$ impurity. NQR parameters of Nb site in $NbAl_3$ is previously reported to be $\nu_{\rm Q,NbAl_3} \approx 2.9$ MHz [12]. Thus, we find the NQR signal at around 3.0 MHz does not originate from $PrNb_2Al_{20}$, but from $NbAl_3$ impurity. Instead, we reproduced an NQR signal at around 1.9 MHz by using $\nu_{\rm Q,Al(3),new}$ and $\eta_{\rm Al(3),new}$ as shown by blue solid lines in Fig. 3. The origin of the NQR

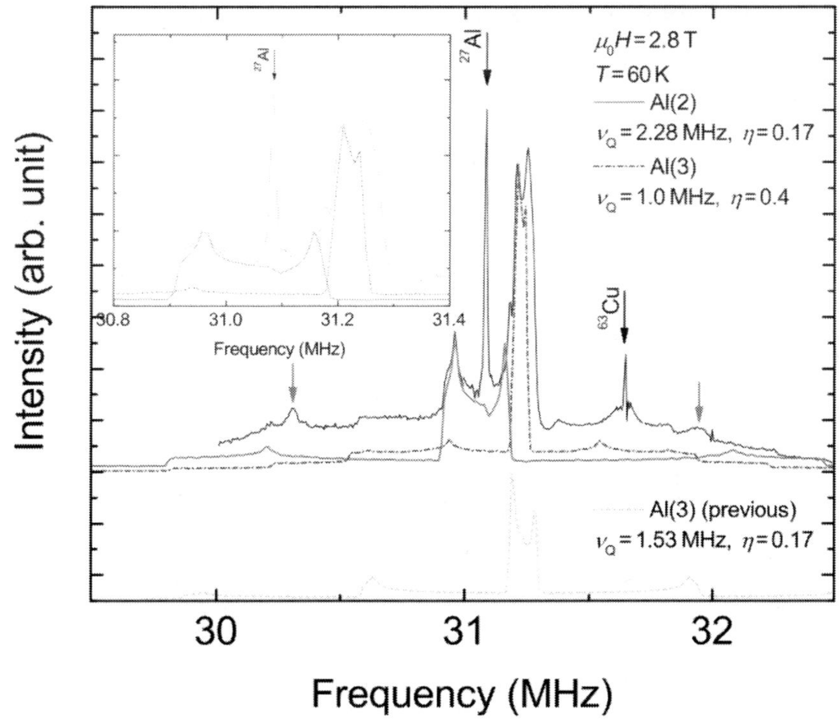

Figure 2. The ^{27}Al-NMR line obtained for a field of 2.8 T and 60 K. Red solid, green dashed, and blue dashed-dotted lines are the results of the simulation described in the text. Al and Cu signals designated by black arrows arise from the impurity and NMR coil. Peaks designated by red arrows are originate from ^{93}Nb in $PrNb_2Al_{20}$. The inset represents an enlarged view of the center of the NMR line.

line at around 1.5 MHz may be an experimental artifact. Further experimental confirmation of NQR lines are needed.

By using $\nu_{Q,Al(3),new}$ and $\eta_{Al(3),new}$, the assignment of ^{27}Al-NMR lines for a single crystal and the determination of the EFG axes are in progress. In addition, NQR signal search at below 2 MHz is also in progress.

4. Summary

We observed ^{93}Nb- and ^{27}Al-NMR signals of a single crystal $PrNb_2Al_{20}$. By using the previously reported NQR parameters, we reproduced the field angle dependence of the ^{93}Nb-NMR lines, while we could not reproduce the Al(3) lines by the previously obtained ^{27}Al(3)-NQR parameters. Thus, we reexamined the powder NMR and NQR spectra and obtained correct NQR parameters for Al(3) site to be $\nu_{Q,Al(3),new} \approx 1$ MHz and $\eta_{Al(3),new} \approx 0.4$. These results indicate the importance of the anisotropic hyperfine interaction and classical dipole field to analyze the temperature and field dependence of NMR lines in $PrNb_2Al_{20}$.

Acknowledgments

We gratefully thank Prof. H. Harima, Dr. S. Kitagawa, and Mr. T. Taniguchi for valuable discussions and helpful advice. This work has been supported by a Grant-in-Aid for Scientific Research on Innovative

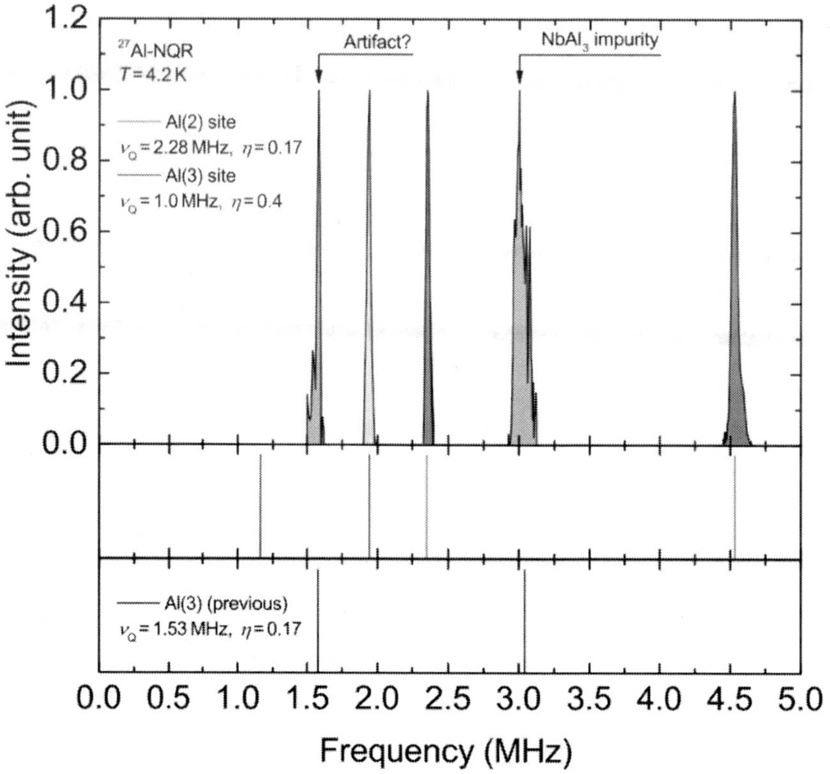

Figure 3. NQR lines obtained by using powder samples at 4.2 K. Intensity is normalized at each peak positions. Black solid lines are the result of the simulations for previously obtained NQR parameters for Al(3) site. This result is inconsistent with NMR results as discussed in the text.

Areas "J-Physics" (No. 15H05885) and Grants-in-Aid for Scientific Research (Nos. 26400359 and 15H05745) of the Ministry of Education, Culture, Sports, Science and Technology, Japan.

References

[1] Onimaru T, Matsumoto K T, Inoue Y F, Umeo K, Saiga Y, Matsushita Y, Tamura R, Nishimoto K, Ishii I, Suzuki T and Takabatake T 2010 *J. Phys. Soc. Jpn.* **79** 033704
[2] Onimaru T, Matsumoto K T, Inoue Y F, Umeo K, Sakakibara T, Karaki Y, Kubota M and Takabatake T 2011 *Phys. Rev. Lett.* **106** 177001
[3] Higashinaka R, Nakama A, Ando M, Watanabe M, Aoki Y and Sato H 2011 *J. Phys. Soc. Jpn.* **80** SA048
[4] Sakai A and Nakatsuji S 2011 *J. Phys. Soc. Jpn.* **80** 063701
[5] Sakai A, Kuga K and Nakatsuji S 2012 *J. Phys. Soc. Jpn.* **81** 083702
[6] Tsujimoto M, Matsumoto Y, Tomita T, Sakai A and Nakatsuji S 2014 *Phys. Rev. Lett.* **113** 267001
[7] Higashinaka R, unpublished.
[8] Kubo T, Kotegawa H, Tou H, Higashinaka R, Nakama A, Aoki Y and Sato H 2015 *J. Phys. Soc. Jpn.* **84** 074701
[9] Kubo T, Kotegawa H, Tou H, Higashinaka R, Nakama A, Aoki Y and Sato H 2015 *J. Phys.: Conf. Ser.* **592** 012093
[10] Tou H, Tsugawa N, Sera M, Harima H, Haga Y and Ōnuki Y 2007 *J. Phys. Soc. Jpn.* **76** 024705; 2007 *J. Phys. Soc. Jpn.* **76** 058001
[11] Kubo T, Kotegawa H, Tou H, Higashinaka R, Nakama A, Aoki Y and Sato H 2014 *JPS Conf. Proc.* **3** 012031
[12] Lue C S, Su T H, Xie B X and Cheng C 2006 *Phys. Rev. B* **74** 094101

Single crystal ^{27}Al-NMR study of the cubic Γ_3 ground doublet system $PrTi_2Al_{20}$

T Taniguchi, M Yoshida, H Takeda, M Takigawa, M Tsujimoto, A Sakai, Y Matsumoto, and S Nakatsuji

Institute for Solid States Physics, University of Tokyo, Kashiwa 277-8581, Japan

E-mail: `taka.taniguchi@issp.u-tokyo.ac.jp`

Abstract. We report results of ^{27}Al-NMR measurements on a single crystal of $PrTi_2Al_{20}$, in which the ground state of Pr^{3+} ions in the crystalline electric field is the nonmagneitc Γ_3 doublet. From the analysis of NMR spectra with the magnetic field applied precisely along the $\langle 111\rangle$ and $\langle 100\rangle$ directions, we determined the electric field gradient tensors for all three inequivalent Al sites (Al(1) $\sim$ Al(3) sites) and the anisotropic Knight shifts for the Al(3) sites. The hyperfine coupling tensor at the Al(3) sites is strongly anisotropic and much larger than the classical dipolar coupling, indicating importance of the anisotropic hybridization between the conduction and f electron states.

1. Introduction

Recently, f electron compounds with a nonmagnetic ground state of the crystalline electric field (CEF) potential have attracted much interest. A number of non-trivial phenomena such as multipole order or quadrupole Kondo effects are expected at low temperatures, when the nonmagnetic CEF ground state has non-zero matrix elements of the quadrupole and octupole momoents [1,2]. Particularly interesting examples are the quadrupole order and unconventional superconductivity reported on PrT_2X_{20} (T:transition metals, X:Zn, Al, $\cdots$), where non-Kramers Pr^{3+} ions ($4f^2$ system) are encapsulated in the Frank-Kasper cage formed by X atoms [3-8]. In a cubic CEF potential, the $4f^2$ ground multiplet of J=4 splits into the Γ_1 nonmagnetic singlet, Γ_3 nonmagnetic doublet, Γ_4 magnetic triplet, and Γ_5 magnetic triplet.

Figure 1 shows the temperature dependence of the magnetic susceptibility M/H of $PrTi_2Al_{20}$ at 6.615 T for $H \parallel \langle 111\rangle$. M/H is nearly independent of temperature below 20 K, indicating a nonmagnetic CEF ground state with Van Vleck paramagnetism. The specific heat measurements show that the $4f$ electronic entropy reaches $R\ln 2$ near 5 K, suggesting that the ground state is the Γ_3 doublet [5]. M/H increases with decreasing temperature above 20 K due to the excited Γ_4 and Γ_5 triplets. This result is consistent with the CEF level scheme deduced from the inelastic neutron scattering experiments, which determined the energies between the ground Γ_3 and the excited Γ_4, Γ_5, and Γ_1 levels to be 65, 108, and 156 K, respectively [9]. A phase transition was detected by an anomaly near 2 K in the specific heat [5]. A clear anomaly in the ultrasonic measurement combined with the absence of anomaly in the magnetic susceptibility indicates that it is a quadrupole transition [10]. At ambient pressure, the superconductivity appears below T_c =200 mK. Application of pressure above 6 GPa causes decrease of the quadrupole transition temperature and increase of the superconducting T_c, which reaches near 1.1 K at 8.7 GPa [8].

The electron effective mass deduced from the upper critical field is also enhanced by pressure exceeding 100 times the bare electron mass [11]. This raises a possibility that the quadrupole fluctuations play a vital role for the formation of Cooper pairs.

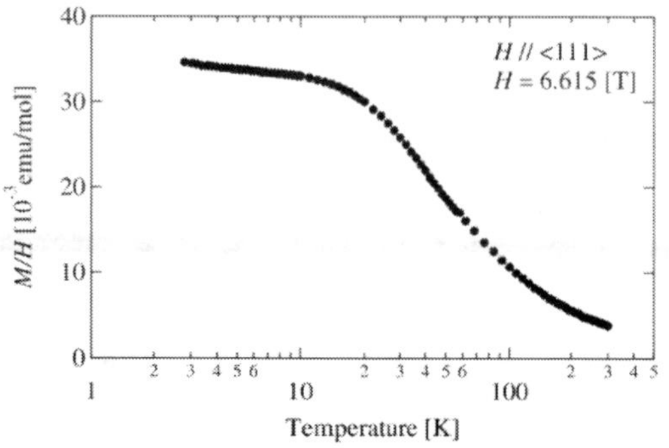

Figure 1. The temperature dependence of the magnetic susceptibility of $PrTi_2Al_{20}$ at 6.615 T for $H \parallel \langle 111 \rangle$

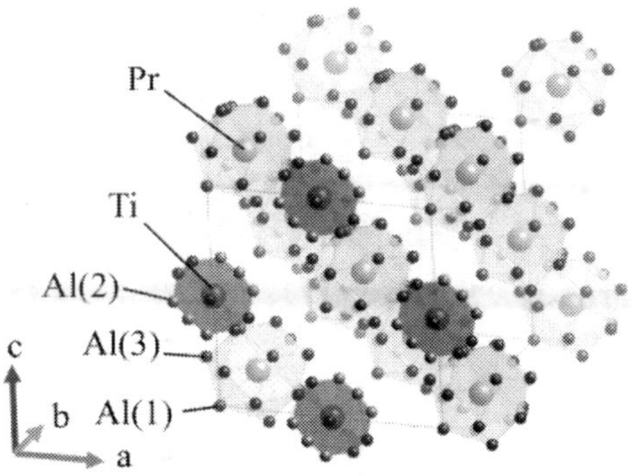

Figure 2. The crystal structure of $PrTi_2Al_{20}$. The Pr and Ti atoms are surrounded by yellow and gray cages, respectivity.

Figure 2 shows the crystal structure of $PrTi_2Al_{20}$. This compound has the cubic $CeCr_2Al_{20}$ type structure with the lattice parameter a=14.723 Å and the space group $Fd\bar{3}m$ [12]. There are three crystallographycally inequivalent Al sites, Al(1), Al(2), and Al(3), occupying the $16c$, $48f$ and $96g$ Wyckoff positions with the point symmetries $\bar{3}m$, $2mm$, and m, respectively. Pr atoms are surrounded by four Al(1) atoms and twelve Al(3) atoms. Ti atoms are surrounded by six Al(2) atoms and six Al(3) atoms.

Tokunaga *et al.* reported the results of NMR Knight shift and nuclear magnetic relaxation rates for selected Al sites and field orientations [13]. In this paper, we report measurements of the electric field gradient (EFG) and the Knight shift for all Al sites. In particular, we have completely determined the EFG tensor for all Al sites and the magnetic hyperfine coupling tensor between Pr-$4f$ electrons and ^{27}Al nuclei at the Al(3) sites.

2. Experimental procedure

A single crystal of $PrTi_2Al_{20}$ was synthesized by the Al flux method [5]. For NMR measurements, the crystal was shaped into a thin plate of the size $2.05 \times 1.14 \times 0.073$ mm^3 with the $\langle 111 \rangle$ direction normal to the plate. There are two reasons for this. First, thinner plates give better signal intensity because the rf penetration depth is much shorter than the thickness. Second, the geometry of a thin plate reduces the distribution of demagnetizing field, thereby resulting in a narrower NMR line width. The NMR spectra were obtained by summing the Fourier transform of the spin-echo signals obtained at equally spaced rf frequencies with a fixed magnetic field of 6.615 T. The orientation of the crystal in the magnetic field was precisely controlled by double-axis goniometer typically within 0.2 degree.

3. NMR spectra and EFG

In general, the NMR resonance frequency is determined by the spacing of eigenvalues of the following Hamiltonian

$$\mathscr{H} = -\gamma\hbar\mathbf{I}\cdot\mathbf{H} + \frac{1}{6}h\nu_Q\left\{\left(3I_Z^2 - I^2\right) + \eta\left(I_X^2 - I_Y^2\right)\right\}. \quad (1)$$

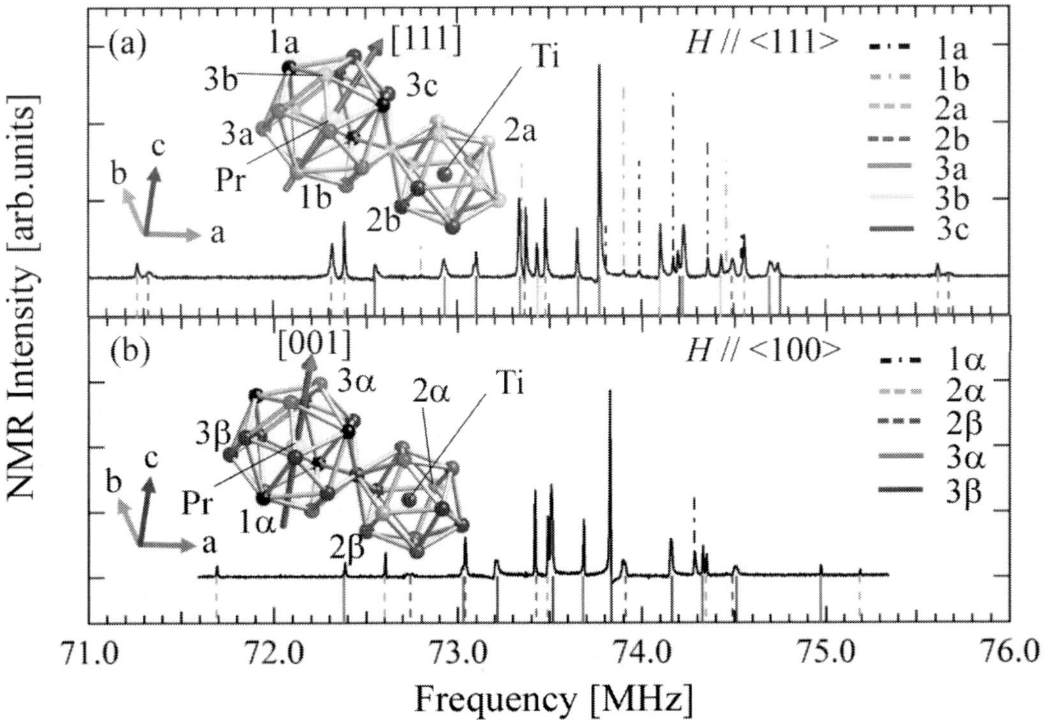

Figure 3. NMR spectra of ^{27}Al nuclei at 6.615 T and 60 K for (a) $\mathbf{H}_{\rm ext} \parallel \langle 111 \rangle$ and (b) $\mathbf{H}_{\rm ext} \parallel \langle 100 \rangle$. The vertical lines show the results of fitting by equation (1) as described in the text. The insets show how the Al(1), Al(2), and Al(3) sites split into inequivalent groups of sites in magnetic fields. The colors of the lines indicate the site assignment.

The first term represents the Zeeman interaction, where h is the Plank constant, $\hbar = h/2\pi$, $\mathbf{I}$ is the nuclear spin and γ is the gyromagnetic ratio (11.09407 MHz/T for ^{27}Al). The effective magnetic field $\mathbf{H}$ acting on nuclei is related to the external magnetic field $\mathbf{H}_{\rm ext}$ as $\mathbf{H} = \mathbf{H}_{\rm ext}(1+K)$, where K is the Knight shift. The second term corresponds to the nuclear quadrupole interaction. The quadrupole coupling constant ν_Q is defined as $\nu_Q = \frac{3eQ}{2I(2I-1)} V_{ZZ}$, where e is the proton's charge, Q is the nuclear quadrupole moment (0.149×10^{-24} cm^{-2} for ^{27}Al) and $V_{ZZ} = \frac{\partial^2 V}{\partial Z^2}$ is the largest principal value of the EFG tensor at the nuclei. Here X, Y, and Z denote the principal axes of the EFG tensor, satisfying the relation $|V_{ZZ}| \geq |V_{YY}| \geq |V_{XX}|$ and the symmetry parameter of EFG is defined as $\eta = \frac{V_{YY} - V_{XX}}{V_{ZZ}}$ ($0 \leq \eta \leq 1$, note that $V_{XX} + V_{YY} + V_{ZZ} = 0$).

Figure 3 shows the ^{27}Al-NMR spectra at 6.615 T and 60 K for (a) $\mathbf{H}_{\rm ext} \parallel \langle 111 \rangle$ and (b) $\mathbf{H}_{\rm ext} \parallel \langle 100 \rangle$. The observed spectra consist of sharp NMR lines, confirming high quality of the single crystal. Since $I = 5/2$ for ^{27}Al nuclei, NMR spectrum of each site should consist of five resonance lines split by the quadrupole interaction. Furthermore, crystallographically equivalent sites generally split into several inequivalent groups of sites in magnetic fields. We have fitted the frequencies of all the observed resonance lines in figure 3 to the calculated values obtained by diagonalizing the Hamiltonian of equation (1) in order to determine the values of K, ν_Q, η, and the directions of the principal axes of EFG as described in detail below. This process also allowed us to make unique site assignment of observed resonance lines as indicated in figure 3.

Journal of Physics: Conference Series **683** (2016) 012016 doi:10.1088/1742-6596/683/1/012016

Al(1)site

The Al(1) site has a 3-fold symmetry along the ⟨111⟩ direction. Therefore the EFG tensor is axially symmetric ($\eta = 0$) and the principal axis with the largest EFG (Z axis) is parallel to the ⟨111⟩ direction. There are four equivalent ⟨111⟩ directions in the cubic crystal. Thus, when $\mathbf{H}_{\text{ext}} \parallel \langle 111 \rangle$, the Z axis is parallel to the field for one quarter of A(1) sites while it is not for the other three quarter of sites. Therefore, Al(1) sites split into two sites with the ratio of 3:1 denoted 1a and 1b in figure 3(a) and the spectrum form the Al(1) sites consists of ten resonance lines. When $\mathbf{H}_{\text{ext}} \parallel \langle 100 \rangle$, all the ⟨111⟩ directions make an identical angle of 54.7 degree with the field, therefore all the Al(1) sites become equivalent (denoted 1α in figure 3(b)). Furthermore, this is the *magic angle*, at which quadrupole splitting vanishes to the first order in the quadrupole interaction. Thus the spectrum form the Al(1) site consists of a single line. By fitting the experimental frequencies of the eleven resonance lines to the calculated values obtained by diagonalizing equation (1), we have determined the four parameters: $\nu_Q = 0.55 \pm 0.02$ MHz, $K\,(1a) = 1.073 \pm 0.005\%$, $K\,(1b) = 0.707 \pm 0.003\%$, and $K\,(1\alpha) = 1.221 \pm 0.011\%$.

Al(2)site

Figure 4 shows a part of the crystal structure, visualizing the $2mm$ point symmetry of the Al(2) sites. The Al(2) site marked by the arrow is located on the two mirror planes, $(0\bar{1}1)$ and (011). Therefore, two of the principal axes of EFG should be perpendicular to these planes, $[0\bar{1}1]$ and $[011]$. The third one then must be along $[100]$, which is the intersection of the two mirror planes. The local symmetry, however, does not determine which one corresponds to the largest principal value of EFG. For $\mathbf{H}_{\text{ext}} \parallel [111]$, the Al(2) sites split into two inequivalent sites with 1:1 ratio denoted 2a and 2b in figure 4 (the red and blue circles). For $\mathbf{H}_{\text{ext}} \parallel [001]$, there are also two inequivalent sites but with 1:2 ratio denoted 2α and 2β (the open and closed circles). In each case, the NMR spectrum of the Al(2) sites consists of ten resonance lines. From the observed frequencies of these lines, we have determined six parameters: $\nu_Q = 2.07 \pm 0.01$ MHz, $\eta = 0.155 \pm 0.001$, $K\,(2a) = 0.091 \pm 0.003\%$, $K\,(2b) = 0.038 \pm 0.005\%$, $K\,(2\alpha) = 0.097 \pm 0.005\%$, and $K\,(2\beta) = 0.130 \pm 0.008\%$. In order to reproduce the experimental results, the Z (X) axis corresponding to the largest (smallest) principal value of EFG must be assigned to the $[011]$ ($[100]$) direction for the Al(2) site marked in figure 4.

Al(3)site

Figure 5 shows a part of the crystal structure, visualizing the m point symmetry of the Al(3) sites that form a cage surrounding a Pr atom. The Al(3) site indicated by the arrow is on the $(1\bar{1}0)$ mirror plane. Therefore one of the principle axes of the EFG should be along the $[1\bar{1}0]$ direction perpendicular to the mirror plane. Other two principle axes are in the mirror plane. We define α to be the angle between one of them and the $[110]$ direction. For $\mathbf{H}_{\text{ext}} \parallel [111]$, the Al(3) sites split into three inequivalent sites with 2:1:1 ratio denoted 3a, 3b and 3c sites in figure 5 (the red, green, and blue circles, respectively), generating fifteen resonance lines. For $\mathbf{H}_{\text{ext}} \parallel [001]$, the Al(3) sites split into two inequivalent sites with 1:2 ratio denoted 3α and 3β (open and closed circles), generating ten resonance lines. From the observed twenty five lines, we have determined eight parameters: $\alpha = 18.1 \pm 1.8$ degree, $\nu_Q = 0.99 \pm 0.01$ MHz, $\eta = 0.594 \pm 0.027$, $K\,(3a) = 0.558 \pm 0.001\%$, $K\,(3b) = 0.519 \pm 0.005\%$, $K\,(3c) = 0.357 \pm 0.012\%$, $K\,(3\alpha) = 0.399 \pm 0.008\%$, and $K\,(3\beta) = 0.608 \pm 0.021\%$. In order to reproduce the experimental results, the Z axis corresponding to the largest principal value of EFG must be assigned to the the $[1\bar{1}0]$ direction and the X axis should be tilted from $[110]$ direction by α as shown in figure 5.

The obtained values of the EFG parameters are summarized in table 1. These are in agreement with the previous results [13] except for minor discrepancies. Our results are based on larger number of data points and should be considered as the refinement.

Table 1. EFG parameters obtained from the NMR spectra at 60 K.

site	ν_Q [MHz]	η	α [deg.]
Al(1)	0.55 ± 0.02	0	-
Al(2)	2.07 ± 0.01	0.155 ± 0.001	-
Al(3)	0.99 ± 0.01	0.594 ± 0.027	18.1 ± 1.8

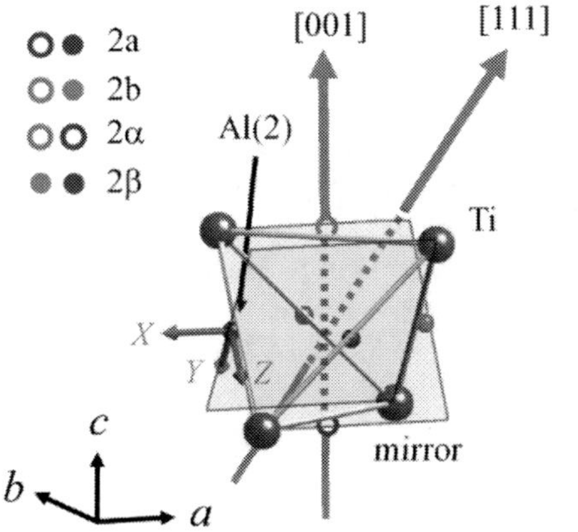

Figure 4. A part of the crystal structure showing the point symmetry the Al(2) site. The semitransparent sheets represent the two mirror planes of the Al(2) sites marked by the black arrow. The principal axes of EFG of this site (X, Y, and Z) are indicated by the red arrows, with Z (X) axis corresponding to the largest (smallest) principal value of EFG. The magnetic field along the [111] direction generates two inequivalent sites denoted 2a and 2b (the red and blue circles), while the field along the [001] direction generates different set of inequivalent sites denoted 2α and 2β (the open and close circles).

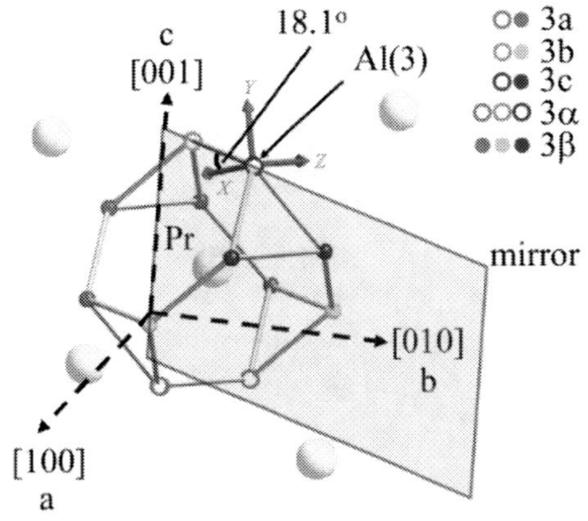

Figure 5. A part of the crystal structure showing the point symmetry of the Al(3) site. The semitransparent sheet represents the mirror plane of the Al(3) site marked by the black arrow. The principal axes of EFG (X, Y, and Z) are indicated by the red arrows, with Z (X) axis corresponding to the largest (smallest) principal value of EFG. The magnetic field along the [111] direction generates three inequivalent sites denoted 3a, 3b, and 3c (the red, green and blue circles), while in the field along the [001] direction, there are two inequivalent sites denoted 3α and 3β. (the open and close circles).

4. Knight shift and hyperfine coupling tensor

The hyperfine field $\mathbf{H}_{\rm hf}$ acting on ^{27}Al nuclei produced by the Pr moment $\mathbf{m}$ can be written as,

$$\mathbf{H}_{\rm hf} = \hat{A}_{hf}\mathbf{m}, \tag{2}$$

where $\hat{A}_{hf}$ is the hyperfine coupling tensor between $\mathbf{m}$ and the nuclei. Because the crystal structure has the cubic symmetry, the magnetic susceptibility χ should be isotropic in the paramagnetic phase. Thus $\mathbf{m}$ is written by

$$\mathbf{m} = \chi \mathbf{H}_{\text{ext}}. \tag{3}$$

Since $|\mathbf{H}_{\text{ext}}| \gg |\mathbf{H}_{\text{hf}}|$ in our case, only the component of $\mathbf{H}_{\text{hf}}$ parallel to $\mathbf{H}_{\text{ext}}$ contributes to the shift K. Therefore, K is described as,

$$K = \frac{\mathbf{H}_{\text{ext}} \cdot \mathbf{H}_{\text{hf}}}{|\mathbf{H}_{\text{ext}}|^2}. \tag{4}$$

From the equations (2), (3), and (4), we obtain

$$K = \frac{\mathbf{H}_{\text{ext}} \cdot \left(\hat{A}_{hf}\mathbf{H}_{\text{ext}}\right)}{|\mathbf{H}_{\text{ext}}|^2}\chi \equiv a_{hf}\chi. \tag{5}$$

The hyperfine coupling constant a_{hf} is defined as the proportionality constant between K and χ.

Figure 6 shows the K-χ plot above 6 K for the 1α site and all the five Al(3) sites. They all show good linear relation. From the slopes of these plots, the values of the hyperfine coupling constants a_{hf} are determined as listed in table 2. We should remark here that the observed Knight shift includes contributions from the demagnetizing field and the Lorentz field, the latter being equal to $4\pi M/3$ (M is the magnetization per volume), in addition to the contribution from the hyperfine field. We have subtracted the contribution from demagnetizing field (estimated from the crystal shape and magnetization data) and the Lorentz field to obtain the values in table 2. The previous study [13] reported that $a_{hf} = 0.170$ T/μ_B for 3a sites, which is consistent with our result.

Table 2. hyperfine coupling constants for the 1α and all Al(3) sites in $PrTi_2Al_{20}$

site	1α	3a	3b
a_{hf} [T/μ_B]	0.458 ± 0.002	0.170 ± 0.001	0.130 ± 0.002
site	3c	3α	3β
a_{hf} [T/μ_B]	0.084 ± 0.001	0.088 ± 0.001	0.170 ± 0.001

Based on these results and the crystal symmetry, we have determined all the components of the hyperfine coupling tensor $\hat{A}_{hf}$. Figure 7 shows a Pr atom at $\left[\frac{1}{8}\frac{1}{8}\frac{1}{8}\right]$ and Al(3) sites surrounding around the Pr atom. Here we explicitly consider the Al(3) site (the open circle) in figure 7, which is on the $(\bar{1}10)$ mirror plane. Because the a and b axes are equivalent for this site, the components of the hyperfine coupling tensor in the crystalline coordinate system can be written as

$$\hat{A}_{hf} = \begin{pmatrix} A_{aa} & A_{ba} & A_{ca} \\ A_{ba} & A_{aa} & A_{ca} \\ A_{ca} & A_{ca} & A_{cc} \end{pmatrix}, \tag{6}$$

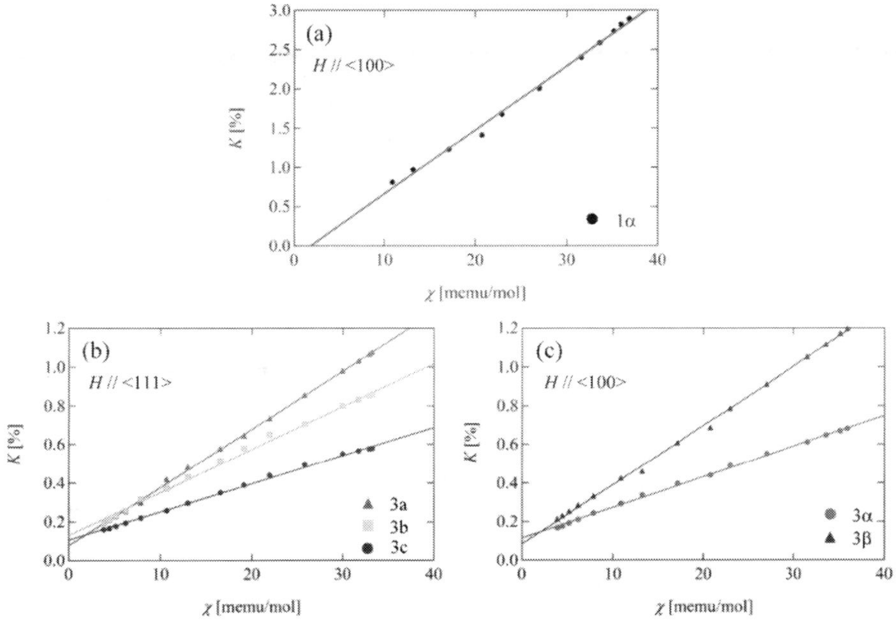

Figure 6. K vs χ plots of (a) the 1α site for $\mathbf{H}_{\mathrm{ext}} \parallel \langle 100 \rangle$, (b) the 3a, 3b, and 3c sites for $\mathbf{H}_{\mathrm{ext}} \parallel \langle 111 \rangle$, and (c) the 3a and 3b sites for $\mathbf{H}_{\mathrm{ext}} \parallel \langle 100 \rangle$.

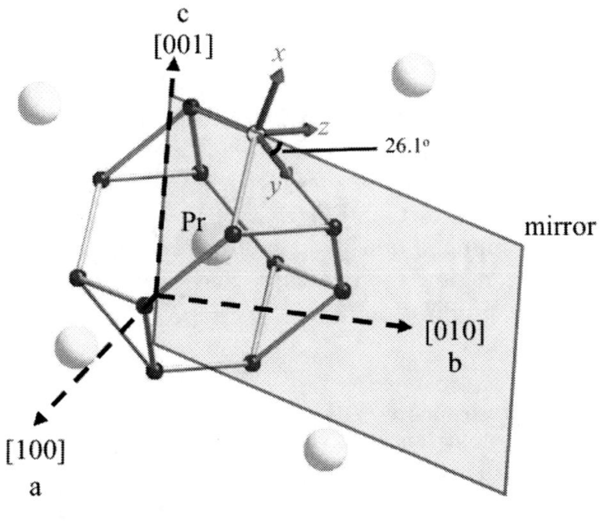

Figure 7. A part of the crystal structure showing the point symmetry of the Al(3) site identical to that shown in figure 6. The x, y, and z axes represent the principal axes of the hyperfine coupling tensor between the Al(3) site (the open circle) and the Pr atom at the center of the cage, chosen to satisfying the condition $|A_{zz}| \geq |A_{yy}| \geq |A_{xx}|$.

which includes four independent parameters, A_{aa}, A_{cc}, A_{ba}, and A_{ca}. The hyperfine coupling tensors at other Al(3) sites can be obtained by appropriate symmetry operations and their components are expressed as linear combinations of these parameters. Therefore, by using the equation (5), the hyperfine coupling constants $a_{\rm hf}$ at any site can be expressed as a linear combination of these four parameters. Since there are five experimental values of a_{hf} listed in table 2, we have five linear equations with four parameters to be determined. We solved four out of the five equations to determine the values of A_{aa}, A_{cc}, A_{ba}, and A_{ca} and used the remaining one equation for the consistency check. The results are satisfactory within the experimental error. We then diagonalized the $\hat{A}_{hf}$ tensor to obtain the principal values and the direction of principal axes. The results are summarized in table 3. Here x, y, and z denote the principle axes chosen to satisfy the condition $|A_{zz}| \geq |A_{yy}| \geq |A_{xx}|$. As shown in figure 7, the z axis is along the $[1\bar{1}0]$ direction and the y axis is tilted from [110] direction by $\beta = 26.1 \pm 3.3$ degree. We also show in table 3 the components of the classical dipole coupling tensor $\hat{A}_d$.

Table 3. The principle values of $\hat{A}_{hf}$ and $\hat{A}_d$

	A_{xx} [T/μ_B]	A_{yy} [T/μ_B]	A_{zz} [T/μ_B]	β [deg.]
$\hat{A}_{hf}$	0.073 ± 0.009	0.135 ± 0.001	0.215 ± 0.005	26.1 ± 3.3
$\hat{A}_d$	-0.035	0.045	-0.011	24.5

The values of $\hat{A}_{hf}$ are several times larger than the values expected from the classical dipolar coupling mechanism, which confirms the dominance of the transferred hyperfine coupling mechanism. The transferred hyperfine coupling arises from the spin polarization of the conduction electrons at the Al site due to the Pr $4f$ electrons. The values of $\hat{A}_{hf}$ thus can be a measure of the strength of the c-f hybridization effects and the anisotropy of $\hat{A}_{hf}$ seem to be associated with the anisotropy of c-f hybridization.

5. Conclusion

We have measured the ^{27}Al-NMR spectra of a single crystal of $PrTi_2Al_{20}$. We have assigned NMR spectra to all Al sites forming the cage lattice and obtained the temperature dependences of the Knight shift at 1α and all Al(3) sites. The hyperfine coupling constants a_{hf} were determined from the slopes of the K-χ plots and we obtained the hyperfine coupling tensor at the Al(3) site. This hyperfine tensor indicates that this system has a strong c-f hybridization effect and the transferred hyperfine coupling is anisotropic. The results provide the quantitative basis to discuss the quadrupole order at low temperatures and effects of multipole fluctuations, which will be the subjects of future publications.

6. Acknowledgement

The work was supported by Grant-in Aids for the Japan Society for the Promotion Science (JSPS) KAKENHI (B) (No. 25287083). T T was supported by the JSPS through the Program for Leading Graduate Schools (MERIT).

References

[1] Cox D L 1987 *Phys. Rev. Lett.* **59** 1240
[2] Cox D L 1994 *Physica B : Condensed Matter* **199-200** 391

[3] Onimal T, Matsumoto K T, Inaue Y F, Umeo K, Sakakibara T, Karaki Y, Kubota M and Takabatake T, *Phys. Rev. Lett.* 2011 **106** 177001
[4] Onimal T, Nagasawa N, Matsumoto K T, Wakiya K, Umeo K, Kittaka S, Sakakibara T, Matsushita Y and Takabatake T, *Phys. Rev. B.* 2012 **86** 184426
[5] Sakai A and Nakatsuji S 2011 *J. Phys. Soc. Jpn.* **80** 063701
[6] Shimura Y, Ohta Y, Sakakibara T, Sakai A and Nakatsuji S 2013 *J. Phys. Soc. Jpn.* **82** 043705
[7] Sakai A, Kuga K and Nakatsuji S 2012 *J. Phys. Soc. Jpn.* **81** 083702
[8] Matsubayashi K, Tanaka T, Sakai A, Kubo Y and Uwatoko Y 2012 *Phys. Rev. Lett.* **109** 187004
[9] Sato T J, Ibuka S, Nambu Y, Yamazaki T, Hong T, Sakai A and Nakatsuji S 2012 *Phys. Rev. B* **86** 184419
[10] Koseki M, Nakanishi Y, Deto K, Koseki G, Kashiwazaki R, Shichinomiya F, Nakamura M, Yoshizawa M, Sakai A and Nakatsuji S 2011 *J. Phys. Soc. Jpn.* **80** SA049
[11] Matsubayashi K, Tanaka T, Suzuki J, Sakai A, Nakatsuji S, Kitagawa K, Kubo Y and Uwatoko Y 2014 *JPS Conf. Proc.* **3** 011077
[12] Niemann S and Jeitschko W 1995 *J. Solid State Chem.* **144** 337
[13] Tokunaga Y, Sakai H, Kambe S, Sakai A, Nakatsuji S and Harima H 2013 *Phys. Rev. B* **88** 085124

Magnetic and thermal properties of NdT_2Al$_{20}$ (T: Ti, V, Cr) single crystals

Takahiro Namiki, Kenta Nosaka, Kyousuke Tsuchida, Qiankun Lei, Ryousuke Kanamori, and Katsuhiko Nishimura

Graduate School of Science and Engineering, University of Toyama, Toyama, 930-8555, Japan

E-mail: `namiki@eng.u-toyama.ac.jp`

Abstract. We have succeeded in growing the caged cubic compounds NdT_2Al$_{20}$ (T: Ti, V, Cr) by the Al-self flux method, and measured magnetic susceptibility χ, magnetization M, and specific heat divided by temperature C/T down to 0.5K. From the measurements, the antiferromagnetic phase transition at T_N = 1.45 K was observed for $NdTi_2Al_{20}$, and ferromagnetic phase transition at T_C = 1.8 K and 1.7 K were observed for NdV_2Al_{20} and $NdCr_2Al_{20}$, respectively.

1. Introduction

Recently, cage-structured compounds such as filled skutterudites have attracted much attention due to the novel properties such as the heavy fermion behavior, unconventional superconductivity and multiple ordering in the Pr- and Sm-based compounds[1, 2, 3, 4, 5]. Among them, the compounds RT_2X_{20} (R: rare earths, T: transition metals, X: Al, Zn, and Cd) also have the caged and cubic structure, also show the novel properties. Pr-based compounds PrT_2Zn$_{20}$ (T = Ir, Rh)[6, 7, 8] and PrT_2Al$_{20}$ (T = Ti, V, Cr, Nb)[9, 10, 11, 12, 13, 14] show the superconductivity, multipole ordering, and the unconventional Kondo effect. These novel properties are coming from a non-Kramers Γ_3 doublet CEF ground state which has nonmagnetic quadrupole moments[9, 7, 15, 16]. For instance, $PrTi_2Al_{20}$ is a heavy fermion superconductor based on strong hybridization between conduction electrons and nonmagnetic quadrupole moments of the Γ_3 ground state[17]. Except the PrT_2X_{20} and SmT_2X_{20}, only a few compounds are studied[12, 18, 19, 20, 21]. To clarify these novel properties, systematic studies are important. We were succeeded in making new NdT_2Al$_{20}$ (T: Ti, V, Cr) polycrystalline samples, and studied magnetic and electronic properties[20]. Recently, we have succeeded in growing high quality NdT_2Al$_{20}$ (T: Ti, V, Cr) single crystals, and studied magnetic and thermal properties.

2. Experimental details

Single crystals of NdT_2Al$_{20}$ (T: Ti, V, Cr) were grown by the Al-self flux method. The quality of the samples was checked using a powder x-ray diffractometer with Cu-Kα radiation. All obtained peaks were indexed with the cubic $CeCr_2Al_{20}$ (Fd$\bar{3}$m) type patterns, suggesting the high quality of samples. This is consistent with the fact that the residual resistivity ratio reaches to 20 for $NdTi_2Al_{20}$ (not shown), which is $\sim$ 4 times higher than the previous report[20]. The evaluated lattice parameters a were 14.77 Å, 14.55 Å, and 14.51 Å, which is consistent with the

Journal of Physics: Conference Series **683** (2016) 012017 doi:10.1088/1742-6596/683/1/012017

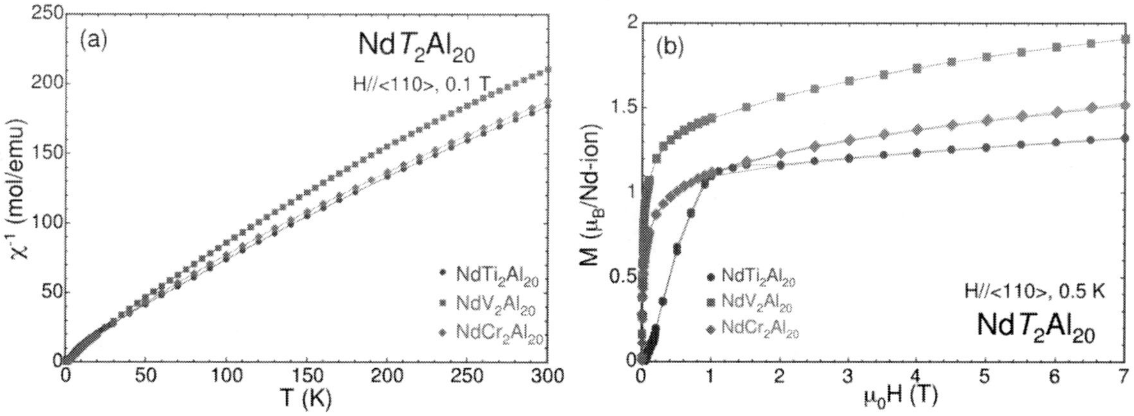

Figure 1. (a) The temperature dependence of the inverse susceptibility ($1/\chi$) for $NdTi_2Al_{20}$ (circle), NdV_2Al_{20} (square), and $NdCr_2Al_{20}$ (diamond) at 0.1 T for $H//\langle 110 \rangle$ direction. (b) The field dependence of the Magnetization (M) for $NdTi_2Al_{20}$ (circle), NdV_2Al_{20} (square), and $NdCr_2Al_{20}$ (diamond) at 0.5 K for $H//\langle 110 \rangle$ direction.

previous report[19]. The crystal axes of the single crystals were determined from Laue pictures. DC magnetization measurements were carried out in a Quantum Design (QD) MPMS down to 1.9 K, and QD MPMS iHelium3 down to 0.5 K and up to 7 T. Specific heat (C) measurements were performed using a quasi-adiabatic method with a QD PPMS down to 0.5 K and up to 9 T.

3. Experimental results and discussions

Figure 1 (a) shows the temperature dependencies of the inverse susceptibility $(\chi)^{-1}$ for NdT_2Al_{20} (T = Ti, V, and Cr) at 0.1 T and $H//\langle 110 \rangle$ direction. At high temperatures, the temperature dependence of the susceptibility (χ) for NdT_2Al_{20} shows Curie-Weiss like temperature dependencies above 200 K, and evaluated the Curie-Weiss temperature Θ_{CW} = -65 K, -79 K, -69 K and the effective magnetic moments of μ_{eff} = 3.98 μ_B, 3.78 μ_B, and 3.95 μ_B,for $NdTi_2Al_{20}$, NdV_2Al_{20}, and $NdCr_2Al_{20}$, respectively. μ_{eff} values are consistent with the theoretical value expected for the full multiplet of Nd^{3+} ($3.64\mu_B$) for these compounds. For $NdCr_2Al_{20}$, the μ_{eff} value becomes smaller than that obtained from the polycrystalline sample[20], and consistent with the no contribution of excess Cr. At low temperatures, χ monotonically increases down to 2 K for NdT_2Al_{20}. Below 2 K, χ^{-1} for $NdTi_2Al_{20}$ shows clear kink due to the antiferromagnetic phase transition at 1.4 K, on the other hand, χ^{-1} for NdV_2Al_{20} and $NdCr_2Al_{20}$ quickly decreases $\sim$ 1.8 K and $\sim$ 1.7 K, indicating the ferromagnetic phase transition. Figure 1 (b) shows the field dependences of the magnetization (M) for $NdTi_2Al_{20}$, NdV_2Al_{20}, and $NdCr_2Al_{20}$ at 0.5 K. The M values at 7 T are below 2 μ_B and similar to the $NdRu_2Zn_{20}$ which has doublet Γ_6 ground state[22]. For $NdTi_2Al_{20}$, M linearly increases at low fields and shows a saturation tendency above 1 T, indicating the antiferromagnetic ground state and consistent with χ^{-1} data. For NdV_2Al_{20} and $NdCr_2Al_{20}$, M increases at very low fields and then shows saturation tendency. The field dependence of M of these two compounds

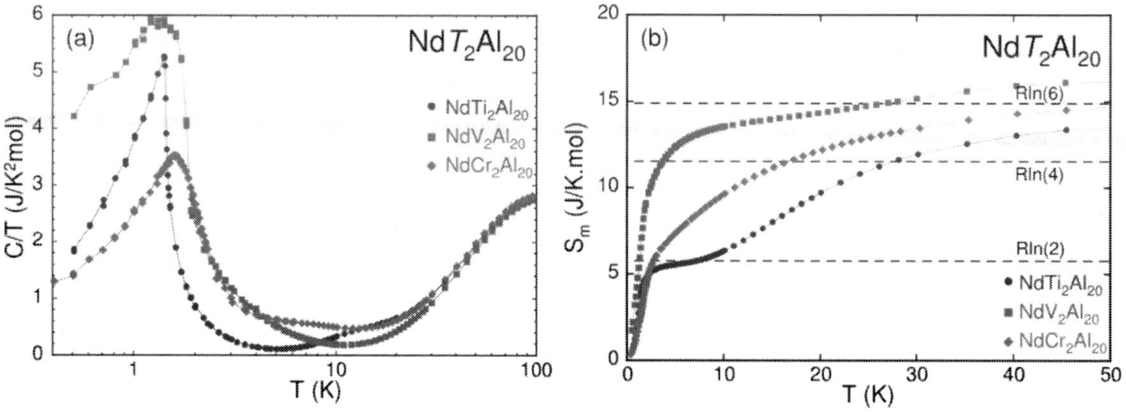

Figure 2. (a) The temperature dependence of the specific heat divided by the temperature (C/T) for $NdTi_2Al_{20}$ (circle), NdV_2Al_{20} (square), and $NdCr_2Al_{20}$ (diamond) at 0 T. (b) the temperature dependences of the magnetic part of the entropy (S_m) for $NdTi_2Al_{20}$, NdV_2Al_{20}, and $NdCr_2Al_{20}$.

shows hysteresis at very low fields (below 0.05 T). These behaviors indicate that NdV_2Al_{20} and $NdCr_2Al_{20}$ have the ferromagnetic ground state and consistent with the χ^{-1} data.

Figure 2 (a) shows the temperature dependences of the specific heat divided by temperature (C/T) for $NdTi_2Al_{20}$, NdV_2Al_{20}, and $NdCr_2Al_{20}$. At high temperatures, the temperature dependences of C/T for these three compounds shows the almost same values of LaT_2Al_{20}, indicating that the $4f$ electrons are well localized. Reflecting the high quality of the sample, no anomaly is observed around 10 K in NdV_2Al_{20} in contrast to the polycrystalline sample[20] which may come from the impurity. At low temperatures, C/T for $NdTi_2Al_{20}$ and $NdCr_2Al_{20}$ shows a broad peak or plateau, which become peak considering only the magnetic part of C/T (C_m/T) obtained from subtracting the phonon part of C/T, at $\sim$ 20 K and $\sim$ 10 K, respectively. For NdV_2Al_{20}, C_m/T shows a peak at higher temperature, $\sim$ 40 K (not shown). These peaks are may come from the Schottky peak due to the CEF. Below 4 K, C/T increases with decreasing temperature, and shows clear anomaly at 1.45 K, 1.8 K, and 1.7 K for $NdTi_2Al_{20}$, NdV_2Al_{20}, and $NdCr_2Al_{20}$, indicating the phase transition. Figure 2 (b) shows the temperature dependencies of the magnetic part of the entropy S_m for $NdTi_2Al_{20}$, NdV_2Al_{20}, and $NdCr_2Al_{20}$. At low temperatures, S_m quickly increases with increasing temperature and reaches $\sim R\ln 2$ at transition temperatures for these compounds, indicating the CEF doublet Γ_6 ground state for $NdTi_2Al_{20}$ and $NdCr_2Al_{20}$. However, because S_m exceeds $\ln 4$ near T_C, it is difficult to decide the CEF ground state.

4. Summary

We have succeeded in growing single crystals, and studied magnetic and thermal properties of $NdTi_2Al_{20}$, NdV_2Al_{20}, and $NdCr_2Al_{20}$. From magnetization and specific heat measurements, we found that NdV_2Al_{20} and $NdCr_2Al_{20}$ shows the ferromagnetic phase transitions at 1.8 K and 1.7 K, and $NdTi_2Al_{20}$ shows antiferromagnetic phase transition at 1.45 K. The CEF ground state should be doublet Γ_6 ground state for $NdTi_2Al_{20}$ and $NdCr_2Al_{20}$.

Acknowledgment

We thanks to Prof. Isikawa for the useful discussions. This work was supported by Takahashi Industrial and Economic Research Fundation and JSPS KAKENHI Grant Number 24740231.

References

[1] Aoki Y, Namiki T, Matsuda T D, Abe K, Sugawara H and Sato H 2002 *Phys. Rev. B* **65** 064446
[2] Aoki Y, Namiki T, Ohsaki S, Saha S R, Sugawara H and Sato H 2002 *J. Phys. Soc. Jpn.* **71** 2098
[3] Matsuhira K, Hinatsu Y, Sekine C, Togashi T, Maki H, Shirotani I, Kitazawa H, Takamasu T and Kido G 2002 *J. Phys. Soc. Jpn. Suppl.* **71** 237
[4] Yoshizawa M, Nakanishi Y, Oikawa M, Sekine C, Shirotani I, Saha S R, Sugawara H and Sato H 2005 *J. Phys. Soc. Jpn.* **74** 2141
[5] Sanada S, Aoki Y, Aoki H, Tsuchiya A, Kikuchi D, Sugawara H and Sato H 2005 *J. Phys. Soc. Jpn.* **74** 246
[6] Onimaru T, Matsumoto K T, Inoue Y F, Umeo K, Saiga Y, Matsushita Y, Tamura R, Nishimoto K, Ishii I, Suzuki T and Takabatake T 2010 *J. Phys. Soc. Jpn.* **79** 033704
[7] Onimaru T, Matsumoto K T, Inoue Y F, Umeo K, Sakakibara T, Karaki Y, Kubota M and Takabatake T 2011 *Phys. Rev. Lett.* **106** 177001
[8] Ishii I, Muneshige H, Suetomi Y, Fujita T K, Onimaru T, Matsumoto K T, Takabatake T, Araki K, Akatsu M, Nemoto Y, Goto T and Suzuki T 2011 *J. Phys. Soc. Jpn.* **80** 093601
[9] Sakai A and Nakatsuji S 2011 *J. Phys. Soc. Jpn.* **80** 063701
[10] Sakai A, Kuga K and Nakatsuji S 2012 *J. Phys. Soc. Jpn.* **81** 083702
[11] Nakatsuji A S S 2012 *J. Phys.: Conf. Ser.* **391** 012058
[12] J Kangas M, C Schmitt D, A Sakai, S Nakatsuji and Y Chan J 2012 *J. Solid State Chem.* **196** 274
[13] Tokunaga Y, Sakai H, Kambe S, Sakai A, Nakatsuji S and Harima H 2013 *Phys. Rev. B* **88** 085124
[14] Kubo T, Kotegawa H, Tou H, Higashinaka R, Nakama A, Aoki Y and Sato H 2015 *J. Phys. Soc. Jpn.* **84** 074701
[15] Higashinaka R, Nakama A, Ando M, Watanabe M, Aoki Y and Sato H 2011 *J. Phys. Soc. Jpn.* **80** SA048
[16] Sato T J, Ibuka S, Nambu Y, Yamazaki T, Hong T, Sakai A and Nakatsuji S 2012 *Phys. Rev. B* **86** 184419
[17] Tsujimoto M, Matsumoto Y, Tomita T, Sakai A and Nakatsuji S 2014 *Phys. Rev. Lett.* **113** 267001
[18] Niemann S and Jeitschko W 1995 *J. Solid State Chem.* **114** 337
[19] Thiede V M T, Jeitschko W, Niemann S and Ebel T 1998 *J. Alloys Compd.* **267** 23
[20] Namiki T, Murata Y, Baba K, Shimojo K and Nishimura K 2014 *JPS Conf. Proc.* **3** 011055
[21] Moze O, Tung L, Franse J and Buschow K 1998 *J. Alloys Compd.* **268**
[22] Isikawa Y, Ejiri J, Mizushima T and Kuwai T 2013 *J. Phys. Soc. Jpn.* **82** 123708

La substitution effect on the magnetic phase transition of $SmTi_2Al_{20}$

Ryuji Higashinaka, Akira Yamada, Tatsuma D. Matsuda, and Yuji Aoki

Department of Physics, Tokyo Metropolitan University, 1-1, Minami-osawa, Hachioji, Tokyo 192-0397, Japan

E-mail: `higashin@tmu.ac.jp`

Abstract. Powder X-ray and speci c heat measurements were performed on $(Sm_xLa_{1-x})Ti_2Al_{20}$ ($x = 0.2$–1) single crystals grown by the Al self- ux method. All samples crystallize in the $CeCr_2Al_{20}$-type structure with $Fd\bar{3}m$ symmetry and the lattice parameter changes almost linearly with respect to the Sm concentration. From a λ-type peak appearing in the speci c heat, it was found that the antiferromagnetic phase transition occurring at 6.5 K for $x = 1$ remains in the large Sm-concentration range. The transition temperature decreases almost linearly depending on the Sm concentration and goes to zero at the nite Sm concentration value of $\sim$ 0.1, indicating the possible existence of a quantum critical point. The robustness of the phase boundary against the magnetic eld remains even in the La-substituted system. This behavior is in marked contrast to the eld-sensitive behavior observed in Ce-based heavy fermion compounds.

1. Introduction

Sm compounds with cage structure have attracted considerable attention as a result of their unique properties, which are attributed to strong c-f hybridization and uctuations of higher-rank multipoles and the rare-earth ion valence. Among the cage-structured compounds containing Sm, $\mathrm{Sm}\mathit{Tr}_2X_{20}$ (Tr = transition metal, X = Zn, Al, Cd), which crystallize in the cubic $CeCr_2Al_{20}$-type structure with the space group $Fd\bar{3}m$ (#227) have been investigated intensively. This is because unusual eld (H)-insensitive phase transitions and heavy fermion (HF) behaviors in the ordered state have been discovered in $\mathrm{Sm}\mathit{Tr}_2\mathrm{Al}_{20}$ (Tr = Ti, V, Cr, and Ta)[1, 2, 3].

The resistivity of these compounds exhibits a log T dependence below 50 K, indicating the occurrence of the Kondo e ect, and exhibits H-insensitive phase transitions at $T_x = 6.5$, 2.9, 1.8, and 2.0 K for $\mathrm{Sm}\mathit{Tr}_2\mathrm{Al}_{20}$ (Tr = Ti, V, Cr, and Ta), respectively. Below T_x, these compounds show unusual H-insensitive HF behaviors with values of 0.15, 0.72, 1.0, and 3.2 J/(K^2 mol-Sm) for $\mathrm{Sm}\mathit{Tr}_2\mathrm{Al}_{20}$ (Tr = Ti, V, Cr, and Ta), respectively[1, 2, 3]. The magnetic susceptibilities of these compounds exhibit weak T dependences, which are evidently di erent from those for both free Sm^{2+} and Sm^{3+} ions. In addition, the X-ray absorption spectra for all Tr show that the Sm ions are in a mixed-valence state, with an average Sm-ion valence of 2.87 0.02 and no signi cant T dependence down to 3 K[5, 6]. These results suggest the possibility that valence uctuation plays a role in the eld-insensitive nature.

From the speci c heat C data, a crystalline-electric- eld (CEF) Γ_8 quartet state of the J =

5/2 multiplet is expected to be the ground state of these compounds[1, 2, 3]. The ordered Sm dipole moments estimated from the upturn in the nuclear speci c heat below T_x are $m_{\mathrm{Sm}} = 0.51$, 0.28, and 0.22 $_{\mathrm{B}}$/Sm for SmTr_2Al$_{20}$ (Tr = Ti, Cr, and Ta), respectively, which are signi cantly suppressed from 0.65 $_{\mathrm{B}}$/Sm expected for a Γ_8 quartet state and 0.84 $_{\mathrm{B}}$/Sm expected for a free Sm^{3+} ion[1, 2]. The suppressed ordered moment has also been estimated from a neutron magnetic structural analysis using a $^{154}\mathrm{SmTi_2Al_{20}}$ single crystal[4]. As a Γ_8 quartet includes octupolar degrees of freedom, octupole contribution to the ordering has been suggested.

Although plural degrees of freedom (charge and higher-rank multipoles) are expected to contribute to the unconventional H-insensitive HF formation judging from various experimental results, the origin of this behavior has not yet been clari ed. For further investigation, it is important to study the e ect of nonmagnetic La ion substitution for the Sm ion on the physical properties of the resultant material and to compare these results with those reported for typical Ce-based HF compounds. In this work, we perform powder X-ray and C measurements using $(\mathrm{Sm}_x\mathrm{La}_{1\ x})\mathrm{Ti_2Al_{20}}$ single crystals.

2. Experimental

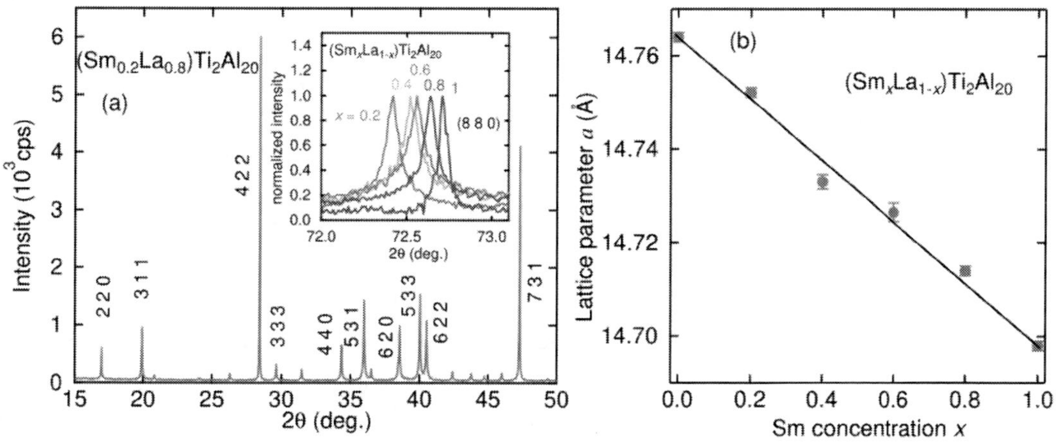

Figure 1. (a) X-ray powder di raction pattern for crushed $(\mathrm{Sm_{0.2}La_{0.8}})\mathrm{Ti_2Al_{20}}$ single crystals obtained using Cu-Kα1 radiation. The Miller indices of the cubic symmetry $Fd\bar{3}m$ for the main peaks are shown. (Inset) Normalized X-ray powder di raction patterns of (8 8 0) peak for $(\mathrm{Sm}_x\mathrm{La}_{1\ x})\mathrm{Ti_2Al_{20}}$ (x = 0.2, 0.4, 0.6, 0.8, and 1). (b) Lattice parameter a as function of nominal Sm concentration x along with that of $\mathrm{LaTi_2Al_{20}}$[7]. The solid line represents Vegard's law.

$(\mathrm{Sm}_x\mathrm{La}_{1\ x})\mathrm{Ti_2Al_{20}}$ (x = 0.2, 0.4, 0.6, 0.8, and 1) single crystals were grown by the Al self- ux method under vacuum. The starting component atomic ratio was 3N (99.9% pure) Sm:3N La:5N Ti:4N Al = x:1-x:2:90. X-ray powder-di raction experiments were carried out using a Rigaku SmartLab powder di ractometer equipped with a Cu-Kα1 monochromator. A silicon standard was used for accurate determination of the lattice parameter a. It was con rmed that all of the $(\mathrm{Sm}_x\mathrm{La}_{1\ x})\mathrm{Ti_2Al_{20}}$ samples crystallize in the $\mathrm{CeCr_2Al_{20}}$-type structure with $Fd\bar{3}m$ symmetry (gure 1(a)). The inset of gure 1(a) shows the normalized X-ray powder di raction pattern of the (8 8 0) peak for various nominal Sm concentration values x. The peak position shifts to smaller 2 with decreasing x. The determined a increases almost linearly with decreasing x, as shown in gure 1(b). As the single crystals were grown by the Al self- ux method, it is possible that the actual Sm concentration of each grown single crystal di ers from the nominal x value given in the starting atomic ratio. Therefore, on the basis of Vegard's law, we determined the value of the corrected Sm concentration x^* for each crystal using $a(x^*) = x^*a(1) + (1 \quad x^*)a(0)$, where $a(1)$ and $a(0)$ represent the lattice parameters of $SmTi_2Al_{20}$ and $LaTi_2Al_{20}$, respectively. We use x^* in the following discussion. C measurements were performed using a quasi-adiabatic heat-pulse method with a physical property measurement system (PPMS; Quantum Design) down to 2 K and up to 9 T.

3. Results and discussion

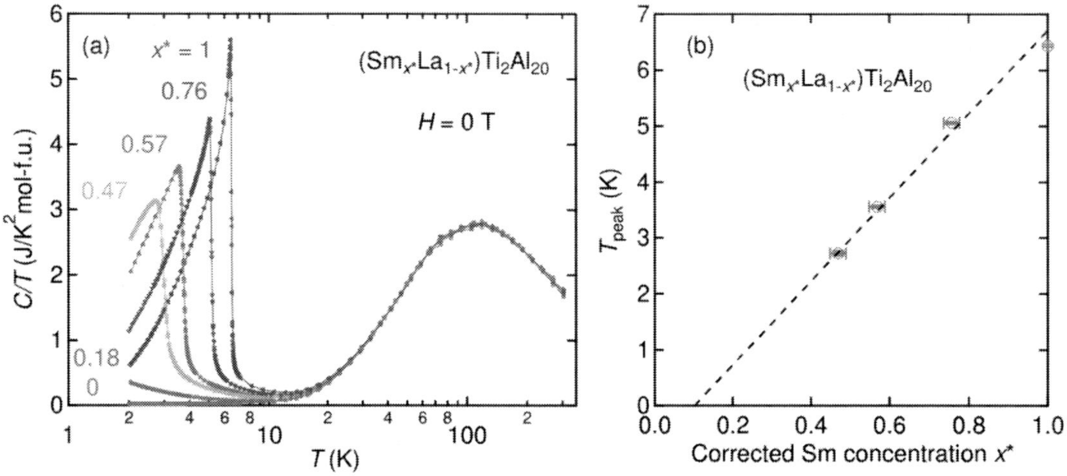

Figure 2. (a) Temperature T dependence of speci c heat C/T of $(Sm_{x^*}La_{1\ x^*})Ti_2Al_{20}$ in zero eld along with that of $LaTi_2Al_{20}$. (b) Corrected Sm concentration x^* dependence of T_{peak} in C/T. The broken line is a guide to the eye.

Figure 2(a) shows the T dependence of C/T of $(Sm_{x^*}La_{1\ x^*})Ti_2Al_{20}$ in zero eld down to 2 K. In the high-T region above 30 K, there is no signi cant di erence in the C/T behavior for all Sm concentrations, which indicates that the phonon contribution to C is not signi cantly a ected by the La substitution. For $x^* = 1$, a clear λ-type peak appears at 6.5 K, indicating the existence of a second-order phase transition. Our recent neutron scattering experiment suggests that this is a magnetic ordering with a $q = 0$ antiferro (AF)-type magnetic structure[4]. The peak temperature T_{peak} shifts to lower values with decreasing x^*. For $x^* = 0.18$, C/T exhibits an increase below 6 K, which is attributable to the Kondo e ect and/or a short-range ordering associated with the phase transition possibly shifted to a temperature below 2 K. The Sm concentration dependence of T_{peak} is shown in gure 2(b). In this gure, T_{peak} varies almost linearly depending on x^* (as shown by the broken line in gure 2(b)) and goes to zero at x^* 0.1. This indicates that a quantum critical point induced by La substitution may exist at x^* 0.1.

As regards the crystal structure of $SmTrAl_{20}$, the Sm ions form a diamond lattice, which has a site percolation threshold of $x_c = 0.43$[8, 9, 10]. Therefore, if the magnetic interaction among the Sm ions were dominated by a nearest-neighbor bond type, T_{peak} would exhibit a signi cant decrease as x approached x_c. Further, no ordering would appear for $x < x_c$. The observed linear

x^* dependence of T_{peak} over the entire Sm concentration range and the clear λ-type (mean- eld type) C anomaly associated with the phase transition indicate that the magnetic interaction among the Sm ions is of long-range type, most likely in line with the strong c-f hybridization.

Figure 3(a) shows T dependence of C/T for $x^* = 0.47$ under various magnetic elds up to

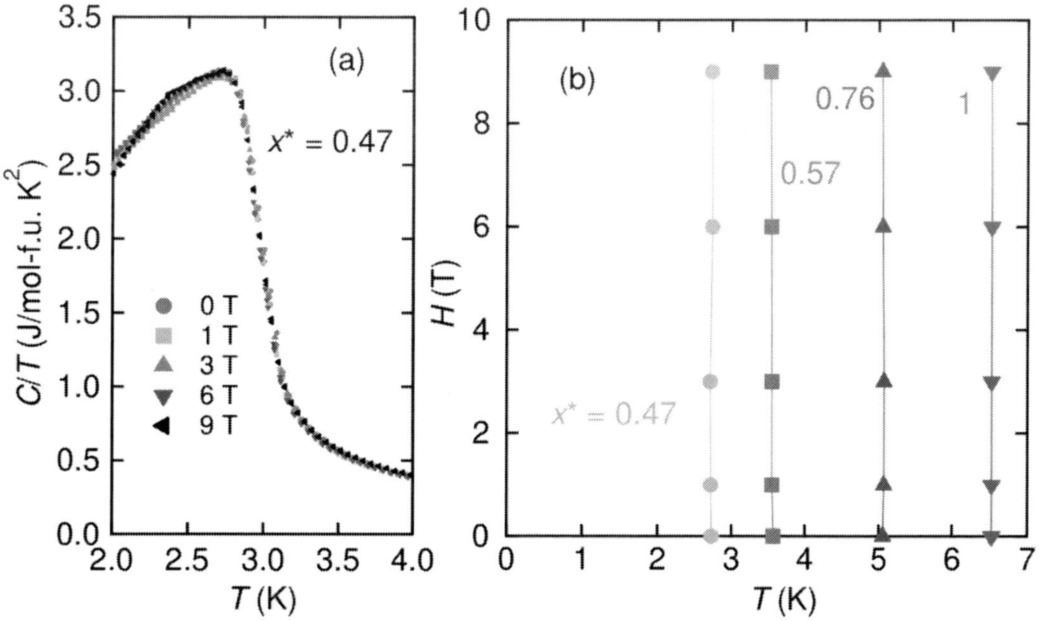

Figure 3. (a) T dependence of C/T in various magnetic elds for $x^* = 0.47$. (b) H-T phase diagram of $(Sm_{x^*}La_{1\ x^*})Ti_2Al_{20}$.

9 T near T_{peak}. This sample shows the robustness of T_x against the eld as observed for $x^* = 1$[1]. The shift of T_{peak} between 0 and 9 T is smaller than 0.05 K. In gure 3(b), we summarize the H-T phase diagram for $(Sm_{x^*}La_{1\ x^*})Ti_2Al_{20}$. For all samples, T_x exhibits almost no H dependence up to 9 T.

The rare-earth element concentration dependence on strongly correlated electronic properties has been investigated for several Ce-based HF compounds, such as $CeCu_6$[11, 12], CeB_6[13, 14, 15, 16] and $CeRu_2Si_2$[17]. Among those compounds, systems with 100% Ce concentration exhibit the Kondo lattice (coherent) regime in thier transport properties at low temperatures. As the Ce concentration decreases, this low-temperature regime continuously changes to the Kondo impurity (incoherent) regime. In the case of Sm compounds, we recently found that the resistivity exhibits a $\log T$ even for $x = 0.01$ in $(Sm_xLa_{1\ x})Ta_2Al_{20}$. The coherence temperature, at which the resistivity exhibits a maximum, shifts to lower values with decreasing x and is not apparent for $x = 0.01$[18]. This fact indicates that the system changes from the Kondo lattice to Kondo impurity regime in a similar manner to typical Ce-based HF compounds. However, strong robustness against H is observed in all x region, which is in marked contrast to the eld-sensitive behavior in Ce-based HF compounds.

In Ce-based HF compounds, the ground state changes with decreasing Ce concentration. For example, the ground state of $(Ce_xLa_{1\ x})B_6$ with $x = 1$ is an antiferromagnetic (AFM) state coexisting with a quadrupolar ordering. With decreasing Ce concentration, the ground state changes to an octupolar ordered phase at x 0.75 and becomes a paramagnetic (PM) state at approximately $x = 0.6$[13, 14, 15, 16]. The existence of these di erent ordered phases indicates competition among the magnetic-dipole, electric-quadrupole, and magnetic-octupole interactions in $(Ce_xLa_{1\ x})B_6$. On the contrary, in $(Sm_{x^*}La_{1\ x^*})Ti_2Al_{20}$, T_{peak} varies linearly depending on x^*. This feature suggests that a single type of Sm ion interaction, which leads to a $q = 0$ AFM ordering for $x^* = 1$, dominates the entire x^* region.

In contrast to the H-dependent ground state in Ce-based HF compounds, the H-T phase diagram of $(Sm_{x^*}La_{1\ x^*})Ti_2Al_{20}$ exhibits almost no H dependence in the phase boundary between the PM phase and the ordered phase for a wide range of x^* values ($0.47 < x^* < 1$), as shown in gure 3(b). This feature indicates that, in the La-substituted Sm 1-2-20 system, the nonmagnetic degrees of freedom possibly play an important role in the robustness of the ordered state against H, in contrast with the Ce-based HF compounds. In order to further investigate the detailed x dependence of physical properties including the possibility of quantum criticality at x^* 0.1 induced by the La substitution, we need to perform C and resistivity experiments at lower temperatures.

4. Summary

We succeeded in growing $(Sm_xLa_{1\ x})Ti_2Al_{20}$ ($x = 0.2$, 0.4, 0.6, 0.8, and 1) single crystals and performed powder X-ray and speci c heat measurements. We con rmed that all of the $(Sm_xLa_{1\ x})Ti_2Al_{20}$ samples crystallize in the $CeCr_2Al_{20}$-type structure and the lattice parameter a increases almost linearly with decreasing x. Using the corrected value of the Sm concentration x^* based on Vegard's law, we found that the transition temperature changes almost linearly as a function of x^* and goes to zero at x^* 0.1, indicating the possible existence of a quantum critical point. Similar to $SmTr_2Al_{20}$, the robustness of the transition temperature against an applied magnetic eld holds even in the La-substituted system. This behavior is in contrast to the eld-sensitive behavior observed in Ce-based HF compounds.

Acknowledgments

The authors are grateful to Professor H. Sato and H. Otsuka for helpful discussions. This work was supported by MEXT/JSPS KAKENHI Grant Numbers 15J07600, 24740239, 15H03693, 15K05178, 15H05884 (J-Physics), 23540421, 23340107, and 20102005.

References

[1] Higashinaka R, Maruyama T, Nakama A, Miyuazaki R, Aoki Y and Sato H 2011 *J. Phys. Soc. Jpn.* **80** 093703
[2] Yamada A, Higashinaka R, Miyazaki R, Fushiya K, Matsuda T D, Aoki Y, Fujita W, Harima H and Sato H 2013 *J. Phys. Soc. Jpn.* **82** 123710
[3] Sakai A and Nakatsuji S 2011 *Phys. Rev. B* **84** 201106(R)
[4] Higashinaka R: in preparation
[5] Higashinaka R, Yamada A, Miyazaki R, Aoki Y, Mizumaki M, Tsutsui S, Nitta K, Uruga T and Sato H 2014 *JPS Conf. Proc.* **3** 011079
[6] Yamada A, Higashinaka R, Fushiya K, Asano T, Matsuda T D, Mizumaki M, Tsutsui S, Nitta K, Ina T, Uruga T and Aoki Y 2015 to appear in *J. Phys. Conf. Ser.*
[7] Niemann S and Jeitschko W 1995 *J. Solid State Chem.* **114** 337
[8] Frisch H. L., Sonnenblick E., Vyssotsky V A and Hammersley J M 1961 *Phys. Rev.* **124** 1021
[9] Kirkpatrick S 1973 *Rev. Mod. Phys.* **45** 574
[10] Stauffer D and Aharony A 1961 *Introduction to Percolation Theory* (2nd ed.) (London: Taylor and Francis)
[11] Sumiyama A, Oda Y, Nagano H, Onuki Y, Shibutani K and Komatsubara T 1986 *J. Phys. Soc. Jpn.* **55** 1294
[12] Satoh K, Kato M, Maeno Y, Fujita T, Onuki Y and Komatsubara T 1987 *Jpn. J. Appl. Phys.* **26** 521
[13] Sato N, Kunii K, Oguro I, Komatsubara T and Kasuya T 1984 *J. Phys. Soc. Jpn.* **53** 3967
[14] Hiroi M, Kobayashi S, Sera M, Kobayashi N and Kunii S 1997 *J. Phys. Soc. Jpn.* **67** 53
[15] Tayama T, Sakakibara T, Tenya K, Amitsuka H and Kunii S 1997 *J. Phys. Soc. Jpn.* **66** 2268
[16] Suzuki O, Goto T, Nakamura S, Matsumura T and Kunii S 1998 *J. Phys. Soc. Jpn.* **67** 4243

[17] Haen P, Lapierre F, Kappler J P, Lejay P, Flouquet J and Meyer A 1988 *J. Magn. Magn. Mater.* **76-77** 143
[18] Yamada A, Higashinaka R, Matsuda T D, Sato H, Miyake A, Tokunaga M and Aoki Y 2015 to appear in *Phys. Procedia*

Low-temperature study of the Sm-based compound $SmOs_2Zn_{20}$

M Tanahashi[1], K Adachi[1], T Sasahara[1], N Kase[1], T Nakano[1] and N Takeda[2]
[1]Graduate School of Science and Technology, Niigata University, Niigata 950-2181, Japan
[2]Department of Materials Science and Technology, Niigata University, Niigata 950-2181, Japan

E-mail: f15b024a@mail.cc.niigata-u.ac.jp

Abstract. We synthesized single crystals of Zn-based Sm-compound $SmOs_2Zn_{20}$. *T*-dependence of magnetization $M(T)$ shows a sudden increase at the ferromagnetic transition temperature $T_C \sim 3$ K. *T*-dependence of electrical resistivity $\rho(T)$ shows typical metallic behavior. At T_C, $\rho(T)$ slightly decreases, which originates from the ferromagnetic transition. Specific heat $C(T)$ shows a clear but broad anomaly at T_C. Magnetic entropy $S_{mag}(T)$ at T_C is about R ln 2, and suggests that a Γ_7 doublet of the $J = 5/2$ multiplet is the ground state.

1. Introduction

RTr_2X_{20} (R = rare earth, Tr = transition metal, X = Al, Zn) crystallizes in the cubic $CeCr_2Al_{20}$-type structure with a space group *Fd-3m* (No.227). In this structure, 16 and 12 X atoms surround R and Tr ions, respectively. Among RTr_2X_{20}, Sm-based compounds have attracted much attention recently. $SmTr_2Al_{20}$ (Tr = Ti, V, Cr, and Ta) shows field-insensitive phase transition and heavy-fermion state [1,2,3]. In order to better understand the $SmTr_2X_{20}$ compounds, we focus on a Zn-based Sm compound and examine the thermodynamic and electrical transport properties of single crystal of $SmOs_2Zn_{20}$.

2. Experimental Details

We prepared single crystals of $SmOs_2Zn_{20}$ by a Zn self-flux method using following starting elements: Sm chips, Os powder, Zn grains in an atomic ratio of 1.5:2:100. The mixture was placed in an alumina crucible and sealed in an evacuated silica tube. The tube was heated up to 900 °C, kept for 2 h, then cooled down to 750 °C for 300 h. The excess Zn was separated by a centrifuge. Small amount of residual Zn-flux on the crystal surfaces were removed by 0.1 % acetic acid. The crystal structure of the single crystal was examined by powder-X-ray diffraction (XRD) implemented using a conventional X-ray spectrometer equipped with Cu-Kα radiation and a monochromator (RAD-2X, Rigaku). The intensity data were collected over a 2θ range of 10-90° with a step width of 0.01° and a counting rate of 4°/minutes. DC-magnetization measurements were carried out by MPMS (Quantum Design) down to 2 K and up to 1 T. Electrical resistivity $\rho(T)$ was measured by a dc-four-probe method in a ^{3}He cryostat (Heliox VL, Oxford) down to 0.28 K. Specific heat $C(T)$ measurements were performed down to 0.28 K in the ^{3}He cryostat using a standard thermal relaxation method.

3. Experimental Results

3.1. Chemical composition and Lattice Constants

Figure 1 shows powder-XRD patterns of $SmOs_2Zn_{20}$ at room temperature. The obtained powder-XRD patterns can be indexed as the space group of *Fd*-3*m* with cubic symmetry. No extra peaks due to impurity phase can be detected. The lattice constant of $SmOs_2Zn_{20}$ is obtained to be a = 1.4482 nm. This value is in a good agreement with that of the previous report [4].

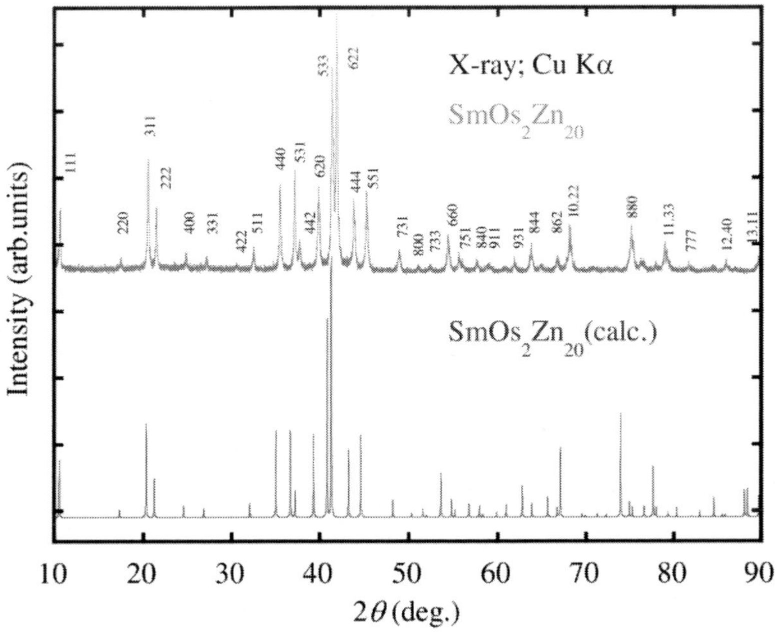

Figure 1. XRD patterns of $SmOs_2Zn_{20}$ at room temperature. (Upper panel is obtained from powder sample and lower panel shows a simulation)

3.2. Magnetic susceptibility and magnetization

T-dependence and *H*-dependence of magnetization $M(T,H)$ were obtained by using powder sample to which single crystal was crushed. Figure 2 shows $M(T)$ for $SmOs_2Zn_{20}$ at a magnetic field of $\mu_0H = 0.1$ T. $M(T)$ increases ferromagnetically at $T_C \sim 3$ K. The inset of Figure 2 shows $M(H)$ at $T = 2$ K for $SmOs_2Zn_{20}$. The theoretical Sm^{3+} free ion magnetic moment is $gJ\mu_B = 0.71\mu_B$, where $g = 2/7$ is the Lande's g factor and $J = 5/2$. From $M(H)$ at 2 K, the saturated magnetic moment of $SmOs_2Zn_{20}$ was estimated to be about 0.77 μ_B. This value is slightly larger than the theoretical value $gJ\mu_B = 0.71\mu_B$.

3.3. Electrical resistivity

Figure 3 shows *T*-dependence of electrical resistivity $\rho(T)$ for $SmOs_2Zn_{20}$. The residual resistivity ratio [RRR = ρ(300 K)/ρ(0.28 K)] is estimated to be approximately 5. The residual resistivity ρ_0 is obtained to be 13.7 μΩ cm. $\rho(T)$ decreases monotonically with decreasing temperature. The data below 10 K is represented in the inset of Fig. 3. At $T_C \sim 3$ K, $\rho(T)$ exhibits a slight decrease, which is originated from the ferromagnetic transition.

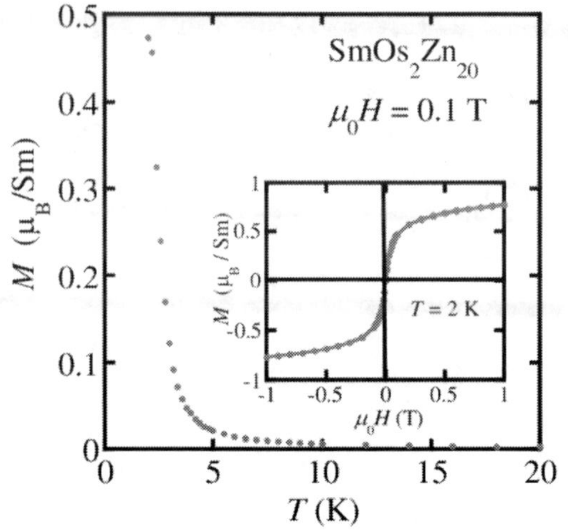

Figure 2. T-dependence of magnetization $M(T)$ for $SmOs_2Zn_{20}$ at a magnetic field of $\mu_0H = 0.1$ T. The inset shows H-dependence of magnetization $M(H)$ at $T = 2$ K for $SmOs_2Zn_{20}$.

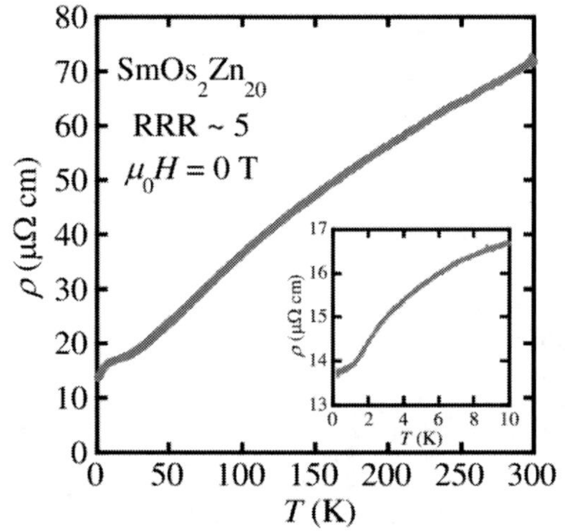

Figure 3. T-dependence of electrical resistivity $\rho(T)$ for $SmOs_2Zn_{20}$. The inset shows $\rho(T)$ below 10 K.

3.4. *Specific heat*

Figure 4 shows T-dependence of specific heat $C(T)$ of $SmOs_2Zn_{20}$. $C(T)$ shows anomaly at $T_C \sim 3$ K, which indicates ferromagnetic transition. The transition is relatively broad, which implies that quality of present sample could be insufficient. The inset of Fig. 4 represents T-dependence of magnetic entropy $S_{mag}(T)$ obtained by subtracting the specific heat of non-magnetic $LaOs_2Zn_{20}$. $S_{mag}(T)$ at T_C is about R ln 2, and shows change of the slope. The change in $S(T)$ near R ln 2 implies that the Γ_7 doublet of the $J = 5/2$ multiplet is the ground state. $S_{mag}(T)$ already reaching 10 J/(K mol) at 6 K, however, suggests the narrow crystalline-electric-field splitting if Γ_7 doublet is the ground state. The large saturated magnetization value close to the theoretical value of the $J = 5/2$ also supports that the crystalline-electric-field effect is weak. The origin of excess moment still remains to be clarified.

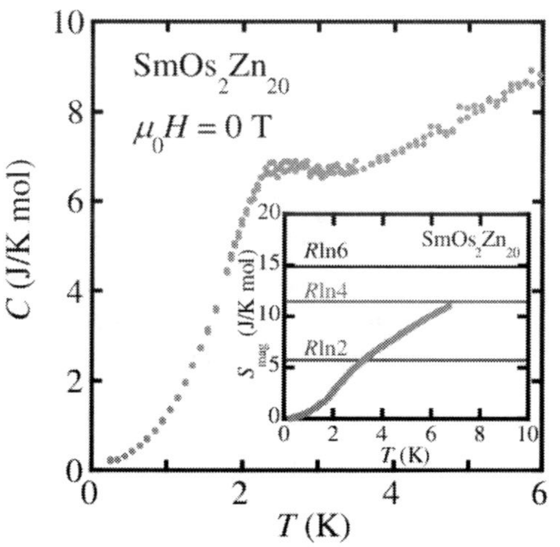

Figure 4. T-dependence of specific heat $C(T)$ for $SmOs_2Zn_{20}$. Inset shows magnetic entropy $S_{mag}(T)$ of $SmOs_2Zn_{20}$.

4. Conclusion

We succeeded in growing the single crystal of $SmOs_2Zn_{20}$ by a Zn self-flux method. By using the single crystal, we performed measurements of magnetization $M(T,H)$, electrical resistivity $\rho(T)$ and specific heat $C(T)$. T-dependence and H-dependence of magnetization $M(T,H)$ suggest that $SmOs_2Zn_{20}$ is ferromagnet with $T_C \sim 3$ K. T-dependences of electrical resistivity $\rho(T)$ and specific heat $C(T)$ show anomalies which originate from the ferromagnetic transition at $T_C \sim 3$ K. Magnetic entropy $S_{mag}(T)$ at T_C is about $R \ln 2$, suggesting that the Γ_7 doublet of the $J = 5/2$ multiplet is the ground state. The magnetic entropy $S_{mag}(T)$, which reaches $R \ln 4$ even at 6 K suggests the narrow crystalline-electric-field splitting if Γ_7 doublet is the ground state.

Acknowledgment

This work was partially supported by the Grant-in-Aid for Young Scientist (B) (No. 26800183) and Innovation Area “Heavy Electrons” (No.20102004) from the Ministry of Education, Culture, Sports, Science, and Technology, Japan, the Sasakawa Scientific Research Grant (26-207) from the Japan Science Society, and the Uchida Energy Science Promotion Foundation (26-1-22).

References

[1] Higashinaka R, Maruyama T, Nakama A, Miyazaki R, Aoki Y and Sato H 2011 *J. Phys. Soc. Jpn.* **80** 093703
[2] Sakai A and Nakatsuji S 2011 *Phys. Rev.* B **84** 201106(R)
[3] Yamada A, Higashinaka R, Miyazaki R, Fushiya K, Matsuda T, Aoki Y and Sato H 2013 *J. Phys. Soc. Jpn.* **82** 123710
[4] Chen Y and Shen J 2011 *Mater. Sci. Forum* **689** 204-210

Mixed valence states in (Sm_xLa_{1-x})Tr_2Al_{20} (Tr = Ti and Ta) studied using X-ray absorption spectroscopy

Akira Yamada[1], Ryuji Higashinaka[1], Kengo Fushiya[1], Takuya Asano[1], Tatsuma D. Matsuda[1], Masaichiro Mizumaki[2], Satoshi Tsutsui[2], Kiyofumi Nitta[2], Toshiaki Ina[2], Tomoya Uruga[2], and Yuji Aoki[1]
[1]Department of Physics, Tokyo Metropolitan University, 1-1 Minamiosawa, Hachioji, Tokyo 192-0397, Japan
[2]Japan Synchrotron Radiation Research Institute, SPring-8, Sayo, Hyogo 679-5198, Japan
E-mail: yamada-akira1@ed.tmu.ac.jp

Abstract. X-ray absorption spectra of (Sm_xLa_{1-x})Tr_2Al_{20} (Tr = Ti and Ta), which exhibit unusual magnetic-field-insensitive physical properties, have been measured at the Sm L_3-edge at selected temperatures between 2.5 and 300 K. The spectra show that the Sm ions are in a mixed valence state and that their average valence is approximately +2.85, with almost no significant temperature and x dependences. For $SmTi_2Al_{20}$, the average valence of the Sm ions does not display any change associated with the magnetic ordering below 6.5 K.

1. Introduction

In recent years, compounds with the general chemical formula RTr_2X_{20} (R = rare earth, Tr = transition metal, X = Al, Zn, and Cd) crystallizing in the cubic $CeCr_2Al_{20}$-type structure have attracted considerable interest, as various strongly correlated electronic properties caused by the strong c-f hybridization have been found. Remarkable phenomena found among the RTr_2X_{20} family include heavy fermion (HF) behavior with a large Sommerfeld coefficient γ of $\sim$ 8 J/mol K^2 in $YbCo_2Zn_{20}$ and superconductivity in the quadrupolar ordered state in $PrTi_2Al_{20}$ [1, 2].

In the case of R = Sm, unusually field-insensitive phase transitions at T_x = 6.5, 2.9, 1.8, and 2.0 K for $SmTr_2Al_{20}$ (Tr = Ti, V, Cr, and Ta, respectively) and field-insensitive HF behavior in the ordered states have been discovered [3, 4, 5]. The resistivity ρ of these compounds exhibits a $-\log T$ dependence, which is unusually insensitive to the magnetic field [3, 4, 5, 6]. Recent electronic transport measurements on $SmTa_2Al_{20}$ have revealed that the transverse magnetoresistance (MR) is always positive, with a H^2 dependence in a wide temperature T region between 2 and 300 K. Therefore, the magnitude of the expected negative MR due to the suppression of Kondo scattering by the magnetic field is significantly smaller than that of the ordinary MR due to the cyclotron motion of the conduction electrons [6]. These electronic transport measurements have also revealed a strongly enhanced deviation from Kohler's rule in MR, which indicates development of the wave-vector $\boldsymbol{k}$ dependence of the conduction electron scattering due to $4f$ electrons with decreasing T [6]. In the ordered state below T_x, these compounds exhibit unusually field-insensitive HF behaviors with large Sommerfeld coefficients of γ = 0.15, 0.72, 1.0, and 3.2 J/mol K^2 for $SmTr_2Al_{20}$ (Tr = Ti, V, Cr, and Ta, respectively).

For $SmTi_2Al_{20}$, the quasiparticle mass enhancement has also been confirmed via a de Haas-van Alphen experiment [7]. The field-insensitive Kondo behavior mentioned above differs considerably from that of typical Ce-based HF compounds such as $CeCu_6$ [8], suggesting the occurrence of the Kondo effect with an unconventional mechanism in the $SmTr_2Al_{20}$ family.

The crystalline electric field (CEF) ground state of these Sm compounds is expected to be a $_8$ quartet of the $J = 5/2$ multiplet [3, 4, 5]. The magnitudes of the ordered moments estimated from the nuclear specific heat in zero magnetic field, $m_{\mathrm{Sm}} = 0.51$ (Ti), 0.28 (Cr), and 0.22 μ_{B}/Sm (Ta), are suppressed from the value of 0.65 μ_{B}/Sm expected for a $_8$ quartet state. This suppression can be ascribed to the c-f hybridization or the involvement of higher-rank multipolar degrees of freedom belonging to the $_8$ state [3, 5]. The magnetic susceptibilities exhibit weak temperature dependences, which are evidently different from those for both free Sm^{2+} and Sm^{3+} ions [3, 4, 5]. X-ray absorption spectroscopy (XAS) at the Sm L_3-edge has revealed that the Sm-ion valence in the $SmTr_2Al_{20}$ family (Tr = Ti, V, Cr, and Ta) is approximately +2.87 with no significant T dependence between 7 and 300 K [9]. From these experimental results, it is expected that the multipolar and/or valence degrees of freedom play a role in the formation of the field-insensitive HF states.

In the case of another field-insensitive HF compound $SmOs_4Sb_{12}$, it is believed that the T-dependent behavior of the Sm-ion valence is key to understanding the field-insensitive HF state in $SmOs_4Sb_{12}$ [10]. Recent XAS experiments on $(Sm_xLa_{1-x})Os_4Sb_{12}$ with $x = 0.2$–1 indicate that the Sm ion valence exhibits $\log T$-dependent behavior [11]. The T-dependent variation of the Sm-ion valence is attributed to an unconventional Kondo effect associated with charge degrees of freedom [12]. In the present study, we investigate the T and x dependences of the Sm-ion valence in $(Sm_xLa_{1-x})Tr_2Al_{20}$ (Tr = Ti and Ta) using XAS measurements.

2. Experimental

$(Sm_xLa_{1-x})Ta_2Al_{20}$ (x = 0.01, 0.05, 0.25, and 1) and $(Sm_xLa_{1-x})Ti_2Al_{20}$ (x = 0.1, 0.2, 0.4, 0.6, and 1) single crystals were grown by the Al self-flux method. For this purpose, mixtures with atomic ratios of Sm(3N):La(3N):Ta(3N5):Al(4N) = x:1-x:2:150 and Sm(3N):La(3N):Ti(5N):Al(4N) = x:1-x:2:90 (the purity is shown in parentheses) were used. X-ray powder diffraction patterns collected using a Rigaku SmartLab powder diffractometer equipped with a Cu K $_1$ monochromator indicated that all samples crystallized in the $CeCr_2Al_{20}$-type cubic crystal structure with space group $Fd\bar{3}m$. XAS spectra at the Sm L_3-edge were measured for $(Sm_xLa_{1-x})Tr_2Al_{20}$ (Tr = Ti and Ta) at BL01B1 of SPring-8 in Japan. This beam line features a double-crystal monochromator equipped with a Si 111 crystal and a Si mirror for eliminating the higher harmonics [13]. The XAS spectra for $(Sm_xLa_{1-x})Ti_2Al_{20}$ (x = 0.1, 0.2, 0.4, 0.6, and 1) and $(Sm_xLa_{1-x})Ta_2Al_{20}$ (x = 0.25 and 1) were recorded in the transmission mode using ionization chambers at various T between 3 and 300 K. For the Sm-dilute compounds $(Sm_xLa_{1-x})Ta_2Al_{20}$ (x = 0.01, 0.05, and 0.25), the fluorescence mode was employed using a 19-element Ge solid-state detector. The energy resolution was approximately 1 eV around the Sm L_3-edge. The incident photon energy was calibrated using the Sm L_3-edge XAS spectrum of Sm_2O_3 at 300 K.

3. Results and discussion

The normalized XAS spectra for $SmTi_2Al_{20}$ and $SmTa_2Al_{20}$ are shown in figures 1(a) and (b), respectively. The spectra consist of two components; the lower and higher energy peaks at the photon energies E = 6716 and 6722 eV correspond to the divalent and trivalent components, respectively. The XAS spectra in the X-ray energy range of 6700–6730 eV was fit according to

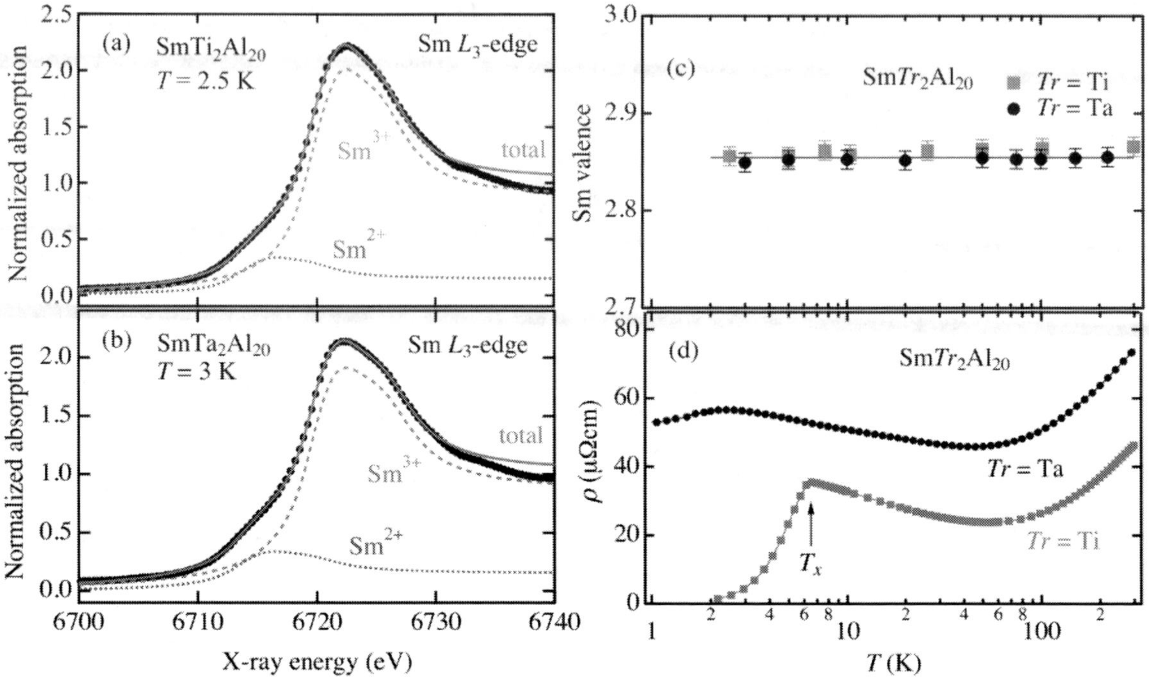

Figure 1. (a) Normalized X-ray Absorption Spectroscopy (XAS) spectra at Sm L_3-edge of $SmTi_2Al_{20}$ measured at $T = 2.5$ K. (b) Normalized XAS spectra at Sm L_3-edge of $SmTa_2Al_{20}$ measured at $T = 3$ K. The closed circles represent the measured spectra, and the broken, dotted, and solid lines indicate the Sm^{3+}, Sm^{2+}, and total components fitted to the experimental data, respectively. (c) T dependence of the average Sm ion valence in $SmTi_2Al_{20}$ and $SmTa_2Al_{20}$. The solid line is a guide for the eye. (d) T dependence of resistivity for $SmTi_2Al_{20}$ and $SmTa_2Al_{20}$ (data taken from Ref. [3, 6]).

$$f(E) = \sum_{i=2}^{3}\left[\sum_{j=0}^{1}\left\{\frac{A_i^j(\ _i/2)}{(E-E_i-j\cdot\Delta)^2+(\ _i/2)^2}\right\} + \frac{(A_i^0+A_i^1)\ _i}{(A_2^0+A_2^1)\ _2+(A_3^0+A_3^1)\ _3}\left\{\frac{1}{2}+\frac{1}{\pi}\arctan\frac{E-(E_i+\delta)}{\ '_i/2})\right\}\right], \qquad (1)$$

where A_i^0, A_i^1, Δ, $\ _i$, $\ '_i$ E_i, and δ are fitting parameters and the subscript $i = 2$ and 3 denotes the divalent and trivalent states, respectively. In the $SmTr_2Al_{20}$ family, the peak structure attributed to the Sm^{3+} states exhibits a noticeably asymmetric shape compared with that of Sm_2O_3, indicating that the $5d$ empty states of Sm are split into two states by the CEF effect [9]. We introduced the superscript $j = 0$ and 1 to equation (1), representing a lower and higher state of the split 5d empty state, respectively. The energy gap Δ and the intensity ratio A_i^0/A_i^1 were assumed to be identical for Sm^{2+} and Sm^{3+}. In the present analysis, we do not consider the oscillatory structure appearing above 6730 eV, which is caused by the extended X-ray absorption fine structure (EXAFS) and which reflects the local crystal structure around the Sm ions. The fitting curve reproduces the experimental data well, confirming the validity of the present model as expressed by equation (1). The average valence of the Sm ions was determined from

Journal of Physics: Conference Series **683** (2016) 012020 doi:10.1088/1742-6596/683/1/012020

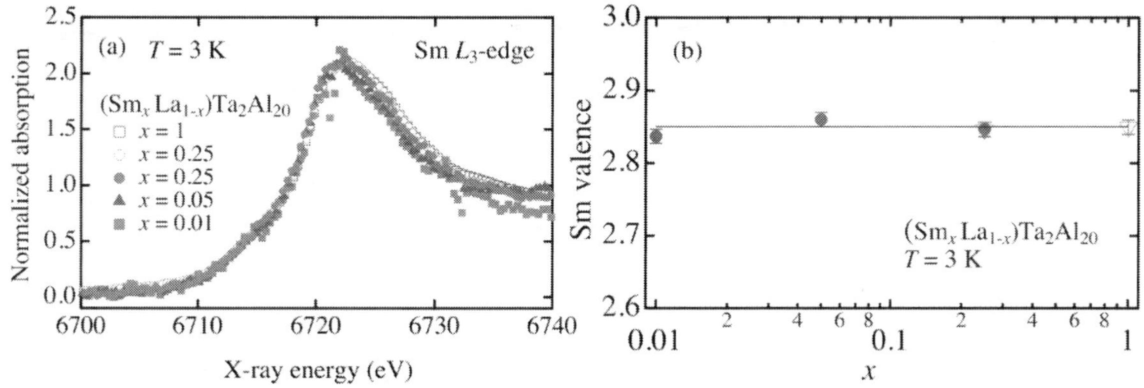

Figure 2. (a) Normalized XAS spectra at Sm L_3-edge of $(Sm_xLa_{1-x})Ta_2Al_{20}$ recorded in transmission (open symbols) and fluorescence (solid symbols) modes, measured at T = 3 K. The small anomalies in the spectra appearing in the energy ranges of 6720.6–6722.6 eV and 6731.5–6733.5 eV for $x = 0.01$ are removed from the fitting. (b) x dependence of the average valence measured in the fluorescence (transmission) mode, represented by solid (open) symbols. The solid line is a guide for the eye.

$$v = 2 + \frac{(A_3^0 + A_3^1)\ _3}{(A_2^0 + A_2^1)\ _2 + (A_3^0 + A_3^1)\ _3}, \tag{2}$$

using the relative intensities of the Sm^{2+} and Sm^{3+} components. The present result clearly demonstrates that the Sm ions are in a mixed valence state with an average valence of approximately +2.85 for both $SmTi_2Al_{20}$ and $SmTa_2Al_{20}$. This value is in agreement with the previous report [9]. The T dependence of the average valence was investigated for both $SmTi_2Al_{20}$ and $SmTa_2Al_{20}$. As shown in figure 1(c), the Sm-ion valence in $SmTr_2Al_{20}$ (Tr = Ti and Ta) displays no significant T dependence between 3 and 300 K. Although ρ exhibits a clear $-\log T$ dependence between 7 and 50 K (between 3 and 40 K) for Tr = Ti (Ta) as shown in figure 1(d), no associated valence change is observed. In $SmTi_2Al_{20}$, no noticeable change appears in the Sm-ion valence below $T_x = 6.5$ K [3].

In order to investigate the x dependence of the Sm-ion valence in the $(Sm_xLa_{1-x})Tr_2Al_{20}$ family, the XAS spectra at the Sm L_3-edge were measured. The normalized spectra for $(Sm_xLa_{1-x})Ta_2Al_{20}$ (x = 0.01, 0.05, 0.25, and 1) measured at 3 K are shown in figure 2(a). As can be seen in figure 2(b), the average Sm-ion valence determined from these XAS spectra exhibits almost no x dependence within the experimental uncertainty. In principle, the Sm-ion valence can shift in response to La substitution because of the following factors: 1) Considering the fact that the radius of a Sm^{2+} ion is larger than that of a Sm^{3+} ion, Sm^{2+} ions become more stable via La substitution, because of the negative chemical pressure caused by the lattice expansion (the lattice parameters of $SmTa_2Al_{20}$ and $LaTa_2Al_{20}$ are 14.750 and 14.811 Å, respecitvely); 2) As La^{3+} substitution increases the conduction electron density (i.e., the Fermi energy), if the phenomenon of the mixed Sm-ion valence is coupled with the conduction electrons in some ways, the Sm-ion valence may shift. For another field-insensitive HF compound $(Sm_xLa_{1-x})Os_4Sb_{12}$, a decrease in the Sm-ion valence by approximately 0.05 from $x = 1$ to $x = 0.2$ has been observed [11]. The present result indicates that these effects are below the detectable levels in the $SmTr_2Al_{20}$ family, even if they exist.

The T dependence of the Sm-ion valence in $(Sm_xLa_{1-x})Tr_2Al_{20}$ (Tr = Ti and Ta) was

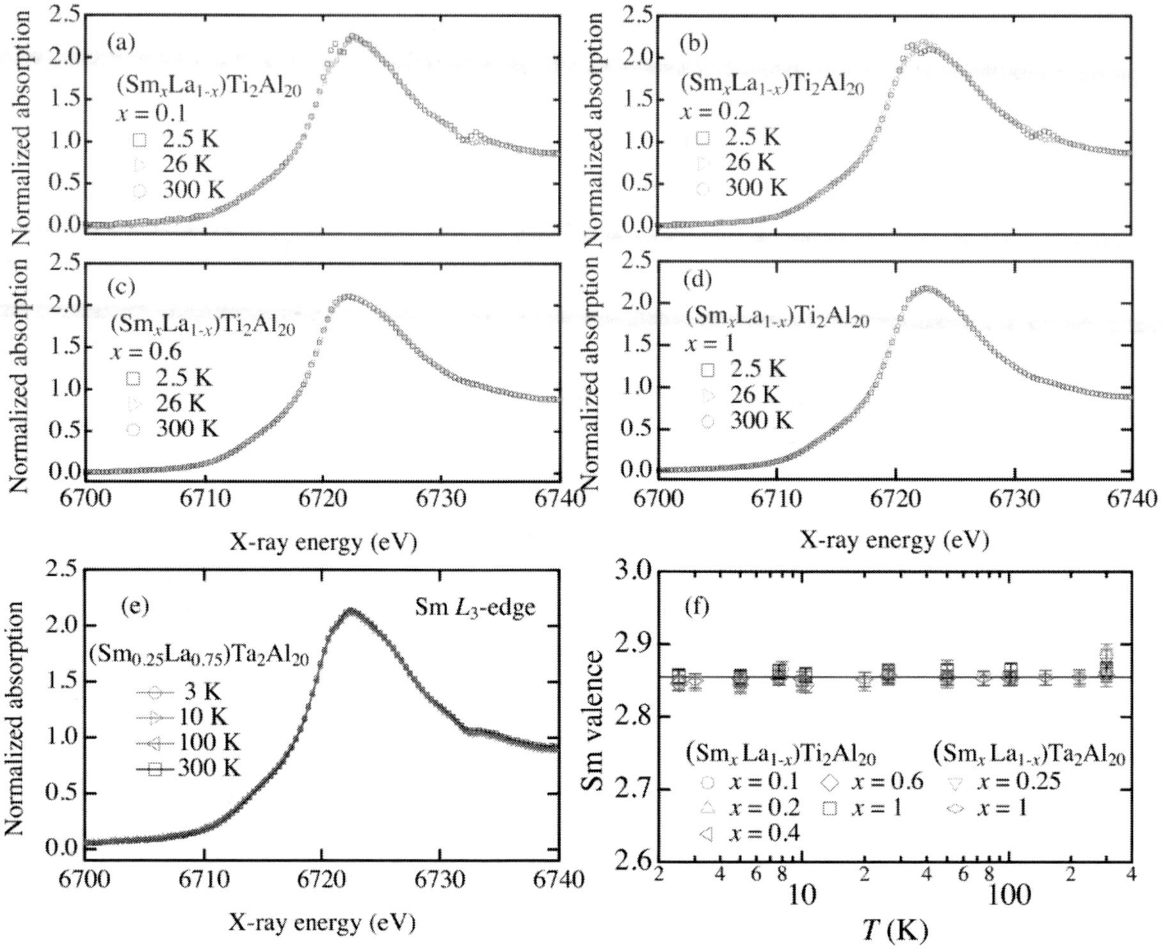

Figure 3. (a)–(d) represent normalized $(Sm_xLa_{1-x})Ti_2Al_{20}$ (x = 0.1, 0.2, 0.6, and 1, respectively) XAS spectra at selected T. (e) Normalized XAS spectra at Sm L_3-edge of $(Sm_{0.25}La_{0.75})Ta_2Al_{20}$ measured at selected T. (f) T dependence of average Sm-ion valence of $(Sm_xLa_{1-x})Tr_2Al_{20}$ (Tr = Ti, Ta). The small anomalies in the spectra appearing in the energy ranges of 6720.6–6722.6 and 6731.5–6733.5 eV for small x samples are removed from the fitting. The solid line is a guide for the eye.

also investigated. Figures 3(a)–(e) show the XAS spectra measured at selected T between 2.5 and 300 K, where no T-dependent change is observed. As summarized in figure 3(f), the valences of the Sm ions in all of the compounds investigated in this study exhibit almost no T dependence. In contrast, it has been reported that the Sm-ion valence displays a clear decrease with decreasing T in $SmOs_4Sb_{12}$ [10] and the Kondo insulator SmB_6 [16]. In $SmOs_4Sb_{12}$, the Sm-ion valence shifts from ~2.83 at 300 K to ~2.76 at 10 K. This T dependence can be attributed to the unconventional Kondo effect associated with charge degrees of freedom with the associated Kondo temperature $\widetilde{T}_K$ = 56 K [11, 12]. For SmB_6, the average valence shifts from ~2.6 at room temperature to 2.5 at 15 K, which is considered to be associated with the development of the Kondo insulating gap with decreasing T [14, 15, 16, 17]. At present, in the $(Sm_xLa_{1-x})Tr_2Al_{20}$ family (Tr = Ti and Ta), the physics of the mixed valence states of the Sm

Journal of Physics: Conference Series **683** (2016) 012020 doi:10.1088/1742-6596/683/1/012020

ions is unclear. Considering the fact that the XAS measurement timescale is $\sim 10^{-12}$–10^{-15} s, systematic studies of Sm-ion valence in Sm mixed-valence compounds are required, combined with complementary techniques for detecting slower dynamics, i.e., Mössbauer ($\sim 10^{-8}$ s) and nuclear quadrupole resonance ($\sim 10^{-7}$–10^{-6} s) experiments.

4. Summary

We have revealed that the Sm ions in the $(Sm_xLa_{1-x})Tr_2Al_{20}$ family (Tr = Ti and Ta) are in a mixed valence state, using X-ray absorption spectroscopy measurements at the Sm L_3-edge. The average Sm-ion valence is approximately +2.85 and is almost independent of x and temperature. Although the electrical resistivity exhibits a clear $-\log T$ dependence between 7 and 50 K (between 3 and 40 K) for Tr = Ti (Ta), no corresponding valence change is observed. For $SmTi_2Al_{20}$, the average valence in the ordered state below T_x = 6.5 K is almost identical to that in the paramagnetic state.

Acknowledgements

The authors are grateful to Professor H. Sato for helpful discussions. The XAS measurements were performed at the facilities of BL01B1 (Proposal No. 2014B1139) in SPring-8, with the approval of the Japan Synchrotron Radiation Research Institute (JASRI). This work was supported by MEXT/JSPS KAKENHI Grant Numbers 23540421, 23340107, 24740239, 20102005, 15H03693, 15J07600, 15K05178, and 15H05884 (J-Physics).

References

[1] Torikachvili M S, Jia S, Mun E D, Hannahs S T, Black R C, Neils W K, Martien D, Bud'ko S L and Canfield P C 2007 *Proc. Natl. Acad. Sci.* U.S.A. **104** 9960
[2] Sakai A, Kuga K and Nakatsuji S 2012 *J. Phys. Soc. Jpn.* **81** 083702
[3] Higashinaka R, Maruyama T, Nakama A, Miyazaki R, Aoki Y and Sato H 2011 *J. Phys. Soc. Jpn.* **80** 093703
[4] Sakai A, Nakatsuji S 2011 *Phys. Rev.* B **84** 201106(R)
[5] Yamada A, Higashinaka R, Miyazaki R, Fushiya K, Matsuda T D, Aoki Y, Fujita W, Harima H and Sato H 2013 *J. Phys. Soc. Jpn.* **82** 123710
[6] Yamada A, Higashinaka R, Matsuda T D, Aoki Y and Sato H 2015 *J. Phys. Soc. Jpn* **84** 103701
[7] Nagashima S, Nishiwaki T, Ohtani A, Sakoda M, Matsuoka E, Harima H and Sugawara H 2014 *JPS Conf. Proc.* **3** 011019
[8] Sumiyama A, Oda Y, Nagano H, Onuki Y, Shibutani K, Komatsubara T 1986 *J. Phys. Soc. Jpn.* **55** 1294
[9] Higashinaka R, Yamada A, Miyazaki R, Aoki Y, Mizumaki M, Tsutsui S, Nitta K, Uruga T and Sato H 2014 *JPS Conf. Proc.* **3** 011079
[10] Mizumaki M, Tsutsui S, Tanida H, Uruga T, Kikuchi D, Sugawara H and Sato H 2007 *J. Phys. Soc. Jpn.* **76** 053706
[11] Fushiya K, Miyazaki R, Higashinaka R, Yamada A, Mizumaki M, Tsutsui S, Nitta K, Uruga T, Suemitsu B, Sato H and Aoki Y 2015 *Phys. Rev.* B **92** 075118
[12] Tanikawa S, Matsuura H and Miyake K 2009 *J. Phys. Soc. Jpn.* **78** 034707
[13] Uruga T, Tanida H, Yoneda Y, Takeshita K, Emura S, Takahashi M, Harada M, Nishihata Y, Kubozono Y, Tanaka T, Yamamoto T, Maeda H, Kamishima O, Takabayashi Y, Nakata Y, Kimura H, Goto S and Ishikawa T 1999 *J. Synchrotron Radiat.* **6** 143
[14] Cohen R L, Eibschtz M and West K W 1970 *Phys. Rev. Lett.* **23** 383
[15] Souma S, Kumigashira H, Ito T, Takahashi T and Kunii S 2002 *Physica* B **312-313** 329
[16] Mizumaki M, Tsutsui S and Iga F 2009 *J. Phys. Conf. Ser.* **176** 012034 *Phys. Rev.* X **3** 041024
[17] SmB_6 is regarded as a topological Kondo insulator. Angle-resolved photoemission spectroscopy results show that the metallic surface states appearing below approximately 110 K are associated with the T-dependent development of the Kondo-insulating gap (see the following references: Xu N *et al* 2013 *Phys. Rev.* B **88** 121102(R) and Frantzeskakis E *et al* 2013 *Phys. Rev.* X **3** 041024).

Journal of Physics: Conference Series **683** (2016) 012021 doi:10.1088/1742-6596/683/1/012021

Low temperature properties of $ZrTr_2Zn_{20}$ (*Tr* = transition metal) with a pyrochlore lattice

S Hasegawa[1], K Morihiro[2], N Kase[1], T Nakano[1] and N Takeda[2]
[1]Graduate School of Science and Technology, Niigata University, Niigata 950-2102, Japan
[2]Department of Materials Science and Technology, Niigata University, Niigata 950-2102, Japan

E-mail: f15b032b@mail.cc.niigata-u.ac.jp

Abstract. Single crystals of $ZrTr_2Zn_{20}$ (*Tr* = Mn, Co, Ru) with a cubic $CeCr_2Al_{20}$-type structure (space group *Fd*-3*m*) were grown by a Zn-flux method. Electrical resistivity $\rho(T)$ of $ZrTr_2Zn_{20}$ shows typical metallic behavior with the residual resistivity ratio (RRR) of 1.1 (Mn), 8.4 (Co) and 10 (Ru). Magnetic susceptibility $\chi(T)$ of $ZrRu_2Zn_{20}$ shows large diamagnetism. $\chi(T)$ of $ZrTr_2Zn_{20}$ (*Tr* = Mn, Co) clearly shows Curie-Weiss behavior, indicating that Mn/Co spins are really active as local magnetic moments or that a large spin fluctuation exists. $\chi(T)$ of $ZrMn_2Zn_{20}$ increases ferromagnetically at around 20 K. Because the anti-ferromagnetic correlation with active localized spins exists in $ZrCo_2Zn_{20}$, geometrical frustration of the pyrochlore lattice of Co is expected. However, specific heat $C(T)$ of $ZrCo_2Zn_{20}$ shows no upturn at low temperatures and Sommerfeld coefficient γ is evaluated to be 24.8 mJ/(mol K^2). We consider that the localized Co spins are insufficient to cause anomaly originated from the geometrical frustration, or the Co spins exist in a disordered arrangement of excess Co.

1. Introduction

Heavy-fermion behavior is one of the central issues in the strongly correlated systems. In this system, 4/5*f* electron plays an important role to form heavy-fermion state ascribed to competition between Kondo effect and RKKY interaction. However, heavy-fermion behavior without 4/5*f*-electron has been discovered in the pyrochlore compound LiV_2O_4 [1], where the heavy-fermion behavior is considered to be ascribed to the geometrical frustration on the pyrochlore lattice.

$YMn_2Zn_{20-x}In_x$ is an itinerant-electron antiferromagnet with magnetic Mn atoms forming a pyrochlore lattice made of corner-sharing tetrahedra [2]. This compound is also known to show heavy-fermion behavior. Specific heat $C(T)$ at low temperatures increases below 2 K and reaches 280 mJ/(mol K^2), suggesting a significant large mass enhancement. However, $YMn_2Zn_{20-x}In_x$ has a possibility that disorder from excess Mn mask intrinsic behavior [2]. Thus, at this stage, it is highly desirable to find a pure single crystal of MTr_2Zn_{20} comprising *Tr* (*Tr* = transition metal) pyrochlore lattice without excess magnetic elements.

We focus on Zr-based compounds $ZrTr_2Zn_{20}$, which can be synthesized without In substitution, and we report the physical properties of the single crystals of $ZrTr_2Zn_{20}$ (*Tr* = Mn, Co, Ru).

2. Experimental details

Single crystals of $ZrTr_2Zn_{20}$ (Tr = Mn, Co, Ru) were grown by a Zn-flux method. Zr chips, Mn, Co, and Ru powder, Zn grains were mixed in a molar ratio of 1:2:80. Each mixture sealed in silica tubes were heated on 1000°C for 24h, and then cooled down to 500°C at a rate of 5°C/h in an electrical furnace. Zn-flux was eliminated by a centrifugation and the remaining one removed by acetic acid (0.1 %). The typical size of the single crystals is about 1 ~ 5 mm as shown in inset of Figure 1 The crystal structure was examined by the powder X-ray diffraction (XRD) technique with Cu-K_α radiation and a graphite monochromator (RAD-2X, Rigaku). The intensity data were collected over a 2θ range of 10-90° with a step width of 0.01° and a counting rate of 4°/minutes. Magnetic susceptibility $\chi(T)$ measurements were performed by a magnetic properties measurements system (MPMS, Quantum Design). Electrical resistivity $\rho(T)$ was measured by an ac-four-probe method in a ^{3}He cryostat (Heliox VL, Oxford) down to 0.3 K. Specific heat $C(T)$ measurements were performed down to 0.3 K in the ^{3}He cryostat using a standard adiabatic heat-pulse method.

3. Experimental results

3.1. Chemical composition and lattice constants

Figure 1 shows the XRD patterns of $ZrTr_2Zn_{20}$ (Tr = Mn, Co, Ru). Remaining Zn-flux is detected in the XRD patterns of $ZrRu_2Zn_{20}$. All peaks except for the Zn peaks can be indexed with a cubic unit cell with a space group Fd-$3m$. The lattice constants obtained from all indexes are 1.4033(2) (Mn), 1.3913(1) (Co) and 1.4038(1) nm (Ru). These values are in a good agreement with those of the previous report [3].

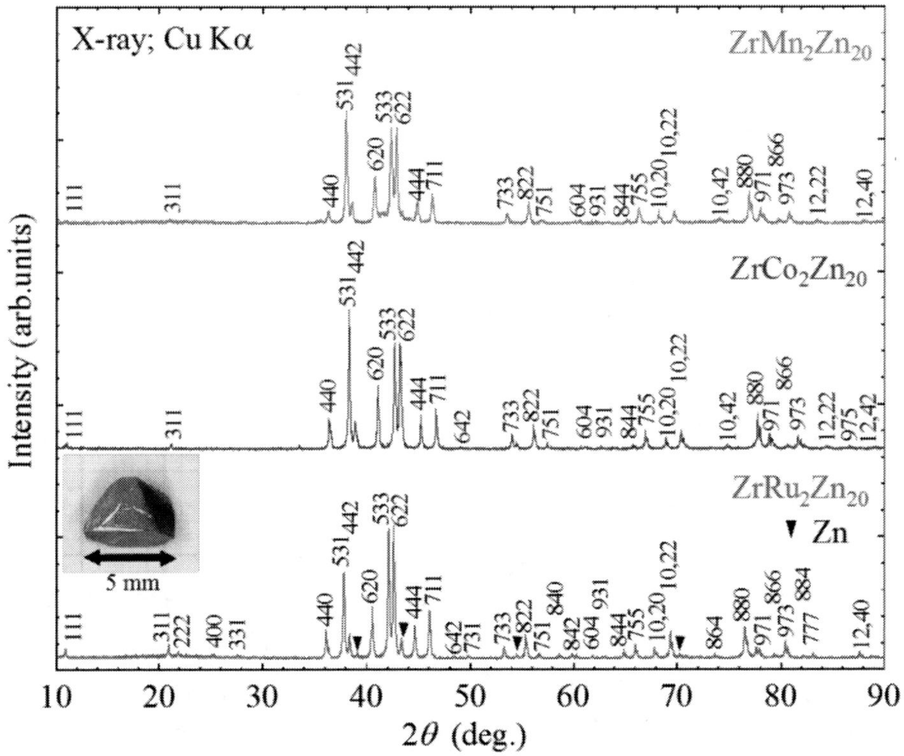

Figure 1. XRD patterns of $ZrTr_2Zn_{20}$ (Tr = Mn, Co, Ru) single crystal. Inset shows image of a single crystal of $ZrRu_2Zn_{20}$.

3.2. Magnetic properties

Figure 2 (a) represents *T*-dependence of χ_0 subtracted magnetic susceptibility $\chi(T) - \chi_0$ of $ZrMn_2Zn_{20}$, where χ_0 is a *T*-independent term of 1.82×10^{-3} emu/mol, and Figure 2 (b) shows $\chi(T)$ of $ZrTr_2Zn_{20}$ (Tr = Co, Ru) in a magnetic field of 0.1 T down to 2 K. $\chi(T)$ of $ZrTr_2Zn_{20}$ (Tr = Mn, Co) clearly shows Curie-Weiss behavior at high temperatures. This result indicates that Mn/Co spins are really active as local magnetic moments or that a large spin fluctuation exists. $\chi(T)$ of $ZrMn_2Zn_{20}$ ferromagnetically increase at $T_C \sim 20$ K. From $\chi^{-1}(T)$ in a temperature range of 100-300 K, Currie-Weiss fits give an effective moment of $\mu_{eff} = 0.68$ μ_B/Mn and Currie temperature $\theta_C = 33$ K for $ZrMn_2Zn_{20}$, $\mu_{eff} = 1.1$ μ_B/Co and Weiss temperature $\theta_w = -96$ K for $ZrCo_2Zn_{20}$. The estimated μ_{eff} values are relatively smaller than that expected when each Mn/Co atom carries an $S = 1/2$ localized spin ($\mu_{eff} = 1.73$ μ_B/Tr), and are comparable to that of paramagnetic compounds RFe_2Zn_{20} (R = Y, Lu) [4]. Because localized spin exists ferromagnetically, geometrical frustration of the pyrochlore lattice of Mn cannot be expected in $ZrMn_2Zn_{20}$. On the other hand, $ZrCo_2Zn_{20}$ has a possibility of heavy-fermion behavior originated from geometrical frustration, because the anti-ferromagnetic correlation with active localized spin exists. $\chi(T)$ of $ZrRu_2Zn_{20}$ shows diamagnetism from room temperature. The diamagnetism is relatively large. Such a large diamagnetism is also seen in AV_2Al_{20} (A = Y, La) [5].

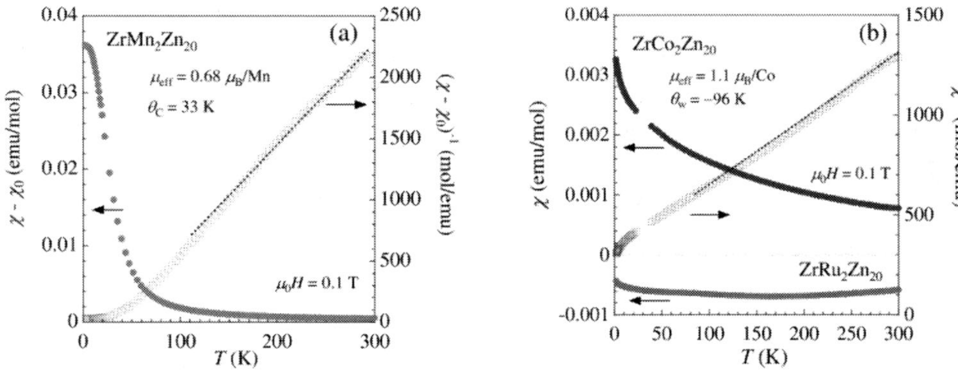

Figure 2. (a) Magnetic susceptibility $\chi(T)$ and inverse susceptibility $\chi^{-1}(T)$ subtracted a temperature-independent term χ_0 of $ZrMn_2Zn_{20}$ in a magnetic field of 0.1 T. (b) $\chi(T)$ and $\chi^{-1}(T)$ of $ZrCo_2Zn_{20}$, and $\chi(T)$ of $ZrRu_2Zn_{20}$ in a magnetic field of 0.1 T. The dotted line on each χ^{-1} data represents a Currie-Weiss fit.

3.3. Specific heat

Figure 3 shows *T*-dependence of specific heat $C(T)$ divided by T of $ZrTr_2Zn_{20}$ (Tr = Mn, Co, Ru) down to 0.3 K. $C(T)/T$ can be fitted as $C/T = \gamma + \beta T^2$; γ is Sommerfeld coefficient and β is lattice specific coefficient. The fittings give $\gamma = 29.6$ and $\beta = 2.90$ (Mn), 24.8 and 1.32 (Co), 13.6 mJ/(mol K^2) and 1.53 mJ/(mol K^4) (Ru). From the simple Debye model of the phonon contribution with the relationship $\Theta_D = (12\pi^4 nR/5\beta)^{1/3}$, gas constant $R = 8.314$ J/(mol K) and $n = 23$ for $ZrTr_2Zn_{20}$, Debye temperatures Θ_D are evaluated to be 249 (Mn), 324 (Co) and 308 K (Ru), respectively. In the heavy fermion compounds originated from geometrical frustration, $C(T)/T$ shows upturn at low temperatures and reaches 207 mJ/(mol K^2) in $YMn_2Zn_{20-x}In_x$ [2]. Because $C(T)/T$ of $ZrCo_2Zn_{20}$ shows no upturn at low temperature and γ is comparable to that of the normal and paramagnetic compounds in 1-2-20 system [5], heavy-fermion behavior is not observed in $ZrCo_2Zn_{20}$. The reason why upturn or large γ is not seen in $ZrCo_2Zn_{20}$ is not clear. One of the reasons is that the localized Co spins are insufficient to cause anomaly originated from the geometrical frustration. In the case of $YMn_2(Zn_{1-x}In_x)_{20}$, μ_{eff} is estimated to be 2.4 μ_B/Mn, which is comparable to that expected when each Mn atom carries an $S = 1$ localized spin [2]. The estimated localized spin of $ZrCo_2Zn_{20}$ (1.1 μ_B/Co) seems to be small to lead to anomaly originated from geometrical frustration. Another reason is that the localized Co spin is not located in the pyrochlore lattice, but exists in a disordered arrangement of excess Co in the compound. It is necessary to determine whether excess Co exists or not by means of detailed crystal structure analysis.

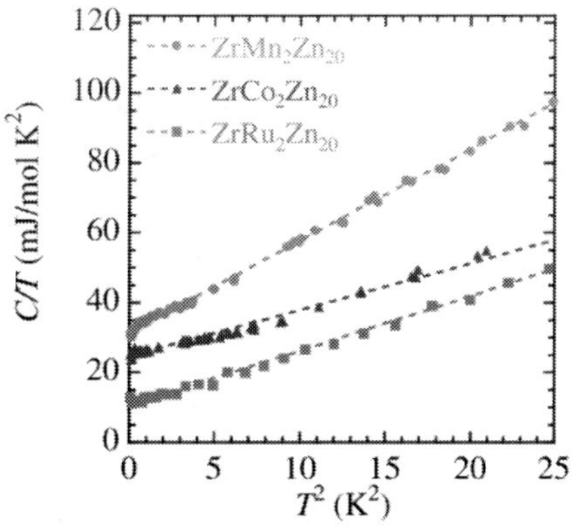

Figure 3. Temperature dependence of specific heat divided by temperature $C(T)/T$ of $ZrTr_2Zn_{20}$ (Tr = Mn, Co, Ru). The dotted lines represent the fitting result (see text).

3.4. Electrical resistivity

Figure 4 shows T-dependence of electrical resistivity $\rho(T)$ of $ZrTr_2Zn_{20}$ (Tr = Mn, Co, Ru) down to 0.29 K. $\rho(T)$ of $ZrTr_2Zn_{20}$ shows normal metallic behavior, although ρ_0 of $ZrMn_2Zn_{20}$ is relatively large. Residual resistivity ratios (RRR) are estimated to be 1.1 (Mn), 8.4 (Co) and 10 (Ru) respectively. The RRR of $ZrMn_2Zn_{20}$ is smaller than those of $ZrTr_2Zn_{20}$ (Tr = Co, Ru). In $\rho(T)$ of $ZrMn_2Zn_{20}$, anomaly ascribed to ferromagnetic transition is not observed. The $\rho(T)$ of $ZrMn_2Zn_{20}$ is quite similar to that of $YMn_2Zn_{20-x}In_x$ (x = 3.44), which is disordered by substitution of In and excess Mn. Large ρ_0 of $ZrMn_2Zn_{20}$ seems to be originated from disorder due to excess Mn [4].

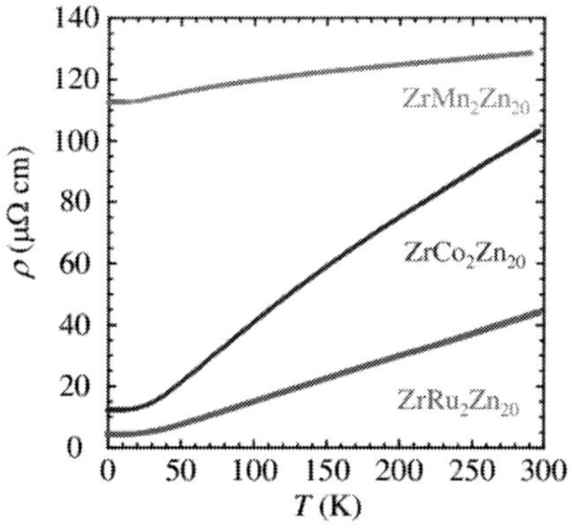

Figure 4. Temperature dependence of electrical resistivity $\rho(T)$ of $ZrTr_2Zn_{20}$ (Tr = Mn, Co, Ru).

4. Conclusion

We succeeded in synthesizing single crystals of ZrTr_2Zn_{20} (Tr = Mn, Co, Ru). $ZrMn_2Zn_{20}$ shows ferromagnetic transition at T_C ~ 20 K. $\rho(T)$ of $ZrMn_2Zn_{20}$ is metallic, but RRR is small and ρ_0 is large despite single crystal. This result indicatives that excess Mn exists. Upturn in $C(T)/T$ at low temperatures is not observed above 0.3 K in ZrTr_2Zn_{20} (Tr = Mn, Co, Ru). Magnetic susceptibility $\chi(T)$ of $ZrRu_2Zn_{20}$ shows large diamagnetism, which is very similar to that of AV_2Al_{20} (A = Y, La). $\chi(T)$ of ZrTr_2Zn_{20} (Tr = Mn, Co) clearly shows Curie-Weiss behavior. The results indicate that Mn/Co spins are really active as local magnetic moments or that a large spin fluctuation exists. $\chi(T)$ of $ZrMn_2Zn_{20}$ ferromagnetically increase at around 20 K. On the other hand, geometrical frustration of the pyrochlore lattice with Co can be expected in $ZrCo_2Zn_{20}$, because the anti-ferromagnetic correlation with active localized spins exists. However, $C(T)/T$ of $ZrCo_2Zn_{20}$ shows normal metallic behavior. We consider that the localized Co spins are insufficient to cause anomaly originated from the geometrical frustration, or the Co spins exist in a disordered arrangement of excess Co.

Acknowledgments

This work was partially supported by the Grant-in-Aid for Young Scientist (B) (No. 26800183) from the Ministry of Education, Culture, Sports, Science, and Technology, Japan, the Sasakawa Scientific Research Grant (26-207) from the Japan Science Society, and the Uchida Energy Science Promotion Foundation (26-1-22).

References

[1] Kondo S, Johnston D C, Swenson C A, Borsa F, Mahajan A V, Miller L L, Gu T, Goldman A I, Maple M B, Gajewski D A, Freeman E J, Dilley N R, Dickey R P, Merrin J, Kojima K, Luke G M, Umemura Y J, Chmaissem O and Jorgensen J D 1997 *Phys. Rev. Lett.* **78** 3729
[2] Okamoto Y, Simizu T, Yamamura J, Kiuchi Y and Hiroi Z 2010 *J. Phys. Soc. Jpn.* **79** 093712
[3] Gross N, Nasch T and Jeitschko W 2001 *J. Solid State Chem.* **161** 288
[4] Jia S, Ni N, Bud'ko S L and Canfield P C 2009 *Phys. Rev.* B **80** 104403
[5] Onosaka A, Okamoto Y, Yamaura J, Hirose T and Hiroi Z 2012 *J. Phys. Soc. Jpn.* **81** 123702

Quadrupole order in the frustrated pyrochlore magnet $Tb_2Ti_2O_7$

H. Takatsu[1,2], T. Taniguchi[1], S. Kittaka[3], T. Sakakibara[3], and H. Kadowaki[1]

[1]Department of Physics, Tokyo Metropolitan University, Hachioji-shi, Tokyo 192-0397, Japan
[2]Department of Energy and Hydrocarbon Chemistry, Graduate School of Engineering, Kyoto University, Kyoto 615-8510, Japan
[3]Institute for Solid State Physics, University of Tokyo, Kashiwa 277-8581, Japan

Abstract. We have studied the hidden long-range order (LRO) of the frustrated pyrochlore magnet $Tb_2Ti_2O_7$ by means of specific-heat experiments and Monte-Carlo (MC) simulations, which has been discussed as the LRO of quadrupole moments inherent to the non-Kramers ion of Tb^{3+}. We have found that the sharp specific-heat peak is collapsed into a broad hump by magnetic fields above 0.3 T for $H//[001]$. This result, qualitatively reproduced by MC simulations, suggests that a field-induced magnetic state overcomes the quadrupolar LRO state, as a similar case of a classical spin ice. The present results support the interpretation that $Tb_{2+x}Ti_{2-x}O_{7+y}$ is a unique material in the boundary between the quadrupolar ($x \geq x_c = -0.0025$) and spin-liquid ($x \leq x_c$) states, where the magnetic field along the [001] axis is a tuning parameter which induces the magnetic ordered state.

1. Introduction

Geometrically frustrated magnetic systems provide us with a rich playground for studying new type of electronic and magnetic behaviors with unconventional order parameters [1, 2]. In particular, the pyrochlore-lattice magnet $Tb_{2+x}Ti_{2-x}O_{7+y}$, a putative candidate of spin liquid (SL) [3, 4], is a unique material showing an unknown long range order (LRO) in the vicinity ($x \geq x_c = -0.0025$) of the SL state [5, 6]. It is found that a clear specific-heat peak appears at $T_c \simeq 0.5$ K for the sample with $x = 0.005$, while no LRO associated with the large magnetic and/or structural phase transitions was observed [5]: the only small Bragg peak (the order of 0.1 μ_B/Tb) appears below T_c but is too small to explain the corresponding entropy change in the specific heat. Therefore, the LRO of $Tb_2Ti_2O_7$ is considered to be a "hidden order" [7], apparently contrast with the magnetic dipole order inferred by earlier theories on the basis of the spin ice (SI) Hamiltonian [8, 9].

Recently, we have investigated the single crystalline sample of $Tb_{2+x}Ti_{2-x}O_{7+y}$ with $x = 0.005$ ($T_c = 0.53$ K), and found experimental key ingredients of the hidden order [10]; i.e., two-step magnetization kink and double peak structure in specific heat, which are induced by magnetic field for $H//[111]$. We have demonstrated those as characteristics of the behaviors of a possible electric quadrupole ordering [10], by using classical Monte Carlo (CMC) and quantum Monte Carlo simulations on the basis of the theoretical model proposed by Onoda and Tanaka [11, 12]. This theory shows that transverse super-exchange interactions between quadrupole moments, which are set into the SI Hamiltonian as an additional term, play a driving

Published under licence by IOP Publishing Ltd

force of the LRO, making $Tb_2Ti_2O_7$ into a type of a quadrupole order in the vicinity of the SL state [11–13]. Following this interpretation, we have also pointed out that the appearance of the inelastic neutron excitation can be understood in terms of the spin-quadrupole wave [10, 14]. These results suggest that the problem of the hidden order in $Tb_2Ti_2O_7$ can now be reconsidered as the novel problem of the quadrupole order in the frustrated pyrochlore magnet with the non-Kramers ion of Tb^{3+}.

The purpose of the present study is to examine the quadrupole order of $Tb_2Ti_2O_7$ by measuring physical quantities in magnetic field along other directions for $H//[111]$, which provide another evidence for the origin of the hidden order and may induce exotic magnetic properties. For this purpose, we performed specific heat experiments under magnetic field for $H//[001]$. CMC simulations are also performed. We found that the sharp peak of the specific heat at $H = 0$ is suddenly suppressed by the magnetic field along [001]. This behavior is understood in terms of the realization of the field-induced magnetic state in weak [001] fields above 0.3 T.

2. Experimental

Single crystals of $Tb_{2+x}Ti_{2-x}O_{7+y}$ were grown by a floating zone method [6]. We used the crystal with $x = 0.005$ in this study. Specific heat $C_P(T, H)$ was measured by a quasi adiabatic method in a dilution refrigerator down to 0.1 K. Magnetic field was applied using a vector magnet system where an accuracy of the field direction to the sample is below 1°. In order to reduce the demagnetization effect, we used a plate-like crystal, cut out from a single crystal rod, along the $\langle 110 \rangle$ plane which includes the [111], [110], and [001] axes. The sample is approximately $0.7 \times 0.9 \times 0.1$ mm^3 and weighs 0.35 mg: this crystal is the same crystal used in the previous study [10]. Since the demagnetization factor for the [001] direction is small enough ($N \simeq 0.09$), demagnetization corrections were not performed in the present study.

CMC simulations were performed using the program of the ALPS package (1024 sites, periodic boundary conditions) on the basis of the effective pseudospin-1/2 Hamiltonian [11, 12] relevant for the non-Kramers magnetic doublets of $Tb_2Ti_2O_7$. We employed the nearest-neighbor (NN) classical SI model for the first three terms of Eq. (1) of Ref. [10] and used following parameters, $J_{\rm nn,eff} = 1.48$ K, $J_{\rm nn} * \delta = 0$, and $J_{\rm nn} * q = 0.85$, so that those correspond to the previously used parameters [10]. These parameters are in the region of the planar antiferropseudospin (PAF) state. Readers are referred to Ref. [10] for details of the Hamiltonian and the parameters. We calculated the specific heat under the [001] field in the present study. We also computed the order parameter of the quadrupolar LRO, which is a type of magnetization associated with the LRO of the xy-components of the Pauli matrices $\boldsymbol{\sigma_r}$: the quadrupolar LRO can be expressed by a pseudospin structure $(\langle \sigma_{\boldsymbol{r}}^x \rangle, \langle \sigma_{\boldsymbol{r}}^y \rangle)$ in the model of Refs. [11, 12].

3. Results and Discussion

Figure 1 shows the temperature dependence of $C_P(T, H)$ measured at several magnetic fields up to 1 T. The sample exhibits a clear sharp peak at $T_c = 0.53$ K in zero field, which is compatible with the behavior of the polycrystalline sample of $x = 0.005$ [5]. Remarkably, once a magnetic field is applied along [001], the sharp peak is collapsed into a broad hump in fields above 0.3 T. This result suggests that the zero-field LRO state is weak for the application of magnetic field for $H//[001]$. In the inset of Fig. 1, we summarized the H–T phase diagram for $H//[001]$. The labels of each phase were determined from the analysis of CMC simulations, which will be discussed later. We found that the quadrupole LRO is replaced by the field-induced magnetic ordered state for $H//[001]$. This result is related to the crystal structure of the pyrochlore lattice and the Ising-like anisotropy of spins along the local $\langle 111 \rangle$ direction. In fact, for magnetic field along the [001] axis, it is expected that magnetic moments are easily induced as those pointing at the inward and outward tetrahedra consisting of the 2-in 2-out spin configurations. It is known for a classical SI material for $H//[001]$ [15, 16].

Journal of Physics: Conference Series **683** (2016) 012022 doi:10.1088/1742-6596/683/1/012022

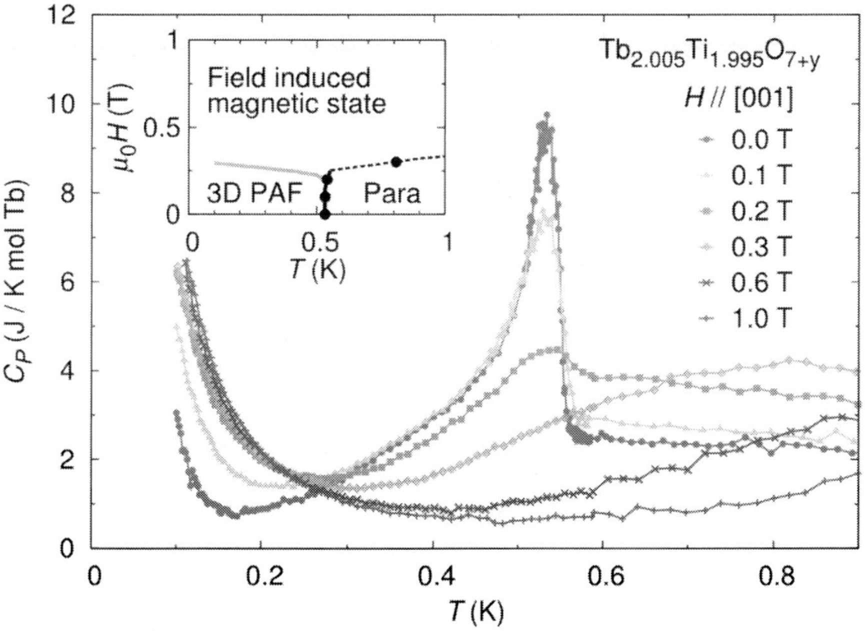

Figure 1. Temperature dependence of the specific heat $C_P(T, H)$ of single-crystalline $Tb_{2+x}Ti_{2-x}O_{7+y}$ with $x = 0.005$ in the applied magnetic field along [001]. Inset shows the H–T phase diagram for $H//[001]$. Filled circles in the inset are peak positions of $C_P(T, H)$. The black solid line up to 0.2 T indicates the phase boundary between the quadrupolar (3D PAF) state and the paramagnetic state. The black dashed line is the line of the crossover between the paramagnetic state and the field-induced magnetic state. The gray solid line is the putative phase boundary between the 3D PAF state and the field-induced magnetic state, which is examined from the analysis of CMC simulations. Labels in the phase diagram are assigned from the analysis of CMC simulations (Fig. 3).

Figure 2 shows the temperature dependence of the calculated specific heat. We confirmed that the experimental behaviors are qualitatively reproduced by CMC simulations, although the slight lowering of T_c is observed in low field. In fact, we observed the abrupt change in the peak height of the specific heat in CMC simulations for $H//[001]$. This result implies that the present model [11, 12] works well for the discussion for the LRO of $Tb_2Ti_2O_7$, where the dominant component for the formation of LRO is the super-exchange interaction between quadrupole moments.

In Figs. 3(a)–(c), we present contour plots of the calculated specific heat, order parameter of the quadrupolar LRO, and magnetization M. The map of specific heat shows the qualitatively similar behavior for experiments, where the abrupt change appears around T_c in low fields and the broad hump shifts to high temperatures when the field is increased. The phase boundary extending to low temperatures is less obvious, however it can be expected by the appearance and disappearance of order parameters as shown in Figs. 3(b) and (c). In fact, we confirmed that the order parameter of quadrupole moments develops in the low-H and low-T region below T_c. Here, this quadrupolar state is found to be the three dimensional (3D) PAF state [10, 14]. However, it is soon destroyed by the large magnetic field for $H//[001]$ (Fig. 3(b)). We realized that M is developed in such field region (Fig. 3(c)). The value of M in the phase boundary is about 70% of the full magnetic moment for the [001] direction ($\sim 2\ \mu_B$/Tb): note that

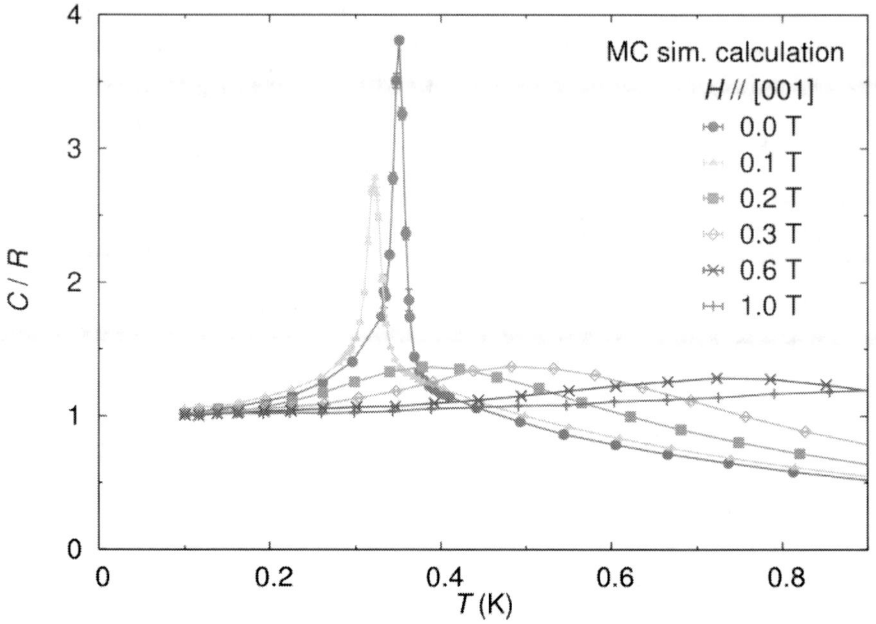

Figure 2. Temperature dependence of the specific heat for several fields calculated by the CMC simulation.

the present CMC simulations can only handle the contribution of the magnetic moment of the ground state doublet ($\mu \simeq 5\ \mu_B$/Tb) [10, 14], and then the fully saturated moment of M along the [001] direction is $\mu/\sqrt{3}$ due to the anisotropy of the [001] direction [15]. Interestingly, we found that the boundary, accompanied by the specific-heat peak and the disappearance of the quadrupole order parameter, emerges at the line of such 70% magnetization (Figs. 3(a)–(c)). This result implies a phase boundary between the quadrupolar LRO state and the field-induced magnetic state or the paramagnetic state ($T < T_c$), and a crossover from the paramagnetic state to the field-induced magnetic state ($T > T_c$) (Fig. 3(d)). This situation is probably realized in the experiments for $H//[001]$. In fact, in the previous magnetization experiment by Legl *et al.* [17], the sharp increase of M has been observed for $H//[001]$. The field where the magnetization reaches 2 μ_B/Tb is 0.3 – 0.4 T at $T \leq 0.3$ K. This is compatible with the present CMC calculations and the experimental results.

Studies of bulk physical properties in magnetic field are indirect observation for the origin of the quadrupole ordering. However, we can observe key signatures associated with the formation and deformation of its ordering. Therefore, the present specific heat experiment for $H//[001]$ and analysis by CMC simulations are important clues to discuss the origin of the LRO of $Tb_2Ti_2O_7$, which should be a quadrupole ordering and is weak for the [001] field . We can now consider that $Tb_2Ti_2O_7$ is a very unique material showing two ground states, the quadrupolar-LRO state and SL state, which are tuned by the minute change in the off-stoichiometric parameter of x for $Tb_{2+x}Ti_{2-x}O_{7+y}$ [5, 6]. Magnetic field is also a tuning parameter which instead induces the magnetic ordered state for $H//[001]$.

4. Conclusion

In conclusion, we have studied the long-range order (LRO) of the frustrated pyrochlore-lattice magnet $Tb_2Ti_2O_7$, which emerges in the vicinity of the spin liquid state by tuning the parameter

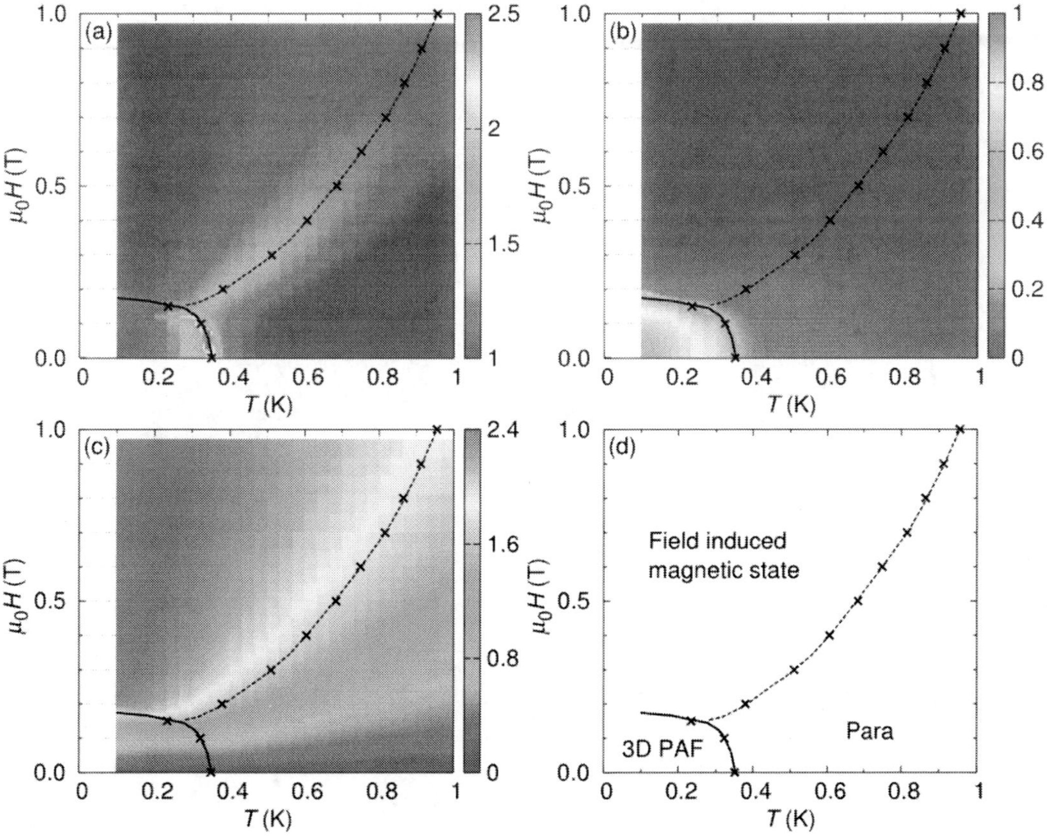

Figure 3. Temperature-field dependence of (a) specific heat, (b) quadrupolar order parameter, and (c) magnetization calculated by the CMC simulation for magnetic field along [001]. (d) H–T phase diagram obtained by the CMC simulation for $H//[001]$. Cross symbols in (a)–(d) are peak positions of the calculated specific heat. The solid lines indicate the phase boundary between the 3D PAF state and the field-induced magnetic state or the paramagnetic state. The dashed lines are the lines of the crossover between the paramagnetic state and the field-induced magnetic state. The lines are guides to the eyes.

x for $Tb_{2+x}Ti_{2-x}O_{7+y}$. We measured the specific heat of the single crystal with $x = 0.005$ ($T_c = 0.53$ K) and found that the sharp specific-heat peak is suddenly suppressed by magnetic field for $H//[001]$. We demonstrated that the experimental behavior is qualitatively reproduced by the classical Monte Carlo simulation on the basis of the Onoda-Tanaka Hamiltonian. These results imply that the field-induced magnetic ordered state overcomes the quadrupolar LRO state at magnetic fields above 0.3 T. This behavior is characteristic for the magnetic field along [001]. The present experimental and calculation results support the scenario that the LRO of $Tb_2Ti_2O_7$ is a quadrupole order, mediated by a quantum phase transition from a spin liquid state, where the magnetic field can be used as a tuning parameter as well for the realization of a magnetic ordered state.

Acknowledgments
We thank S. Onoda, Y. Kato, R. Higashinaka and M. Wakita for useful discussions. This work was supported by JSPS KAKENHI grant numbers 25400345 and 26400336. The specific heat measurement was performed using the facilities of ISSP, Univ. of Tokyo.

References

[1] Lacroix C, Mendels P and Mila F (eds) 2011 *Introduction to Frustrated Magnetism* (Berlin, Heidelberg: Springer)
[2] Gingras M J P and McClarty P A 2014 *Rep. Prog. Phys.* **77** 056501
[3] Gardner J S, Dunsiger S R, Gaulin B D, Gingras M J P, Greedan J E, Kiefl R F, Lumsden M D, MacFarlane W A, Raju N P, Sonier J E, Swainson I and Tun Z 1999 *Phys. Rev. Lett.* **82** 1012
[4] Gardner J S, Keren A, Ehlers G, Stock C, Segal E, Roper J M, Fak B, Stone M B, Hammar P R, Reich D H and Gaulin B D 2003 *Phys. Rev. B* **68** 180401
[5] Taniguchi T, Kadowaki H, Takatsu H, Fak B, Ollivier J, Yamazaki T, Sato T J, Yoshizawa H, Shimura Y, Sakakibara T, Hong T, Goto K, Yaraskavitch L R and Kycia J B 2013 *Phys. Rev. B* **87** 060408(R)
[6] Wakita M, Taniguchi T, Edamoto H, Takatsu H and Kadowaki H arXiv:1509.04583
[7] Santini P, Carretta S, Amoretti G, Caciuffo R, Magnani N and Lander G H 2009 *Rev. Mod. Phys.* **81** 807
[8] Gingras M J P, den Hertog B C, Faucher M, Gardner J S, Dunsiger S R, Chang L J, Gaulin B D, Raju N P and Greedan J E 2000 *Phys. Rev. B* **62** 6496
[9] Kao Y J, Enjalran M, Maestro A D, Molavian H R and Gingras M J P 2003 *Phys. Rev. B* **68** 172407
[10] Takatsu H, Kittaka S, Kasahara A, Kono Y, Sakakibara T, Kato Y, Onoda S, Fåk B, Ollivier J, Lynn J W, Taniguchi T, Wakita M, and Kadowaki H, arXiv:1506.04545
[11] Onoda S and Tanaka Y 2010 *Phys. Rev. Lett.* **105** 047201
[12] Onoda S and Tanaka Y 2011 *Phys. Rev. B* **83** 094411
[13] Lee S B, Onoda S and Balents L 2012 *Phys. Rev. B* **86** 104412
[14] Kadowaki H, Takatsu H, Taniguchi T, Fak B and Ollivier J 2015 *SPIN* **5** 1540003
[15] Fukazawa H, Melko R G, Higashinaka R, Maeno Y and Gingras M J P 2002 *Phys. Rev. B* **65** 054410
[16] Hiroi Z, Matsuhira K and Ogata M 2003 *J. Phys. Soc. Jpn.* **72** 3045
[17] Legl S, Krey C, Dunsiger S R, Dabkowska H A, Rodriguez J A, Luke G M and Pfleiderer C 2012 *Phys. Rev. Lett.* **109** 047201

Journal of Physics: Conference Series **683** (2016) 012023 doi:10.1088/1742-6596/683/1/012023

Quantum spin liquid and electric quadrupolar states of single crystal $Tb_{2+x}Ti_{2-x}O_{7+y}$

M Wakita, T Taniguchi, H Edamoto, H Takatsu and H Kadowaki

Department of Physics, Tokyo Metropolitan University, Hachioji-shi, Tokyo 192-0397, Japan

Abstract. The ground states of the frustrated pyrochlore oxide $Tb_{2+x}Ti_{2-x}O_{7+y}$, sensitively depending on the small off-stoichiometry parameter x, have been studied by specific heat measurements using well characterized samples. Single crystal $Tb_{2+x}Ti_{2-x}O_{7+y}$ boules grown by the standard floating zone technique are shown to exhibit concentration (x) gradient. This off-stoichiometry parameter is determined by precisely measuring the lattice constant of small samples cut from a crystal boule. Specific heat shows that the phase boundary of the electric quadrupolar state has a dome structure in the x-T phase diagram with the highest $T_c \simeq 0.5$ K at about $x = 0.01$. This phase diagram suggests that the putative U(1) quantum spin-liquid state of $Tb_{2+x}Ti_{2-x}O_{7+y}$ exists in the range $x < x_c \simeq -0.0025$, which is separated from the quadrupolar state via a first-order phase-transition line $x = x_c$.

1. Introduction

Magnetic systems with geometric frustration have been intensively studied experimentally and theoretically for decades [1]. Spin systems on networks of triangles or tetrahedra, such as triangular [2], kagomé [3], and pyrochlore [4] lattices, play major roles in these studies. Subjects fascinating many investigators in recent years are quantum spin liquid (QSL) states [5, 6], where conventional long-range orders (LRO) are suppressed to very low temperatures.

Among frustrated magnetic pyrochlore oxides [4], $Tb_2Ti_2O_7$ (TTO) has attracted much attention because it does not show any conventional LRO down to 50 mK [7], suggesting that it is a candidate for a QSL state. Although many experimental studies of TTO have been performed to date, the problem why TTO does not show any magnetic LRO remains very difficult [8, 9]. This is partly because TTO shows strong sample dependence [10], extremely strong for single crystals. And accordingly, simple interpretation of experimental data is precluded.

Recently, we investigated polycrystalline samples of off-stoichiometric $Tb_{2+x}Ti_{2-x}O_{7+y}$, and showed that a very small change of x induces a quantum phase transition between a spin liquid state ($x < -0.0025 = x_c$) and a LRO state with a hidden order parameter ($x_c < x$) [11]. The x-T phase diagram of $Tb_{2+x}Ti_{2-x}O_{7+y}$ suggested in Ref. [11] has a dome-shape LRO phase boundary. More recently, we study the hidden LRO using an x-controlled single crystal, which shows a very sharp peak in specific heat at $T_c = 0.53$ K ($x \simeq 0.005$) [12]. By using semi-quantitative analyses, we propose [12–14] that the LRO of $Tb_{2+x}Ti_{2-x}O_{7+y}$ is an electric multipolar (or quadrupolar) state. This LRO state was theoretically predicted [15] using electronic superexchange interactions for non-Kramers ions, including Tb^{3+}, which have both magnetic dipole and electric quadrupole (16-pole, and 64-pole) moments. In addition, quite intriguingly, the estimated parameter set [12] of the effective pseudospin-1/2 Hamiltonian is

Journal of Physics: Conference Series **683** (2016) 012023 doi:10.1088/1742-6596/683/1/012023

located very close to a theoretical phase boundary between the electric quadrupolar and U(1) quantum spin-liquid states [15, 16], which could naturally explain the spin liquid state of TTO.

The purpose of this investigation is to extend our study of polycrystalline $Tb_{2+x}Ti_{2-x}O_{7+y}$ [11] to single crystals in the hope that the above scenario for the TTO problem is reinforced. We grow single crystals of $Tb_{2+x}Ti_{2-x}O_{7+y}$ by the standard floating zone (F) technique [17] and have found that very precise measurements of the lattice constant are useful to characterize the single crystals. Specific heat of these samples with different off-stoichiometry parameters (x) have been measured down to 0.1 K to obtain an x-T phase diagram.

2. Experimental methods and results

Polycrystalline samples of $Tb_{2+x}Ti_{2-x}O_{7+y}$ were prepared by the standard solid-state reaction as described in Ref. [11]. The two starting materials, Tb_4O_7 and TiO_2, were heated in air at 1350 °C for several days with periodic grindings to ensure a complete reaction. The value of x was adjusted by changing the mass ratio of the two materials, and is nominal with an offset about ±0.002. The resulting $Tb_{2+x}Ti_{2-x}O_{7+y}$ powder samples were used for single crystal growth by the standard F technique [17]. Crystal growth was carried out in an Ar gas flow atmosphere using a double ellipsoidal image furnace (NEC SC-N35HD).

-ray powder-diffraction experiments were carried out using a RIGAKU-SmartLab diffractometer equipped with a Cu $K_{\alpha 1}$ monochromator. To precisely measure the lattice constant we performed θ-2θ scans on powder mixtures of polycrystalline or crushed-crystalline $Tb_{2+x}Ti_{2-x}O_{7+y}$ and Si [11, 18]. Absolute values of lattice constants are normalized by using the certified lattice parameter for a temperature of 22.5 °C of the SRM-640d Si powder, $a = 5.43123$ Å [19], being further corrected for the temperature dependence [20].

Temperature dependence of the lattice constant $a(T, x)$ of $Tb_{2+x}Ti_{2-x}O_{7+y}$ was measured using a polycrystalline sample with $x = -0.0075$, and the result is shown in Fig. 1(a). The x dependence of $a(T = 26.0°C, x)$ of polycrystalline samples is plotted in Fig. 1(b), where we converted the published lattice constants (Fig. 1 in Ref. [11]) to those at 26.0 °C [18].

Figure 2 shows a single crystal $Tb_{2+x}Ti_{2-x}O_{7+y}$ boule that was grown from a feed rod of $x = -0.005$ powder and was post-annealed for about 7 days at 1000 °C in air. Lattice constants

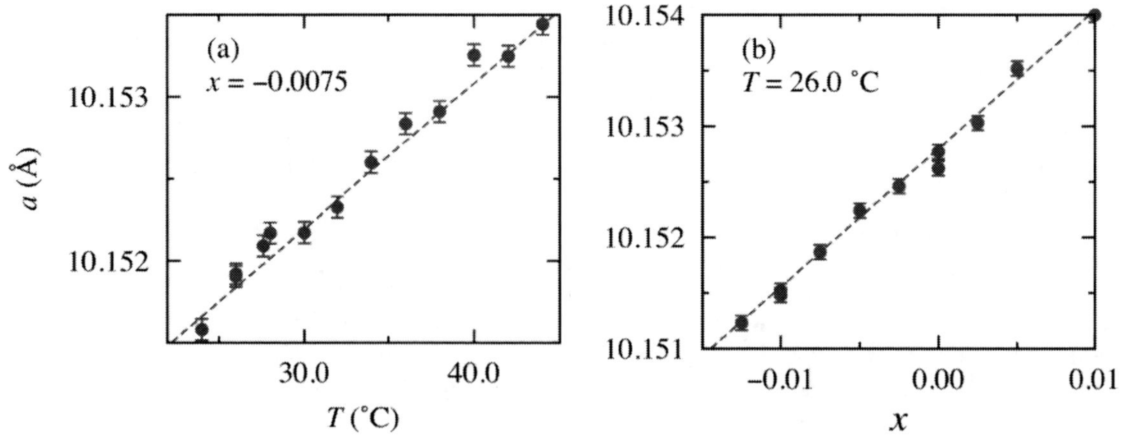

Figure 1. Lattice constant $a(T, x)$ of polycrystalline $Tb_{2+x}Ti_{2-x}O_{7+y}$ samples. (a) Temperature dependence of $a(T, x = -0.0075)$. (b) Off-stoichiometry parameter dependence of $a(T = 26.0°C, x) = 0.124418x + 10.15280$ from Ref. [18].

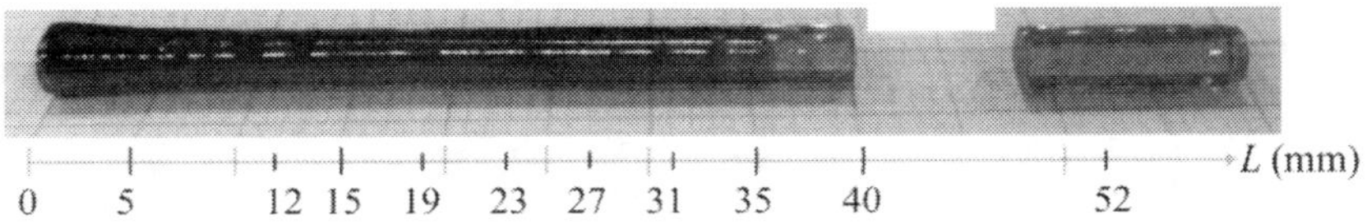

Figure 2. Single crystal $Tb_{2+x}Ti_{2-x}O_{7+y}$ boule grown by the F method, where the missing part ($40 < L < 48$ mm) was cut before taking this photograph. The numbers represent distances L along the growth direction, where small crystals are cut at these L values. Lattice constant and specific heat of these crystals are shown in Figs. 3 and 4.

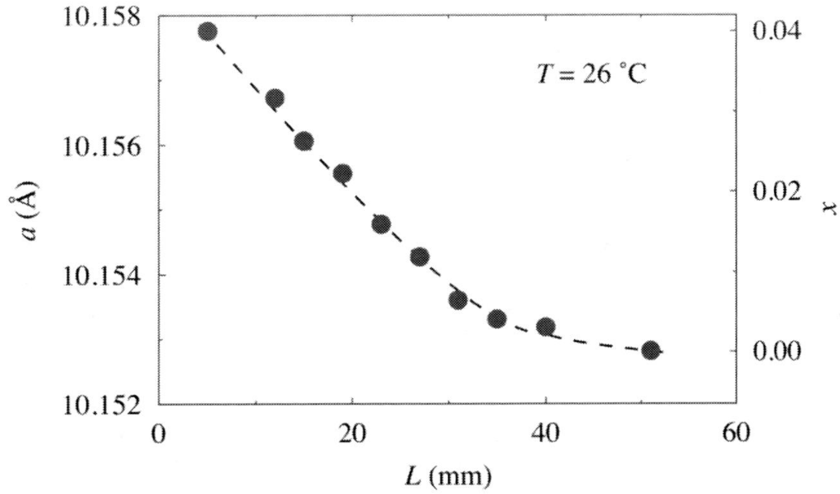

Figure 3. Lattice constants of small $Tb_{2+x}Ti_{2-x}O_{7+y}$ crystals cut from the boule shown in Fig. 2. These lattice constants are converted to x using the polycrystalline curve, i.e., $a(T = 26.0°\text{C}, x)$ of Fig. 1(b) and are shown on the right vertical axis.

of small crystals cut from this boule were measured at 26.0 °C and are plotted as a function of the distance along the growth direction L shown in Fig. 3. We assume that $a(T = 26.0°\text{C}, x)$ of polycrystalline samples (Fig. 1(b)) and its linear extension to the range $x > 0.01$ can be used to estimate the off-stoichiometry parameter (x) of the small crystals. These x values are shown on the right vertical axis of Fig. 3. One can see that the boule has a systematic x gradient. During the crystal growth the off-stoichiometry parameter starts from $x \simeq 0.04$ ($L = 1 - 5$ mm), then decreases linearly as a function of L, and finally varies more slowly ($L > 40$ mm).

To characterize crystal samples we also measured specific heat $C_P(T)$ at low temperatures using a ^{3}He or an adiabatic demagnetization refrigerator. In Fig. 4(a) we show specific heat as a function of temperature for the several crystals cut from the boule (Fig. 2) and a few from another boule. Based on these $C_P(T)$ data we draw a tentative x-T phase diagram for the single crystals in Fig. 4(b). We note that these $C_P(T)$ data and the x-T phase diagram for the single crystals are quite consistent with those of polycrystalline $Tb_{2+x}Ti_{2-x}O_{7+y}$ [11]. This indicates that our trial method of estimating small x ($|x| < 0.01$) for single crystals using the precise measurement of the lattice constant is probably reliable.

The x-T phase diagram (Fig. 4(b)) implies that one has to take a special care of very small change of the off-stoichiometry existing even in a single crystal boule to investigate

Journal of Physics: Conference Series **683** (2016) 012023 doi:10.1088/1742-6596/683/1/012023

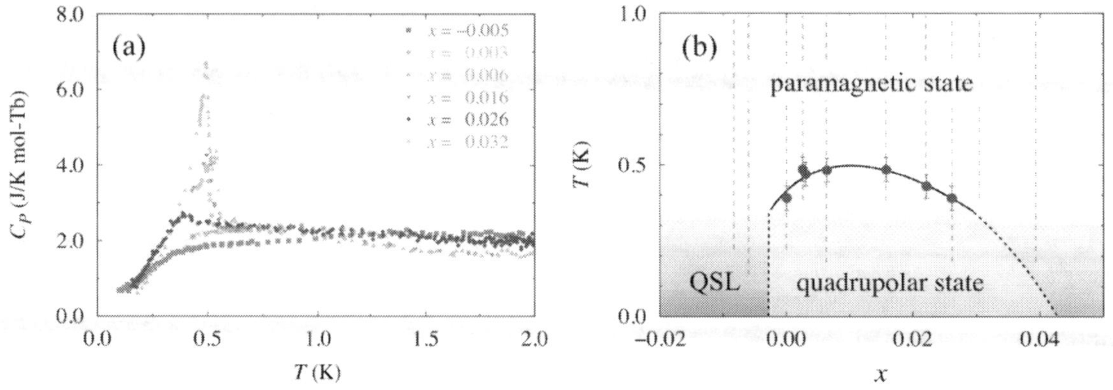

Figure 4. (a) Temperature dependence of specific heat of several single crystals. The x values are estimated by the method shown in Fig. 3. (b) x-T phase diagram determined from the specific heat measurements of single crystals. Temperature ranges of the specific heat measurements are shown by vertical blue dashed lines.

$Tb_{2+x}Ti_{2-x}O_{7+y}$ (or nominal $Tb_2Ti_2O_7$). Previous experimental investigations using small TTO crystals will have to be reinterpreted as investigations on different $Tb_{2+x}Ti_{2-x}O_{7+y}$ crystals. In particular, previous experiments using large crystals, especially inelastic neutron scattering for example Refs. [21–24], require special caution in their interpretation, because the crystals may not be sufficiently homogeneous.

3. Discussion and summary

The x-T phase diagram shows that around $x = x_c \simeq -0.0025$ the transition temperature T_c of the quadrupolar state [12] disappears abruptly in a small x range. This suggests that the neighboring putative QSL state is separated by a first-order phase-transition line $x = x_c$ [11, 12]. It is interesting that this type of first-order phase transition between U(1) QSL and quadrupolar states is predicted by a gauge mean-field theory [16], presumably relevant to TTO [12]. One may naturally expect that $Tb_{2+x}Ti_{2-x}O_{7+y}$ with $x = x_c$ is on the theoretical border of U(1) QSL and quadrupolar states [16], and that the spin liquid state of $Tb_{2+x}Ti_{2-x}O_{7+y}$ with $x < x_c$ is U(1) QSL of Ref. [16]. This is a very intriguing hypothesis for further studies.

On the other hand, in a larger x range of $x > 0.01$ the transition temperature of the quadrupolar state seems to decrease gradually and the specific heat peak gradually becomes smaller as x is increased. These suggest that an effect of randomness controls the system. A possible scenario of the randomness effect may be as follows. Most of excess Tb atoms reside on the Ti^{4+} site and become Tb^{4+} ions. These magnetic Tb^{4+} ions behave as magnetic impurities in the system, where local magnetic short-range order is restored around each Tb^{4+} ion. The quadrupolar state is completely suppressed in $x > 0.04$.

In summary, we have investigated single-crystalline samples of the frustrated pyrochlore oxide $Tb_{2+x}Ti_{2-x}O_{7+y}$ by growing single crystals using the standard floating zone technique and by characterizing them using -ray diffraction techniques and specific heat measurements down to 0.1 K. We show that a precise determination of the lattice constant is useful for estimating the small off-stoichiometry parameter x. Small crystals cut from single crystal rods, exhibiting x gradient, show three different low temperature behaviors: a paramagnetic QSL ($x < x_c$), a long range quadrupolar, and possibly a randomness dominating state. The phase boundary of the quadrupolar state shows a dome structure in the x-T phase diagram with the highest $T_c \simeq 0.5$

K at $x = 0.01$ and suggests existence of a first-order phase-transition line separating the QSL and quadrupolar states.

Acknowledgments

We thank S. Onoda and Y. Kato for useful discussions. This work was supported by JSPS KAKENHI grant numbers 25400345 and 26400336.

References

[1] Lacroix C, Mendels P and Mila F (eds) 2011 *Introduction to Frustrated Magnetism* (Berlin, Heidelberg: Springer)
[2] Wannier G H 1950 *Phys. Rev.* **79** 357
[3] Syôzi I 1951 *Prog. Theor. Phys.* 306
[4] Gardner J S, Gingras M J P and Greedan J E 2010 *Rev. Mod. Phys.* **82** 53
[5] Lee P A 2008 *Science* **321** 1306
[6] Balents L 2010 *Nature* **464** 199
[7] Gardner J S, Dunsiger S R, Gaulin B D, Gingras M J P, Greedan J E, Kiefl R F, Lumsden M D, MacFarlane W A, Raju N P, Sonier J E, Swainson I and Tun Z 1999 *Phys. Rev. Lett.* **82** 1012
[8] Gingras M J P and McClarty P A 2014 *Rep. Prog. Phys.* **77** 056501
[9] Petit S, Guitteny S, Robert J, Bonville P, Decorse C, Ollivier J, Mutka H and Mirebeau I 2015 *EPJ Web of Conferences* **83** 03012
[10] Takatsu H, Kadowaki H, Sato T J, Lynn J W, Tabata Y, Yamazaki T and Matsuhira K 2012 *J. Phys. Condens. Matter* **24** 052201
[11] Taniguchi T, Kadowaki H, Takatsu H, Fåk B, Ollivier J, Yamazaki T, Sato T J, Yoshizawa H, Shimura Y, Sakakibara T, Hong T, Goto K, Yaraskavitch L R and Kycia J B 2013 *Phys. Rev. B* **87** 060408
[12] Takatsu H, Kittaka S, Kasahara A, Kono Y, Sakakibara T, Kato Y, Onoda S, Fåk B, Ollivier J, Lynn J W, Taniguchi T, Wakita M and Kadowaki H arXiv:1506.04545
[13] Kadowaki H, Takatsu H, Taniguchi T, Fåk B and Ollivier J 2015 *Spin* **5** 1540003
[14] Takatsu H, Taniguchi T, Kittaka S, Sakakibara T and Kadowaki H 2016 *J. Phys.: Conf. Ser.* **683** 012022
[15] Onoda S and Tanaka Y 2011 *Phys. Rev. B* **83** 094411
[16] Lee S, Onoda S and Balents L 2012 *Phys. Rev. B* **86** 104412
[17] Gardner J, Gaulin B and Paul D 1998 *J. Crystal Growth* **191** 740
[18] Taniguchi T, Kadowaki H, Takatsu H, Fåk B, Ollivier J, Yamazaki T, Sato T J, Yoshizawa H, Shimura Y, Sakakibara T, Hong T, Goto K, Yaraskavitch L R and Kycia J B 2015 *Phys. Rev. B* **92** 019903
[19] Kaiser D L and Watters R L 2010 NIST Certificate of SRM 640d, `https://www-s.nist.gov/srmors/view_cert.cfm?srm=640D`
[20] Izumi F and Nakai I (eds) 2009 *Funmatsu Xsen kaiseki no jissai* 2nd ed (Tokyo: Asakurashoten)
[21] Yasui Y, Kanada M, Ito M, Harashina H, Sato M, Okumura H, Kakurai K and Kadowaki H 2002 *J. Phys. Soc. Jpn.* **71** 599
[22] Fennell T, Kenzelmann M, Roessli B, Mutka H, Ollivier J, Ruminy M, Stuhr U, Zaharko O, Bovo L, Cervellino A, Haas M K and Cava R J 2014 *Phys. Rev. Lett.* **112** 017203
[23] Guitteny S, Robert J, Bonville P, Ollivier J, Decorse C, Steffens P, Boehm M, Mutka H, Mirebeau I and Petit S 2013 *Phys. Rev. Lett.* **111** 087201
[24] Fritsch K, Ross K A, Qiu Y, Copley J R D, Guidi T, Bewley R I, Dabkowska H A and Gaulin B D 2013 *Phys. Rev. B* **87** 094410

Experimental exploration of novel semimetal state in strong anisotropic Pyrochlore iridate $Nd_2Ir_2O_7$ under high magnetic field

Z M Tian[1], Y Kohama[1], T Tomita[1], J Ishikawa[1], H Mairo[1], K Kindo[1], S Nakatsuji [1,2]
[1] Institute for Solid State Physics, University of Tokyo, Kashiwa, Chiba 277-8581, Japan
[2] PRESTO, Japan Science and Technology Agency (JST), Kawaguchi, Saitama, 332-0012, Japan
E-mail: tianzhaoming@issp.u-tokyo.ac.jp

Abstract. We report the anisotropic magnetotransport of $Nd_2Ir_2O_7$ single crystal under high magnetic field (B) up to 50 T with B along various directions. Only for B // [001] direction, a novel semimetal state is realized under high magnetic field evidenced by a field-induced insulating-semimetalic phase transition with critical field B_{MI}~12 T related to the destruction of all-in-all-out (AIAO) state. In contrast, abnormal magnetotransport hysteresis behavior is observed for B // [111] direction below the metal-insulator transition temperature (T_{MI}), and magnetotransport reveals the Ir spin structure keep in the AIAO state under high magnetic field with temperature just below T_{MI}, in prospect to realize Weyl semimetal state.

1. Introduction

Transition metal oxides (TMOs) of 5d elements have attracted a great deal of attention due to the interplay between a strong spin-orbit (SO) coupling and moderate electron correlation, providing a new platform for emergent electronic-or magnetic collective phenomena. Among these materials, pyrochlore iridates $R_2Ir_2O_7$ (R = Y, rare-earch element) are particular interest because various novel topological phases have been predicted to exist in certain correlation regimes, such as topological Mott insulator, Weyl semimetal (WSM) and axion insulator [1-4]. Experimental detection of this emergent topological phase should display anomalous Hall effect (AHE) and chiral magnetic effect [5,6], while these are still waiting for its experimental verification in spite of various theoretical proposals.

Most of $R_2Ir_2O_7$ compounds except for R = Pr exhibit metal-insulator transition (MIT), as successive increasing ion radius of R ions causes the change from magnetic ordered insulators to a spin-disordered metal [7,8]. Near the transition from Mott insulating phase to the metallic ground state within the phase diagram, $Nd_2Ir_2O_7$ (Nd-227) shows significantly smaller MIT temperature with T_{MI}~32 K, supporting a picture of Nd-227 residing at the boundary between correlated metal and insulator. Recent experiment on Nd-227 polycrystals suggests the nontrivial gapless metallic state localized at the AF domain wall [9]. Neutron diffraction experiments suggest the Ir^{4+} moments of Nd-227 consisting of an all-in-all-out (AIAO) magnetic structure as the ground state [10], where all the moments are pointing toward or outward from the center of each tetrahedron, leading to the coexistence of two kinds of antiferromagnetic (AF) domains namely all-in-all-out (A domain) and all-out-all-in (B domain) magnetic order [11]. These two configurations are not equivalent and are related by time-reversal symmetry, as the key ingredients to realize the predicted Weyl semimetal state [2]. Experimental investigation on single crystal would be highly desirable to unveil the emergent state due to the masked anisotropy effect of polycrystals, this will further lead us to realize the anisotropic control of spin texture and electronic structure by chemical (ionic radius of R) or physical pressure and external magnetic field.

In this study, anisotropic magnetotransport on $Nd_2Ir_2O_7$ single crystals is reported with magnetic field along different directions including [111], [110] and [001] axis. Specially, a novel semimetal state under high magnetic field are suggested to be realized for B // [001] direction by observation of MIT related to the destruction of AIAO state. For [111] direction, abnormal magnetoresistance hysteresis is observed for temperature below T_{MI}.

2. Experiment details

Single crystals of $Nd_2Ir_2O_7$ have been grown by a potassium fluoride-flux method. Powder and single crystal X-ray diffraction analyses show single phase and pyrochlore structure with lattice constant a = 10.386(3) Å. Electron-probe microanalysis (EPMA) shows a slight deviation within 2% from stoichiometry in the Ir/Nd ratio for the single crystals.

The magnetotransport were measured by a standard four-probe method with current path along the [110] direction and magnetic field was always oriented perpendicular to the current direction with different typical magnetic field direction. DC resistivity measurement in a DC magnetic field were carried out using a commercial system (PPMS, Quantum Design) which has a base temperature of 2 K and highest magnetic field of 14 T. Electrical transport measurements up to 50 T in a pulsed high magnetic field were performed with frequency f = 100 kHz and electrical current I = 280 μA at the international megagauss laboratory at ISSP, the University of Tokyo. During the measurements, the electrical current was applied to [1-10] direction within the (100) plane. Field dependence of the magnetization in a field up to 14 T was measured using a commercial vibrating sample magnetometer system (PPMS, Quantum Design). Temperature dependence of magnetization was measured using a commercial SQUID (superconducting quantum interface device) magnetometer (MPMS, Quantum Design) in zero-field cooling and field cooling (FC) procedure from 2 to 300 K.

3. Results and discussion

Temperature dependence of resistivity for $Nd_2Ir_2O_7$ single crystal were measured under different magnetic field for B along [001] direction, as shown in Figure 1(a). The Metal-insulator (MI) transition temperature with T_{MI} ~ 20 K under zero field is lower than the value ~ 32 K observed in polycrystalline samples, most likely due to a carrier doping by slight off-stoichiometry around 2 %. The low resistivity of the zero-field cooling (ZFC) compared with the field cooling (FC) with 14T is attributed to the multi-domain wall state. Under magnetic field, the resistivity below T_{MI} was suppressed and become metallic state with B = 14T, this is consistent with field sweeping results at 2 K as shown in Figure 1(b), confirming the existence of field induced MIT behavior.

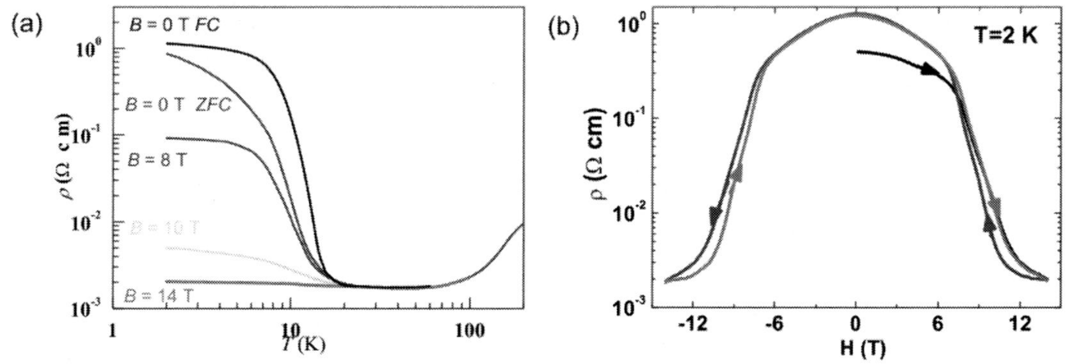

Figure 1 (a) Temperature dependence of resistivity measured under different magnetic field for B // [001] direction, zero field cooling (ZFC) and field cooling under 14 T for B = 0 T; B = 8T, 10T and 14 T. (b) Field dependence of resistivity measured at 2 K for ZFC procedure.

To further clarify the field induced MI transition behavior, we measured the angle dependence of magnetoresistivity under pulsed magnetic field up to 50 T, presented in Figure 2(a). The field B was rotated within the (100) plane, perpendicularly to the current direction [1-10] with the angle (θ) between the field and the [001] direction as schematically shown in Figure 2(b) inset. In this definition, $\theta = 35^{o}$, $\theta = 55^{o}$ and $\theta = 90^{o}$ correspond to the field direction of [112], [111] and [110] axes, respectively. For field along these three axis directions, only a gradual decrease of the resistivity was found, revealing that the insulating state is stable up to 50 T. In sharp contrast, in the [001] direction, a rapid jump of the resistivity was observed at a critical field of $B_c = 12$ T above which, the resistivity becomes nearly constant. The magnetoresistance $[R(0)-R(B)]/R(B)$ amounts to 60,000 %. This sharp resistivity drop characterizes the field-induced phase transition to the high-B semimetal state. The narrow but clear hysteresis (~1 T) seen at the critical field indicates that the transition is first-order.

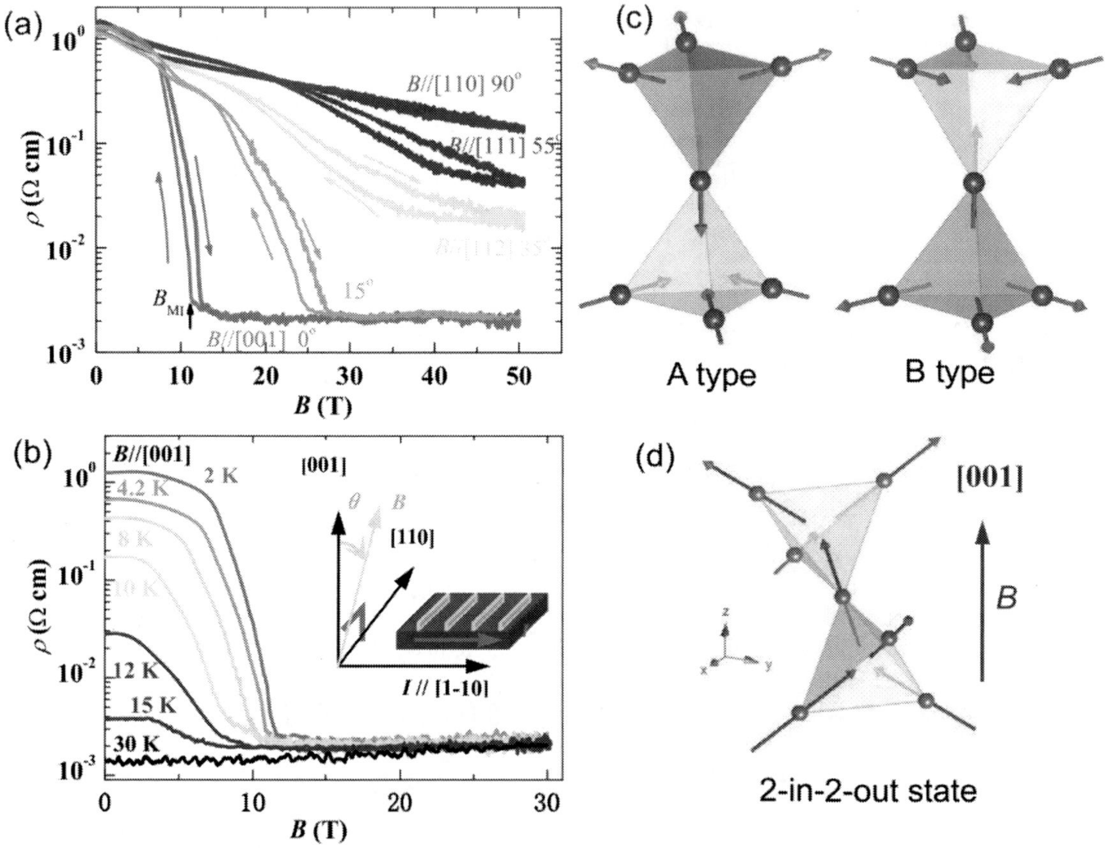

Figure 2 (a) Angle dependence of magnetoresistivity measured under pulsed magnetic field up to 50 T at 2 K with I // [1-10] direction as shown in the inset of Figure 2 (b), (b) magnetoresistance measured at different temperatures from 2 K to 30 K; (c) All-in all-out domain state with A-type (AIAO) of domain(left) and B-type (AOAI) of domain (right); (d) 2-in-2-out domain with B // [001] direction of Nd spin moments.

To map out the field-induced MIT, we show the resistivity versus field at various temperatures in Figure 2(b). The enhanced resistivity is confined to a region of small B and T, and outside this region, the resistivity is only weakly B and T dependent. Thus, there are simply two phases: an insulating antiferromagnetic state at small B and T, and a semimetallic state outside the ordered region. The strong anisotropic magnetoresistance is also evident in the isothermal magnetization measured at 2 K shown in Figure 3(a). The magnetization for *B* // [001] direction monotonically increases and then becomes saturated at 14 T, revealing the formation of 2-in-2-out state for Nd spin structure under high field as shown in Figure 2(d). Based on the neutron results [10,12], strong 4f-5d exchange interaction exists which set the Nd magnetic moment below 15 K to AIAO arrangement. Driven by the 4f-5d exchange interaction, the Nd spins polarization in the 2-in-2-out state will lead to the Ir magnetic moments to order in the same 2-in-2-out magnetic arrangement. In this case, the all-in-all-out (AIAO) AF order shown in Figure 2(c) becomes unstable, which lead to the field induced semi-metallic state at low field for the [001] orientation. This semimetal state corresponds to the destruction of AIAO spin configuration.

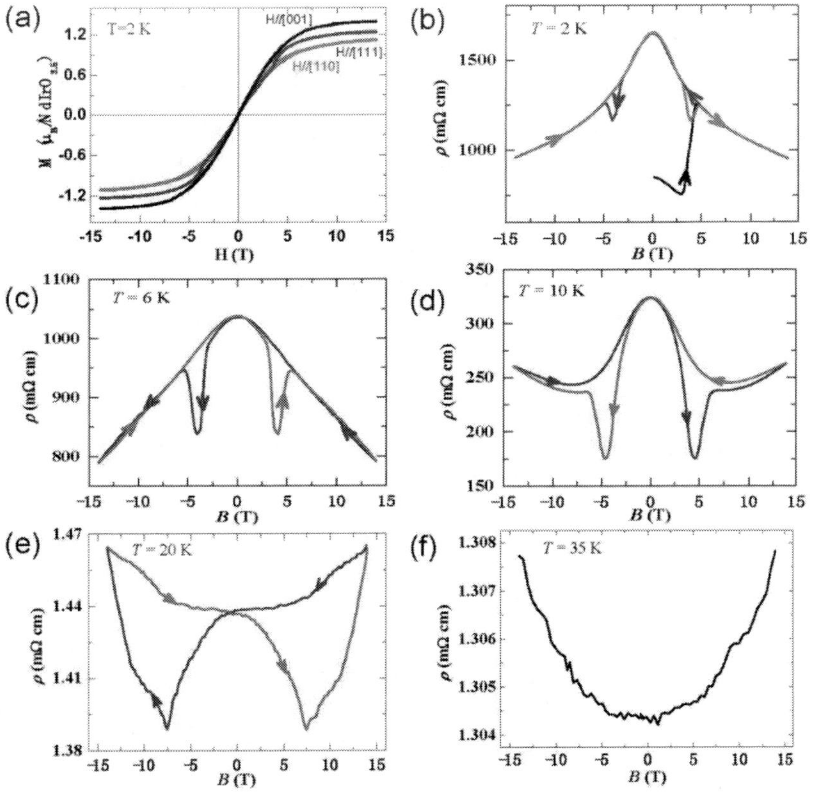

Figure 3 (a) Magnetization as a function of magnetic field at 2 K with magnetic field along the [111], [110] and [001] direction. (b-f) Field dependence of magnetoresistance for *B* // [111] direction measured at different temperatures under ZFC cooling.

For *B* // [111] seen from Figure 3(a), magnetic hysteresis is observed, while *M*(*B*) is smooth for *B* // [100] and [110]. In particular, the field derivative of the [111] magnetization curve reveals two distinct kinks around B_{c1}=2.0 T and B_{c2}=3.9T (not shown here). Naively, these are suggestive of Nd^{3+} spin flops (the Ir moments are negligibly small on this scale [13]), which would arise in a classical "spin-ice" like Ising model [14]. Such a model (using the single-ion moment 2.4μ_B/Nd) predicts saturation at 1.4, 1.2, and 0.98 μ_B/Nd for *B* along [001], [111], and [110], consistent with 1.4, 1.2, and 1.1 μ_B/Nd in experiment. However, the smoothness of the sweeps for the [100] and [110] field orientations, as well as the fact that the down-sweep sequence shows no anomaly and reaches zero magnetization at zero field, are inconsistent with classical Ising spins.

Finally, we focus on magntoresistance with *B* // [111] direction at different temperatures in Figure 3(b-f). Compared to the [001] direction, no dramatic MIT is seen for *B* // [111] at 2 K. Instead, the

[111] MR shows complex hysteresis corresponding to changes in the Nd spin structure [Figure 3(a)]. The results in the initial up-sweep process after ZFC shows a gradual decrease with field and spikes sharply, reaching a minimum value at 3.4 T. In the consecutive field cycles, the resistance exhibits dips at ±3.9T where the derivative of *M* peaks, revealing the close correlations between the MR and magnetization. Notably, the MR after the initial sweep forms a smooth curve interrupted only near the spin flop transition by a temporary dip. These observations can be understood in terms of two Ising domains of the AIAO order [Figure 2(c)] which can be aligned by a [111] field. Without such alignment, the initial zero-field state has multi-domains, and hence its lower resistivity can be attributed to more conductive domain walls [15]. This is also consistent with the dip in the spin flop region of the hysteresis loop, where again multi-domains are generated at this field.

For *B* // [111] direction, we also note that positive magnetoresistance response is displayed under high magnetic field for T > 10 K compared with negative magnetoresistance at low temperatures (T < 10 K), Since the Nd moments freeze below 15 K [10], this difference can be attributed to the coupling energy difference between the 4f-5d magnetic moments. Below 10 K, the 3-in-1-out state of Nd spin structure is formed under high magnetic field, which will destruct the AIAO state of Ir spin structure leading to the negative magnetoresistance. As temperature increases up to 10 K, the 3-in-1-out state of Nd spin structure becomes unstable and 4f-5d exchange interaction become weak. Above 15 K, Nd spin moment keeps in the paramagnetic state, in this case, the AIAO spin structure of Ir moment only can switch between the all-in-all-out (A-type) domain and all-out-all-in (B-type) domain with the reversal of magnetic field as seen at 20 K, which preserve the all-in-all-out spin structure as a key ingredient to realize Weyl semimetal sate. So, we argue that weyl semimetal state is expected to be existed only for *B* // [111] direction.

4. Conclusion

The anisotropic magnetotransport of $Nd_2Ir_2O_7$ single crystal reveals that two ordered spin structures can be realized for *B* along [001] and [111] directions, suggesting the existence of two different semimetal state. Compared with [001] direction, weyl semimetal is suggested to be existed for *B* // [111] direction just below MIT temperature due to the stability of AIAO state under high magnetic field.

Acknowledgement

We thank L. Balents, B. Kamran, Y Matsuo and T. Higo for useful discussions. This work was partially supported by a Grant-in-Aid for Scientific Research (No. 25707030) from the Japan Society for the promotion of Science (JSPS and by PRESTO, JST, Japan). Z.M. Tian acknowledges the financial support of the JSPS fellowship for foreign researchers (No. P14026).

References:

[1] Pesin D and Balents L 2010 *Nat. Phys.* **6** 376
[2] Wan X, Turner A M,Vishwanath A and Savrasov S Y 2011 *Phys. Rev B* **83** 205201
[3] Burkov A A, Hook M D and Balents L 2011 *Phys. Rev B* **84** 235126
[4] Witczak-Krempa W and Kim Y B 2012 *Phys. Rev B* **85** 045124
[5] Machida Y, Nakatsuji S, Onoda S, Tayama T and Sakakibara T 2010 *Nature* **463** 210
[6] Udagawa M and Moessner M 2013 *Phys. Rev. Lett* **111** 036602
[7] Matsuhira K, Wakeshima M, Hinatsu Y and Takagi S 2011 *J. Phys. Soc. Jpn* **80** 094701
[8] Nakatsuji S, Machida Y, Maeno Y, Tayama T, Sakakibara T, Van Duijin J, Balicas L, Millican J N, Macaluso R T, and Chan J Y 2006 *Phys. Rev. Lett* **96** 087204
[9] Ueda, K, fujioka J, Takahashi Y, Suzuki T, Ishiwata S, Taguchi Y, Kawasaki M and Tokura Y 2014 *Phys. Rev B* **89** 075127
[10] Tomiyasu, K. Matuhira K, Iwasa K, Watahiki M, Takagi S, Wakeshima M, Hinatsu Y,Yokoyama M, Ohoyama K and Yamada K 2012 *J. Phys. Soc. Jpn* **81** 034709
[11] Arima, T 2013 *J. Phys. Soc. Jpn* **82** 013705

[12] Ma E Y, Cui Y T, Ueda K, Tang SJ, Chen K, Tamura N, Wu P M, Fujioka J, Tokura Y and Shen Z X 2015 *Science* **350** 538
[13] Higashinaka R, Fukazawa H and Maeno Y 2003 *Phys. Rev B* **68** 014415
[14] Shapiro M C, Riggs S C, Stone M B, De la Cruz C R, Chi S, Podlesnyak A A and Fisher I R 2012 *Phys. Rev B* **85** 214434
[15] Yamaji Y and Imada M 2014 *Phys. Rev X* **4** 021035

Frustrated magnetism in a Mott insulator based on a transition metal chalcogenide

S Kawamoto[1], T Higo[1], T Tomita[1], S Suzuki[1], Z M Tian[1], K Mochitzuki[1], A Matsuo[1], K Kindo[1] and S Nakatsuji[1]

[1]Institute for Solid State Physics, University of Tokyo, Kashiwa, Chiba, Japan

E-mail: a.kawamoto@issp.u-tokyo.ac.jp

Abstract. A three dimensional geometrically frustrated Mott insulator of tetrahedral cluster compound $GaTa_4Se_8$ has been studied by X-ray diffraction, susceptibility and specific heat measurements. Temperature dependence of the magnetic susceptibility shows an anomaly peak at 53 K, which is supposed to be the structural transition from cubic to tetragonal structure on the analogy of $GaNb_4S_8$. Moreover, small g factor compared with $GaNb_4S_8$ compound revealed strong spin orbit coupling (SOC) which is inherent in this $5d$ compound.

1. Introduction

Recently, geometrically frustrated Mott insulators have attracted intensive attentions. The geometrical frustration can suppress the formation of long range magnetic order state like antiferromagnetic order due to the symmetry of the lattice and allow for the fascinating phenomena, such as quantum spin liquid state and spin quadrupole moment [1-2]. For examples, the organic Mott insulator κ-(BEDT-TTF)$_2$Cu$_2$(CN)$_3$ [3-5], which has a dimer spin $s = 1/2$ located on a geometrically frustrated triangular lattice, exhibits spin liquid state. Exotic two dimensional (2D) magnetism is found to be the triangular lattice antiferromagnet $NiGa_2S_4$ [2,6-7], where long range interaction plays a role due to the proximity to the Mott transition.

$GaTa_4Se_8$ belongs to lacunar spinel compounds GaM_4X_8 with $M =$ V, Nb or Ta and $X =$ S, Se or Te [8,9]. The crystal structure can be divided into two unit on geometrically frustrated face centered cubic (fcc) lattice : GaM_4 and M_4X_4 including M_4 cluster. Considering that valence of this compound is $Ga^{3+}(M^{3.25+})_4(X^{2-})_8$, the M_4 cluster has seven valence electrons. M ions are located on the center of the octahedron formed by X ions, so that the $3d$ band splits into e_g and t_{2g} bands. According to molecular orbital theory, by making cluster, six electrons fill the lower energy state and one electron occupies three-fold degenerate t_2 state [10]. Due to strong SOC inherent in $4d$ and $5d$ transition metals, $s = 1/2$ is entangled with effective orbital angular moment $l_{\rm eff} = 1$. Thus, total angular moment $j_{\rm eff} = 3/2$ states are proposed to be induced [11]. Then the ground state of this system can be considered as quarter filled $j_{\rm eff} = 3/2$ Mott state, which is similar with iridiate compounds [12].

Lacunar spinel compounds GaM_4X_8 can exhibit a variety of unusual transport and magnetic behavior. For examples, the insulating states were observed at ambient pressure for $GaNb_4S_8$, $GaNb_4Se_8$, and $GaTa_4Se_8$ [9,13]. Under high pressure, the insulating states can transfer into a metallic or superconductor state in Ga(Nb,Ta)$_4$Se$_8$ system [13]. But this insulating state theoretically is not suggested at room temperature according to LDA + U calculation result [9].

Journal of Physics: Conference Series **683** (2016) 012025 doi:10.1088/1742-6596/683/1/012025

By considering SOC additionally, this insulating state could be theoretically explained. Thus these compounds can be considered as unconventional Mott insulator [11] where the electronic conduction originates from the electrons hopping among the widely separated M_4 clusters. A sudden drop of the susceptibility is observed around 31 K and effective moment is 0.86 μ_B [14] for $GaNb_4S_8$. However μSR and NMR measurements show there is no long range magnetic order down to 4.5 K [15,16]. Moreover, these measurements reveal that spin singlet state is formed and Nb_8 octamer consisting of two Nb_4 clusters is proposed. Above 53 K, from the NMR measurement dynamical Jahn-Teller effect is suggested.

In this work, we report geometrically frustrated Mott insulator $GaTa_4Se_8$ which would realize strong SOC since $5d$ transition metal Ta ($s = 1/2$) cluster is included. Here we performed the susceptibility and heat capacity measurements to investigate the physical properties.

2. Experimental

Polycrystalline samples of $GaTa_4Se_8$ were synthesized by a solid state reaction method sealed in a quartz tube. Then it is heated at 900°C. The structural properties were checked by -ray diffraction (RINT2100, RIGAKU). Temperature dependences of magnetic susceptibility were measured by a commercial SQUID magnetometer (MPMS, Quantum Design) from 2 K to 300 K. Heat capacity was checked by a commercial system (PPMS, Quantum Design) from 2 to 70 K under 0 and 9 T.

3. Results and Discussion

3.1. X-ray diffraction

Room temperature RD shows the sample is single phase with no impurities. All diffraction peaks can be indexed by the cubic $GaTa_4Se_8$ structure type ($F\bar{4}3m$) with an fcc unit cell, and the lattice constant $a = 10.385(2)$ Å.

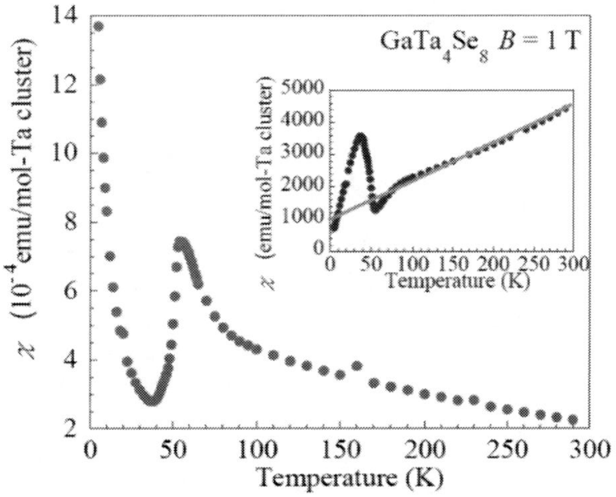

Figure 1. Temperature dependence of the magnetic susceptibility measured with magnetic field $B = 1$ T. The inset shows the inverse susceptibility $1/\chi$. The green lines are linear fit from 70 K to 300 K.

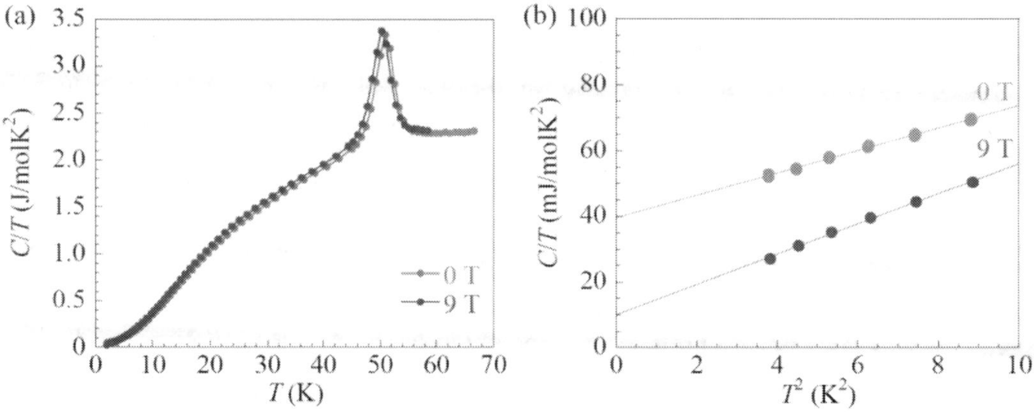

Figure 2. (a)Temperature dependences of heat capacity C/T under 0 T and 9 T. (b) Heat capacity is fitted by $C/T = \gamma + \beta T^2$.

3.2. Susceptibility
The magnetic susceptibility $\chi = M/B$ shows a peak at 53 K shown in Fig 1, which is presumably related to the structural distortion as observed on the analogy of $GaNb_4S_8$ [14]. A more sharp peak is observed for our sample around 53 K indicating that high quality polycrystals were synthesized. By fitting high temperature $1/\chi$ using Curie-Weiss Law, Weiss temperature and effective moment of $GaTa_4Se_8$ is $\theta_W = -86.2$ K and $0.86\mu_B$/Ta cluster, respectively.

The large negative Weiss temperature suggested the strong antiferromagnetic interaction between inter Ta cluster with $s = 1/2$. Considering that this compound has geometrically frustrated fcc lattice, strong magnetic frustration might be realized. The effective moment is much smaller than the theoretical value 1.73 μ_B calculated by $g = 2$ and $s = 1/2$. On the other hand, g factor of $GaTa_4Se_8$ is around 0.99 calculated with $s = 1/2$ and $p_{\rm eff} = 0.86$ μ_B. This value is much smaller compared with g factor of $GaNb_4S_8$, which is $g = 2.03$ by using the value of previous report [14,17]. Thus, $GaTa_4Se_8$ is thought to be one of best candidates to demonstrate large SOC and $j_{\rm eff}$ physics.

In the intermediate temperature between 53 K and 70 K, the inverse magnetic susceptibility deviates from linear fit. This is similar to a prediction made by theoretical calculations for double perovskite compounds which also have $j_{\rm eff} = 3/2$ and magnetic ions are located on the fcc lattice and in this compounds spin nematic state is discussed [18]. Actually, the origin of this behavior is unclear, but the unconventional behavior in this temperature region is possibly related to quadropole state [18].

3.3. Heat capacity
The temperature dependence of specific heat is shown in Fig 2 (a). We estimated the amount of entropy change ΔS which is obtained by peak integration between 42 K and 56 K. ΔS is about 5 J/K close to Rln2. But, when both orbital and magnetic moment degrees of freedom are released, Rln4 should be obtained. If only orbital or magnetic degrees of freedom were released, ΔS would give Rln2. Considering the case of $GaNb_4S_8$, there is possibility that the ΔS comes from the formation of spin singlet state proximity to 53 K. Then, it is speculated that the orbital degrees of freedom might be already released above 53 K and magnetic degrees of freedom remain. In this case, dynamic Jahn-Teller effect, which doesn't involve structural distortions, is expected to occur at high temperature to release the orbital degree.

Figure 2 (b) is fitted by $C/T = \gamma + \beta T^2$ and γ value means residual entropy at the low temperature. By applying magnetic field, γ value decreased. We can interpret this decreasing on two ways. One is the spin singlet breaking and the other interpretation is residual spins. But in the high magnetic field measurement, there are no sign of singlet breaking like magnetic moment jump. In this case, the decreasing of γ value is not from spin singlet breaking but residual moment.

4. Summary

We succeeded in synthesizing single phase polycrystalline samples of $GaTa_4Se_8$. Magnetic susceptibility shows an anomaly at 53 K and deviation from Currie Weiss fit in the range of 53–70 K. Our magnetism investigation revealed that this compound exhibits strong SOC. Moreover from both magnetic and specific heat measurement results spin singlet state might realize in the compound. To confirm spin singlet state, further measurements such as μSR should be carried out to detect a long range ordering.

5. Acknowledgment

This work is partially supported by PRESTO, Japan Science and Technology Agency, Grants-in-Aid for Scientific Research (No. 25707030), by Grants-in-Aids for Scientific Research on Innovative Areas (15H05882, 15H05883) and Program for Advancing Strategic International Networks to Accelerate the Circulation of Talented Researchers (No. R2604) from the Japanese Society for the Promotion of Science.

6. Reference

[1] Anderson P W 1973 *Mater. Res. Bull.* **8** 153
[2] Tsunetsugu H and Arikawa M 2006 *J. Phys. Soc. Jpn.* **75** 083701
[3] Komatsu T, Matsukawa N, Inoue T and Saito G 1996 *J. Phys. Soc. Jpn.* **65** 1340
[4] Shimizu Y, Miyagawa K, Kanoda K, Maesato M and Saito G 2003 *Phys. Rev. Lett.* **91** 107001
[5] S S Lee and Lee P A , 2005 *Phys. Rev. Lett.* **95** 036403-1
[6] Nakatsuji S, Nambu Y, Tonomura H, Sakai O, Jonas S, Broholm C, Tsunetsugu H, Qiu Y and Maeno Y, 2005 *Science* **309** 1697
[7] Nakatsuji S, Nambu Y, and Onoda S 2010 *J. Phys. Soc. Jpn.* **79** 011003
[8] Pocha R, Johrendt D and Pöttgen R S 2000 *Chem. Mater.* **12** 2882
[9] Pocha R, Johrendt D, Ni B and Abd-Elmeguid M M 2005 *J. Am. Chem. Soc.* **127** 8732
[10] Nakamura H, Chudo H and Shiga M 2005 *J. Phys.: Condens. Matter* **17** 6015
[11] Kim H S, Im J, Han M J and Jin H 2014 *Nat. Commun.* **17** 3988
[12] Kim B J, Jin H, Moon S J, Kim J Y, Park B G, Leem C S, Yu J, Noh T W, Kim C, Oh S J, Park J H, Durairaj V, Cao G and Rotenberg E 2008 *Phys. Rev. Lett.* **101** 076402
[13] Abd-Elmeguid M M, Ni B, Khomskii D I, Pocha R, Johrendt D, Wang X and Syassen K 2004 *Phys. Rev. Lett.* **93** 126403-1
[14] Jakob S, Müller, Johrendt D, Altmannshofer S, Scherer W, Rayaprolc S and Pöttgen 2007 *J. Mater. Chem.* **17** 3833
[15] Tabata Y, Kajinami Y, Waki T, Watanabe I and Nakamura H 2010 *J. Phys.: Conf. Ser.* **513** 042014
[16] Waki T, Kajinami Y, Tabata Y, Nakamura H, Yoshida M, Takigawa M and Watanabe I 2010 *Phys. Rev. B* **81** 020401(R)
[17] Matsuura H and Miyake K 2013 *J. Phys. Soc. Jpn.* **82** 073703-1
[18] Chen G, Pereira R and Balents L 2010 *Phys. Rev. B* **82** 174440

Journal of Physics: Conference Series **683** (2016) 012026 doi:10.1088/1742-6596/683/1/012026

Magnetic and Transport Properties of Frustrated γ-MnPd alloys

T Higo[1], N Kiyohara[1], K Iritani[1], A A Nugroho[2], T Tomita[1] and S Nakatsuji[1,3]

[1] Institute for Solid State Physics, University of Tokyo, Kashiwa, Chiba 277-8581, Japan
[2] Faculty of Mathematics and Natural Sciences, Bandung Institute of Technology, Bandung, Jawa Barat 40132, Indonesia
[3] PRESTO, Japan Science and Technology Agency, Kawaguchi, Saitama 332-0012, Japan

E-mail: tomoya@issp.u-tokyo.ac.jp

Abstract. We have synthesized γ-$Mn_{1-x}Pd_x$ alloys with $x = 0.22$ and 0.24 and carried out the susceptibility and resistivity measurements. The γ-$Mn_{1-x}Pd_x$ shows the Néel transition to the non-coplanar antiferromagnetic 3Q phase at $\sim$ 290 K for $x = 0.22$ and at $\sim$ 250 K for $x = 0.24$. Below the Néel temperature, two characteristic temperatures, which support the existence of the magnetic and structural phase transitions, were observed. These results were used to construct a phase diagram of the γ-$Mn_{1-x}Pd_x$ alloys with high x concentrations where non-collinear and/or non-coplanar spin structures are expected.

1. Introduction

Geometrically frustrated magnets have attracted much interest because of the possible emergence of novel magnetic phases. Competing magnetic interactions imposed by the geometrical arrangement of magnetic moments suppress the formation of a conventional magnetic order [1–3]. The concept of geometrical frustration is often discussed by treating an antiferromagnet consisting of a two-dimensional triangular lattice of Ising spins. In this case, the disordered spin state is associated with a macroscopic degeneracy in the ground state of the spin arrangements. On the other hand, in the XY-type or Heisenberg-type spin systems, non-collinear or non-coplanar spin arrangements are expected in order to minimize the energy. Recently, an increasing interest has been devoted to studying the properties of non-collinear and non-coplanar spin configurations in geometrically frustrated magnets because the associated vector and/or scalar spin chirality may induce exotic phenomena such as multiferroicity [4], unconventional anomalous Hall transport [5–8] and chiral spin liquid states [5, 9]. Especially, non-collinear and non-coplanar antiferromagnets have been intensively investigated because of zero or vanishingly small magnetization [8, 10–13].

Mn-based alloys generally have antiferromagnetic interaction due to the occupation of 3*d* orbitals [14]. Mn-based alloys consisting of the face-centered-cubic (fcc) lattice, which is one of three dimensional geometrically frustrated lattices, are considered to have various spin structures depending on the synthesis condition [15]. Here, we focus on γ-Mn disordered alloys, which are known to have complex magnetic and crystal structures including non-collinear and non-coplanar antiferromagnetic spin arrangements. The magnetic and crystal structures of γ-Mn

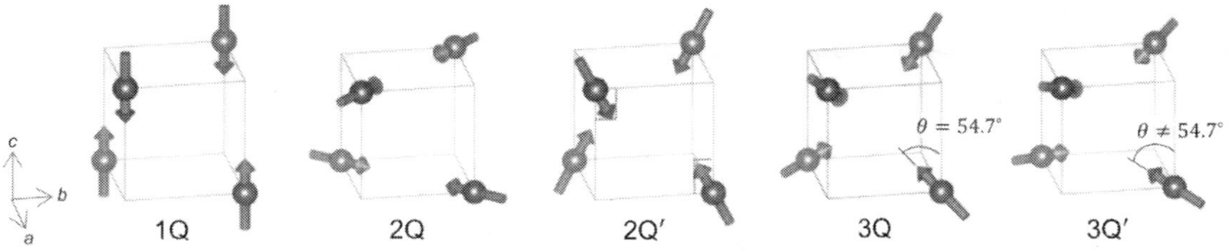

Figure 1. Spin structures of γ-Mn disordered alloys, illustrating a magnetic primitive cell composed of four atoms in the face-centered structures. The gray spheres represent randomly disorered ions composed of Mn and other elements. θ is the angle between the magnetic moment and the c-axis.

alloys have been investigated by neutron scattering, X-ray diffraction, Mössbauer, susceptibility, Young's modulus measurements and theoretical calculations [14, 16–24]. These experimental and theoretical studies for γ-Mn alloys found that the 1Q (collinear), 2Q (non-collinear), 2Q′ (non-collinear) or 3Q (non-coplanar) spin structures basically become stable in the case of $a = b > c$ (face-centered-tetragonal : fct), $a = b < c$ (fct), $a > b > c$ (face-centered-orthorhombic : fco) or $a = b = c$ (fcc), respectively [16, 17, 19, 23, 25] (Fig. 1). As the results of the previous measurements, the lattice distortions seemed to be connected with the spin structures. However, recent detailed investigations of γ-Mn alloys revealed that the distortions are not always accompanied by the magnetic phase transition. For instance, γ-MnNi alloys show that the fct-fcc structural transition is not accompanied by the 2Q-3Q magnetic transition and exhibits another type of magnetic structure 3Q′ ($54.7° < \theta < 90°$), which has spins tilted along the ab-plane from the 3Q ($\theta = 54.7°$) spin structure, in the concentration range of $a = b < c$ (fct) [25]. It is expected that a new kind of spin structures induced by geometrical frustration such as the 3Q′ spin structure appears in other γ-Mn alloys. To understand the essence of the physics of non-collinear and non-coplanar antiferromagnetic spin textures, it is significant to investigate the magnetic and crystal structures in γ-Mn disordered alloys.

Hori *et al.* have studied the magnetic and crystal structures of γ-$Mn_{1-x}Pd_x$ alloys ($0.08 \leq x \leq 0.17$), which have a platinum group metal similarly to γ-MnNi alloys, by X-ray diffraction, neutron scattering and susceptibility measurements [22]. However, there are few results of the fcc phase with $x > 0.15$ where the non-collinear 2Q and/or non-coplanar 3Q spin structures may appear. In this paper, we have investigated the magnetic and transport properties for γ-$Mn_{1-x}Pd_x$ disordered alloys with high x concentrations.

2. Experimental

Polycrystalline samples of γ-$Mn_{1-x}Pd_x$ alloys were prepared by melting mixtures of Mn (4N) and Pd (4N) using a mono-arc furnace filled with Ar gas at 1 atm. Before the melting, the surface oxides on Mn flakes were removed by etching with 5% nitric acid solution. Single crystals of the γ-$Mn_{1-x}Pd_x$ were grown by the Czochralski method using a tetra-arc furnace (TAC-5100, GES) filled with Ar gas at 1 atm. The polycrystalline samples were used as starting materials. From powder X-ray diffraction measurements using a X-ray diffractometer (RINT-2100, Rigaku), the polycrystals and single crystals were confirmed to be the single phase. A scanning electron microscopy with energy dispersive X-ray (SEM-EDX) analysis determined that the Pd concentrations of the samples are $x = 0.22$ and 0.24. The estimated error of our SEM-EDX analysis is within ± 0.5 at.% Pd. The magnetic susceptibility was measured between 50 and 350 K under both zero-field-cooled (ZFC) and field-cooled (FC) conditions using a commercial SQUID magnetometer (MPMS, Quantum Design). The electrical resistivity was

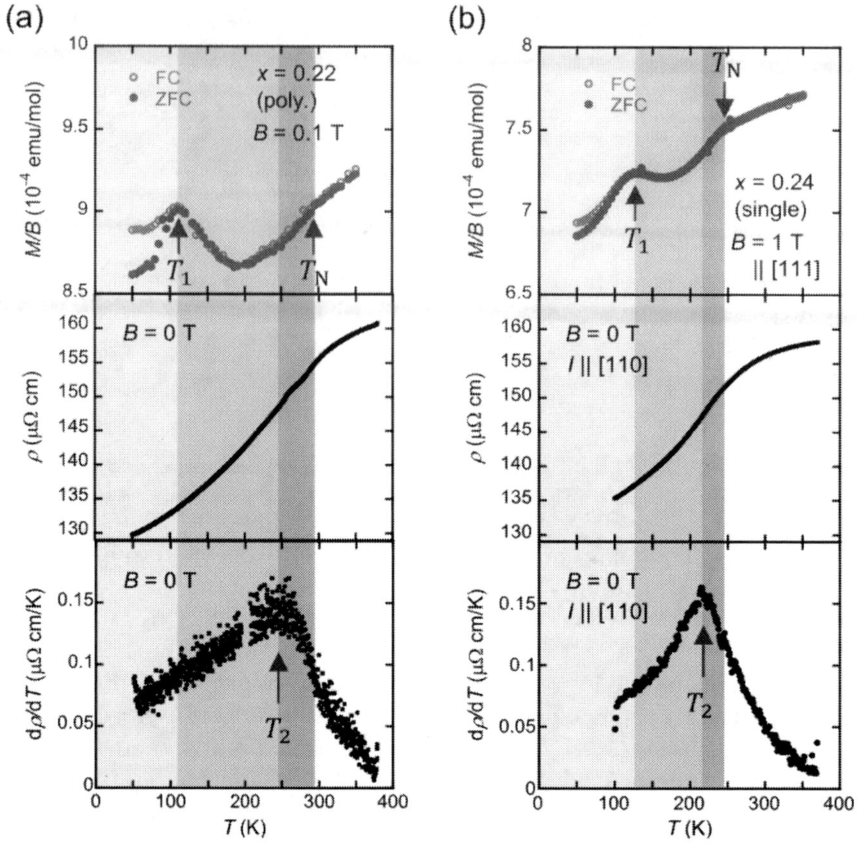

Figure 2. Temperature dependence of the susceptibility M/B, the resistivity ρ and the temperature derivative of the resistivity $d\rho/dT$ of the γ-$Mn_{1-x}Pd_x$ disordered alloys with (a) $x = 0.22$ and (b) $x = 0.24$.

measured by a standard four-probe method between 50 and 380 K using a commercial system (PPMS, Quantum Design). We carried out the X-ray diffraction, SEM-EDX, susceptibility and resistivity measurements using the samples cut from the same ingot for each composition.

3. Results and discussion

Results of magnetic and resistivity measurements for the γ-$Mn_{1-x}Pd_x$ alloys with $x = 0.22$ and 0.24 are shown in Fig. 2(a) and 2(b), respectively. The temperature dependence of the susceptibility exhibits a kink at $\sim$ 290 K for $x = 0.22$ and at $\sim$ 250 K for $x = 0.24$. Similar kink was observed in the previous study for γ-MnPd and other γ-Mn alloys [14, 19, 22, 23, 25]. Neutron scattering and Young's modulus measurements confirmed that this kink is related to the Néel transition [14, 19, 22, 24]. Therefore, the Néel temperature T_N of the γ-$Mn_{1-x}Pd_x$ alloys is determined by the kink in the susceptibility at $\sim$ 290 K for $x = 0.22$ and at $\sim$ 250 K for x $= 0.24$. Below T_N, the non-coplanar 3Q spin structure should be realized as observed in other γ-Mn alloys [14, 16, 17, 23, 25].

Furthermore, far below T_N, another kink in the susceptibility was observed at $\sim$ 110 K for $x = 0.22$ and at $\sim$ 130 K for $x = 0.24$, similarly observed in γ-MnNi and γ-MnIr alloys

Journal of Physics: Conference Series **683** (2016) 012026 doi:10.1088/1742-6596/683/1/012026

[23, 25]. Fishman *et al.* pointed out that the magnetic phase transition from one Q-phase to another Q-phase can be marked by a small jump in the susceptibility of γ-Mn alloys [16]. Susceptibility, neutron scattering measurements and theoretical calculations for γ-MnNi alloys have revealed that the kink, which lies well below T_N, is ascribed to the 2Q-3Q(3Q′) magnetic phase transition [25]. These results suggest that the magnetic phase transition between two different magnetic phases occurred at a characteristic themperature T_1, which is defined by the kink in the susceptibility observed at $\sim$ 110 K for $x = 0.22$ and at $\sim$ 130 K for $x = 0.24$ in the γ-$Mn_{1-x}Pd_x$ alloys. The hysteresis between the ZFC and FC susceptibility was also seen just below T_1. There are several possible origins for this hysteresis, which include spin glass freezing, spin canting or weak ferromagnetic moments arisen from the non-collinear or non-coplanar antiferromagnetic domain wall. To determine the origin of this hysteresis, more detailed measurements such as the AC and DC susceptibility measurements under various magnetic fields are needed.

The temperature dependence of the resistivity of the γ-$Mn_{1-x}Pd_x$ alloys shows metallic behavior, similarly observed in the previous study for γ-MnPd alloys [24]. The temperature derivative of the resistivity $d\rho/dT$ exhibits an obvious peak at $\sim$ 250 K for $x = 0.22$ and at $\sim$ 220 K for $x = 0.24$. According to the investigations for γ-MnNi and γ-MnPd alloys, two anomalies which have different origins can be observed in the resistivity measurements [24, 25]. One is an anomaly derived from the Néel transition. The other is an anomaly originating from the fct-fcc structural phase transition well below T_N [25]. Since the peak temperature T_2 observed in $d\rho/dT$ is $\sim$ 40 K ($x = 0.22$) or $\sim$ 30 K ($x = 0.24$) lower than T_N, the peak would not be derived from the Néel transition. However, it is also difficult to conclude that the fct-fcc phase transition occurs at T_2 from our results and the previous study [22]. As another possibility, the other magnetic and/or structural phase transitions can be assumed. However, to confirm the detailed phase, further experiments with a microscopic probe are necessary, such as X-ray and neutron diffraction measurements.

As a result of the susceptibility and resistivity measurements in the γ-$Mn_{1-x}Pd_x$ disordered alloys with high x concentrations, three characteristic temperatures, T_1, T_2 and T_N were obtained. Here, we added these characteristic temperatures in the magnetic and structural phase diagram given by Hori *et al.* [22], which is presented in Fig. 3.

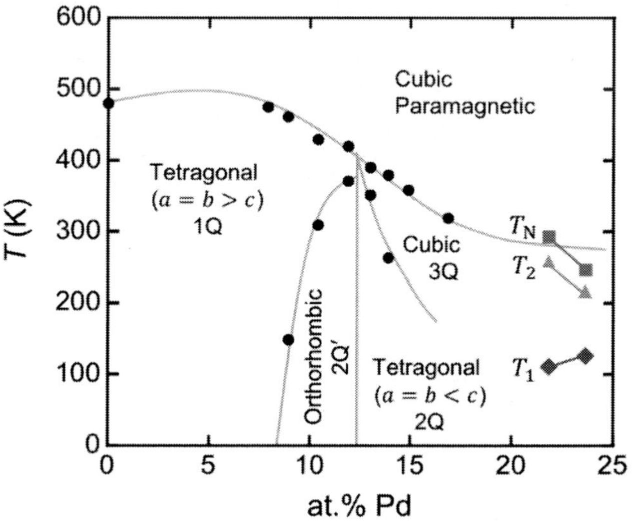

Figure 3. Magnetic and structual phase diagram of the γ-$Mn_{1-x}Pd_x$ disordered alloys, together with the previous study [22] (black circles). The purple diamonds represent the characteristic temperature T_1 determined by the low-temperature kink in the susceptibility. The light blue triangles represent the charcteristic temperature T_2 determined by the anomaly in the resistivity. The red squares represent the Néel temperature T_N.

4. Summary

We have synthesized γ-$Mn_{1-x}Pd_x$ alloys with $x = 0.22$ and 0.24, and carried out the susceptibility and resistivity measurements. The susceptibility shows the Néel transition to the 3Q phase at $T_N \sim 290$ K for $x = 0.22$ and at $T_N \sim 250$ K for $x = 0.24$. Below T_N, moreover, the temperature derivative of the resistivity exhibits a peak at ~ 250 K for $x = 0.22$ and at ~ 220 K for $x = 0.24$, which is presumably ascribed to the magnetic and/or structural phase transition. In addition, the low-temperature susceptibility suggests that the magnetic phase transition occurred at ~ 110 K for $x = 0.22$ and at ~ 130 K for $x = 0.24$. To investigate the detailed low-temperature magnetic and crystal structures of the γ-$Mn_{1-x}Pd_x$ alloys with high x concentrations, further experiments with a microscopic probe are needed.

Acknowledgements

We would like to thank M. Ikhlas for contributing the discussions. This work is partially supported by PRESTO, Japan Science and Technology Agency, Grants-in-Aid for Scientific Research (No. 25707030) and Program for Advancing Strategic International Networks to Accelerate the Circulation of Talented Researchers (No. R2604) from the Japanese Society for the Promotion of Science (JSPS), by Grants-in-Aids for Scientific Research on Innovative Areas (15H05882 and 15H05883) from the Ministry of Education, Culture, Sports, Science, and Technology of Japan and by the JSPS Research Fellowship for Young Scientists. The use of the facilities of the Materials Design and Characterization Laboratory at the Institute for Solid State Physics, the University of Tokyo, is gratefully acknowledged.

References

[1] Ramirez A P 1994 *Annu. Rev. Mater. Sci.* **24** 453
[2] Moessner R and Ramirez A P 2006 *Phys. Today* **59** 24
[3] Collins M F and Petrenko O A 1997 *Can. J. Phys.* **75** 605
[4] Tokura Y, Seki S and Nagaosa N 2014 *Rep. Prog. Phys.* **77** 076501
[5] Machida Y, Nakatsuji S, Onoda S, Tayama T and Sakakibara T 2010 *Nature* **463** 210
[6] Neubauer A, Pfleiderer C, Binz B, Rosch A, Ritz R, Niklowitz P G and Böni P 2009 *Phys. Rev. Lett.* **102** 186602
[7] Taguchi Y, Oohara Y, Yoshizawa H, Nagaosa N and Tokura Y 2001 *Science* **291** 2573
[8] Sürgers C, Fischer G, Winkel P and Löhneysen H V 2014 *Nature Commun.* **5** 3400
[9] Nakatsuji S, Machida Y, Maeno Y, Tayama T, Sakakibara T, Duijn J V, Balicas L, Millican J N, Macaluso R T and Chan J Y 2006 *Phys. Rev. Lett.* **96** 087204
[10] Ishikawa J J, O'Farrell E C T and Nakatsuji S 2012 *Phys. Rev. B* **85** 245109
[11] Ueda K, Fujioka J, Takahashi Y, Suzuki T, Ishiwata S, Taguchi Y, Kawasaki M and Tokura Y 2014 *Phys. Rev. B* **89** 075127
[12] Yamaura J, Ohgushi K, Ohsumi H, Hasegawa T, Yamauchi I, Sugimoto K, Takeshita S, Tokuda A, Takata M, Udagawa M, Takigawa M, Harima H, Arima T and Hiroi Z 2012 *Phys. Rev. Lett.* **108** 247205
[13] Higo T and Nakatsuji S 2015 *J. Phys. Soc. Jpn.* **84** 053702
[14] Fukamichi K, Umetsu R Y, Sakuma A and Mitsumata C 2006 *Handbook of Magnetic Materials* vol 16 ed Buschow K H J (Amsterdam: North-Holland/Elsevier) p 209
[15] Krén E, Kádár G, Pál L, Sólyom J, Szabó P and Tarnóczi T 1968 *Phys. Rev.* **171** 574
[16] Fishman R and Liu S 1999 *Phys. Rev. B* **59** 8672
[17] Fishman R and Liu S 1998 *Phys. Rev. B* **58** R5912
[18] Endoh Y and Ishikawa Y 1971 *J. Phys. Soc. Jpn.* **30** 1614
[19] Yamauchi R, Hori T, Miyakawa M and Fukamichi K 2000 *J. Alloys Compd.* **309** 16
[20] Hori T, Morii Y, Funahashi S and Niida H 1995 *Physica B* **213&214** 354
[21] Etheridge G, Cussen L and Kennedy S 1995 *Physica B* **213&214** 351
[22] Hori T, Tsuchiya Y, Ishii Y and Hojou K 2002 *Mater. Trans.* **43** 436
[23] Sasao K, Umetsu R, Fukamichi K and Sakuma A 2003 *J. Alloys Compd.* **352** 21
[24] Hicks T J, Pepper A R and Smith J H 1968 *J. Phys. C: Solid State Phys.* **1** 1683
[25] Fishman R S and Lee W 2000 *Phys. Rev. B* **61** 12159

NMR studies on polyphosphide $Ce_6Ni_6P_{17}$

T Koyama[1], H Yamada[1], K Ueda[1], T Mito[1], Y Aoyama[2], T Nakano[2], and N Takeda[3]

[1]Graduate School of Material Science, University of Hyogo, Kamigori, Hyogo 678-1297, Japan
[2]Graduate School of Science, Niigata University, Ikarashi 2, Niigata 950-2181, Japan
[3]Faculty of Engineering, Niigata University, Niigata 950-2181, Japan

E-mail: t-koyama@sci.u-hyogo.ac.jp

Abstract. We report the result of ^{31}P nuclear magnetic resonance (NMR) studies on $Ce_6Ni_6P_{17}$. The observed NMR spectra show a Lorentzian-type and an asymmetric shapes, reflecting the local symmetry around each P site in the cubic unit cell. We have identified the observed NMR lines corresponding to three inequivalent P sites and deduced the temperature dependence of the Knight shift for each site. The Knight shifts increase with decreasing temperature down to 1.5 K, indicating a localized spin system of $Ce_6Ni_6P_{17}$. Antiferromagnetic correlation between $4f$ spins is suggested from the negative sign of the Weiss-temperature.

1. Introduction

Rare-earth polyphosphides $Ln_6Ni_6P_{17}$ (Ln=La, Ce, Pr) crystallize in the cubic structure (space group $I\bar{4}3m$, No. 217) with 58 atoms per unit cell [1]. A subject of interest in this structure is the coordination of Ln atoms which occupy a crystallographically single site ($12e$) and form an octahedral cage at the corner and the body-centered position in the cubic unit cell as shown in Fig. 1. The nearest-neighbor (NN) bond of Ln-Ln is the edge of octahedron and the next-nearest-neighbor (NNN) is Ln-Ln dumbbell along the cubic edge. When Ln is a magnetic ion and satisfies some particular conditions, unusual magnetic properties possibly emerge. In the case that the NN interaction is antiferromagnetic and much stronger than the NNN one, typical 3D frustration in the octahedron would be expected. On the other hand, if the NNN interaction exceed the NN one, dimerization between the NN pairs takes place [2]. For Ce-based polyphosphide $Ce_6Ni_6P_{17}$, the distances between NN Ce atoms (4.139 Å) and NNN (4.262 Å) are so close that the competition between frustration and dimerization is expected.

Magnetic susceptibility $\chi(T)$ of $Ce_6Ni_6P_{17}$ increases monotonically with decreasing temperature in the measured temperature range of 2 - 350 K [3]. The low temperature specific heat exhibits broad peak at around 1.4 K followed by two sharp peaks at 0.5 and 0.9 K. These two peaks were considered to associate with magnetic transitions [3]. The magnetic entropy is evaluated to be about a half of Rln2 implying a frustrated system of $Ce_6Ni_6P_{17}$ [3]. In order to investigate the magnetic properties microscopically, we performed ^{31}P nuclear magnetic resonance (NMR) measurement on $Ce_6Ni_6P_{17}$.

2. Experimental method

Single crystals of $Ce_6Ni_6P_{17}$ were grown by molten Sn-flux method. The obtained single crystals were less than 500μm. The x-ray diffraction pattern shows that the sample contains a small

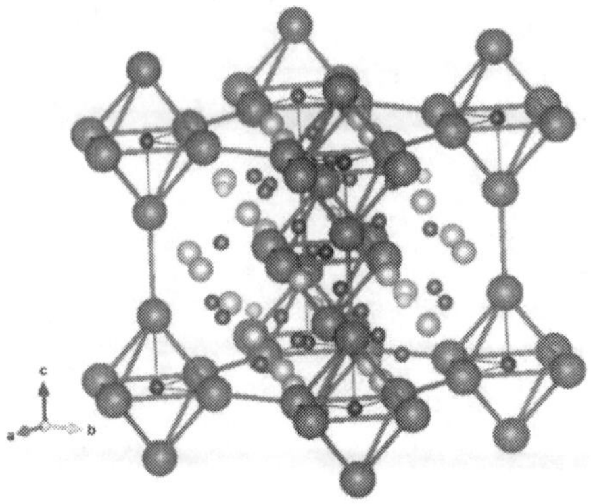

Figure 1. Crystal structure of $Ce_6Ni_6P_{17}$.

amount of paramagnet NiP_3 as a impurity phase [3]. For the NMR measurement, powered single crystals were immersed into daphne oil 7373 to avoid partial orientation of particles in a magnetic field. The NMR measurement was carried out using a phase-coherent spin-echo spectrometer. Field-swept NMR spectra were obtained by recording the integrated spin-echo intensity at a constant frequency.

3. results and discussion

Figure 2 shows the temperature dependence of the ^{31}P NMR spectra (gyromagnetic ratio $^{31}\gamma$=17.235 MHz/T, nuclear spin I=1/2) at 190 MHz. P atoms occupy crystallographically three different sites 2a, 8c, and 24g, as shown in Fig. 1. Generally, in the absence of a nuclear quadrupole moment, the powder pattern of a NMR spectrum reflects the anisotropy of Knight shift. Its components are written as K_x, K_y, and K_z for the external field parallel to the local x, y, and z axes, respectively. When a nucleus is located in an environment of cubic symmetry, these values satisfy the relation of $K_x = K_y = K_z$ resulting in the Lorentzian-type NMR line shape. When the nuclei are in a noncubic environment, these components are $K_x = K_y \neq K_z$ ($K_x \neq K_y \neq K_z$) for axial symmetry (lower than axial). In these cases, the resonance is not symmetrical any more [4].

By considering the local symmetry around each P site, we assigned corresponding sites by following procedure. Since 2a site has cubic symmetry, the Lorentzian shape of the resonance line painted in blue as shown in Fig. 2 can be assigned to 2a. Next, the spectrum painted in green can be considered to be consisting of 8c and 24g sites. Taking the intensity ratio and T_2 correction into account, this spectrum can be reproduced by the superposition of two powder-pattern signals with axial and lower than axial symmetries as shown in Fig. 3. Here T_2 is the spin-echo decay time. The deduced components of the Knight shift at 150 K are listed in Table 1. The resonance lines pained by blue and green shift to lower field with decreasing temperature. On the other hand, we observed the temperature-independent line indicated by the arrow in Fig. 2. This line is probably the signal of the paramagnetic impurity of NiP_3.

Figure 4 shows the temperature dependence of the Knight shift K_{2a}, K_{8c}, and K_{24g} for 2a, 8c, and 24g, respectively. K_{2a} is defined at the peak position of the Lorentizan-type of spectra and corresponds to an isotropic component of the Knight shift $K_{\rm iso}$. Moreover, K_{8c}, and K_{24g} are defined at the peak positions of the asymmetric spectra and roughly correspond to K_y. We could not obtain reliable shift values of K_{8c}, and K_{24g} below 40 K. This is because the overlapping of the signals of intrinsic and impurity phases, and the broadening of the spectra make the peak positions ambiguous. In rare-earth compounds, the temperature dependence of

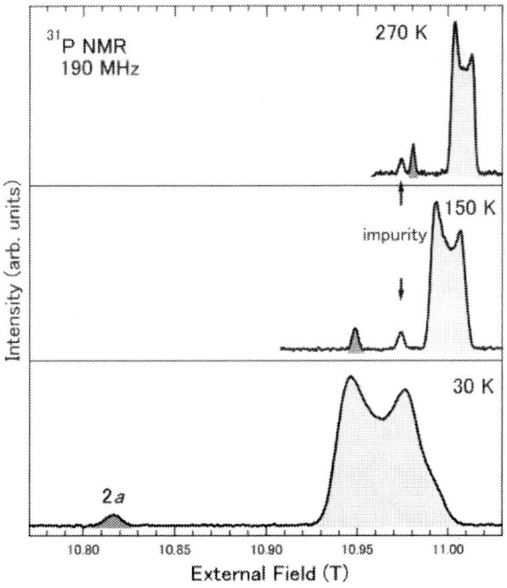

Figure 2. Temperature dependence of the ^{31}P NMR spectra.

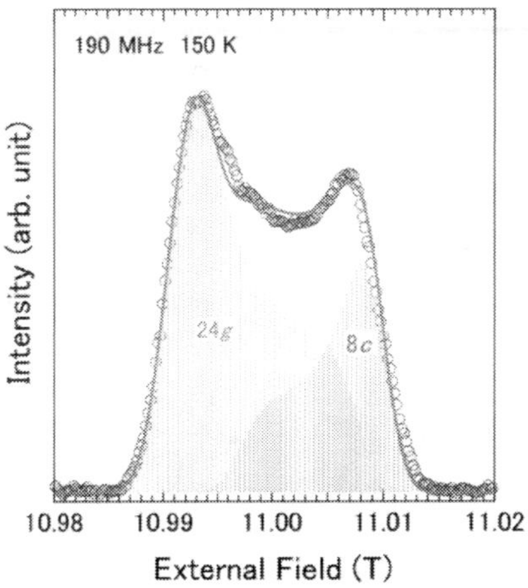

Figure 3. ^{31}P NMR spectrum at 150 K. The solid red curve corresponds the simulated overlap of two spectra drawn as shadows.

Knight shift $K(T)$ of ligand nuclei is related to the spin susceptibility χ_s of rare-earth ions with a transferred hyperfine field H_{hf} as

$$K(T) = K_0 + \frac{H_{hf}}{N_A \mu_B} \chi_s(T), \tag{1}$$

where K_0 is an on-site contribution consisting of a conduction electron and a field-induced Van Vleck orbital terms, N_A is Avogadros number, and μ_B is the Bohr magneton. As shown in Fig. 4, K_{2a}, K_{8c}, and K_{24g} show similar temperature dependence: increase with decreasing temperature. It is apparent that K_{2a} shows a larger temperature dependence than K_{8c} and K_{24g}. This is attributed to large H_{hf} at $2a$ site. H_{hf} originates from the transferred hyperfine couplings with 4f spin of neighboring Ce ions. P at $2a$ site is surrounded octahedrally by six Ce atoms, while the number of NN Ce atom of $8c$ and $24g$ are two and three, respectively. Moreover, $2a$ site is closer to the NN Ce atoms than the others. These explain the larger temperature dependence of K_{2a}. Since P at $2a$ site is the most sensitive to the magnetism of Ce-4f electron, we focus on the temperature variation of K_{2a}.

In most of rare-earth compounds, it is often observed that magnetization data at low

Table 1. Deduced ^{31}P Knight shift parameters of each crystallographic site at 150 K. The local axes of x and y are tentatively defined as the direction of $K_x > K_y$. K_{iso} is defined as $K_{iso} \equiv (K_x + K_y + K_z)/3$.

Position	Site Symmetry	K_x (%)	K_y (%)	K_z (%)	K_{iso} (%)
$2a$	$\bar{4}3m$	0.67	0.67	0.67	0.67
$8c$	$.3m$	0.12	0.12	0.24	0.16
$24g$	$..m$	0.3	0.27	0.13	0.23

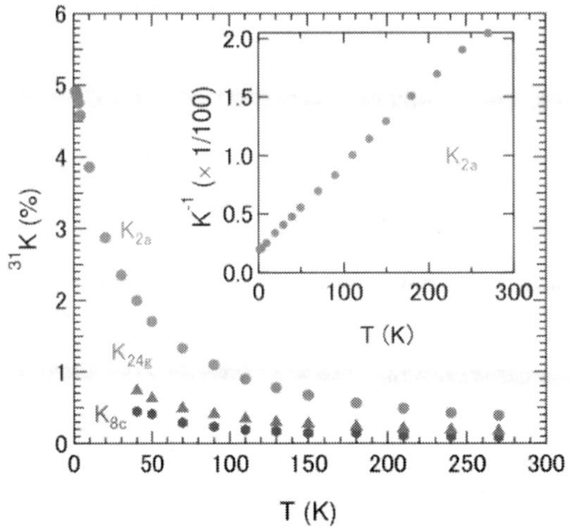

Figure 4. Temperature dependence of the Knight shifts (see text). Inset shows the inverse of K_{2a}.

temperatures show the departure from straight-line against magnetic field. Then we employed data above 90 K and estimated $H_{\rm hf}$ from a slope in the K_{2a} versus χ plot. The obtained value is roughly $H_{\rm hf} \simeq 0.78$ T/$\mu_{\rm B}$. In order to deduce the Weiss-temperature, K_{2a} was analyzed using the equation of $K(T) = K_0 + C/(T - \theta)$, where C and θ are fitting parameters. We found that K_0 is negligible regardless of the temperature range above 90 K. Then we plotted the inverse of K_{2a} as shown in the inset of Fig. 4 and estimated the value of θ. The obtained varies from -15 K to -10 K depending on the fitting temperature range. The negative sign indicates an antiferromagnetic correlation between 4f spins. NMR measurement at low magnetic field and low temperatures is in progress to clarify the ground state of $Ce_6Ni_6P_{17}$.

4. Conclusion

^{31}P NMR measurement has been performed on $Ce_6Ni_6P_{17}$. NMR spectra show the Lorentzian-type and an asymmertric shapes. Taking the local symmetry around each site into account, we could assign three inequivalent P sites ($2a$, $8c$, and $24g$) and obtained each Knight shift. The large temperature dependence of K_{2a} could be understood by considering the close NN distance and the octahedral coordination of Ce atoms around $2a$ site. The temperature dependence of this K_{2a} indicates a well-localized 4f electron state and antiferromagnetic correlation between local moments in $Ce_6Ni_6P_{17}$.

Acknowledgments

This work was supported by JPSJ KAKENHI Grant Number 26800193.

References

[1] Braun D J and Jeitschko W, 1978 *Acta. Cryst. B* **34** 2069
[2] Koyama T, Yamashita H, Takahashi Y, Kohara Y, Watanabe I, Tabata Y, and Nakamura H, 2008 *Phys. Rev. Lett.* **101** 126404
[3] Takeda N, Izumi K, Ono H, Yodono S, and Nakano T, 2012 *J. Phys: Conf. Ser.* **391** 012071
[4] Carter G C, Bennett L H, and Kahan D J, *Metallic Shifts in NMR, Part I, Progress in Materials Science, Vol. 20 (Pergamon, Oxford, 1977)*

Journal of Physics: Conference Series **683** (2016) 012028 doi:10.1088/1742-6596/683/1/012028

Noncentrosymmetric superconductivity in a clean crystal of type II superconductor Bi-Pd

Srinivasan Ramakrishnan, Bhanu Pratap and A. Thamizhavel

Department of Condensed Matter Physics and Materials Science, Tata Institute of Fundamental Research, Mumbai-400005, India

E-mail: `ramky07@gmail.com`

Abstract. In this work, we present the bulk superconductivity of a high-quality single crystal of monoclinic BiPd (α-BiPd, space group $P2_1$) below 3.8 K by studying its electrical resistivity, magnetic susceptibility, and heat capacity. This is the cleanest noncentrosymmetric superconductor (NCS)that display anisotropy due to spin-orbit scattering and also exhibits unusual superconducting properties due to s and p wave mixing as evidenced by the observation of Andreev bound state and multiple energy gaps via point contact measurements. In addition, Fermi surface studies suggest multiband superconductivity in this compound. Penetration depth studies and NQR investigations support mixing of s and p wave Copper paring in this crystal. Moroever, Muon spin rotation measurements indicate strong field dependence of the Ginzburg-Landau coefficient of this superconductor. Unusual pairing and multiband superconductivity are extremely sensitive to disorder and they can be observed only in cleanest (RRR $>$ 170) single crystals.

1. Introduction

Ever since the discovery of the non-centrosymmetric heavy fermion superconductor $CePt_3Si$ [1], there is widespread research activity to understand the nature of superconductivity in such unconventional superconductors. The term non-centrosymmetricEcharacterizes the symmetry of a crystal lattice without inversion center. In such materials the standard classification in even-parity spin-singlet and odd-parity spin triplet superconducting phases is obsolete, because the electrons are exposed to antisymmetric spin-orbit coupling, e.g. Rashba-type of coupling [2] which arises due to electric field gradient in the crystal which has no inversion symmetry. An inherent feature is then the mixing of spin-singlet and spin-triplet Cooper pairing channels which are otherwise distinguished by parity. This mixing of pairing states is expected to cause a two-component order parameter. New forms of pairing appear giving rise to unusual temperature field dependence of the superconducting parameters. During recent years we have witnessed rapid developments on the side of experiments and synthesis of novel materials as well as in the theoretical understanding of this type of superconductors. Indeed many new materials have been found, in particular, among the heavy fermions for which unconventional Cooper pairing is expected. Exotic mixed-state phases (vortex matter) have been predicted in theory and are under experimental investigations. For example in the case of $CePt_3Si$ where superconductivity occurs at ambient pressure, whereas, transition to such a state happens in UIr[3], $CeRhSi_3$[4] and $CeIrSi_3$[5] only under pressure. However, the study of superconductivity in non-centrosymmetric materials which do not exhibit heavy fermion features is also important since it avoids additional complication that arises due strong f-electron correlations. Discovery of such materials also continues to increase, starting from binary carbides ($R_2C_{3-\delta}$ with R=La or Y)[6], $Cd_2Re_2O_7$ [7] , $Li_2(Pd,Pt)_3B$ [8, 9], Mg_2Al_3[10] and the recently found $BaPtSi_3$[11]. However, many of them (except Li_2Pt_3B) exhibit conventional BCS-like super-conductivity due to small spin-orbit scattering. One of the common features in these compounds which show conventional superconductivity with small spin-orbit scattering is the absence of the large density of states at the Fermi level. Hence, it will be of interest to study a superconducting material which have conduction electrons with high density of states at the Fermi level (not from high f-electron correlations) but has no inversion symmetry. In this work we report our investigations in one such material, namely α-BiPd which has no inversion symmetry in its monoclinic crystal structure (space group $P2_1$) and it shows bulk superconductivity below 3.8 K as determined from resistivity, magnetization and heat capacity studies. The unconventional nature of the superconductivity is probed also by point contact spectroscopy, NMR and penetration depth measurements.

2. Experimental details

The compound BiPd undergoes polymorphic transformation from α-BiPd (monoclinic, $P2_1$) [12, 13] to β-BiPd (orthorhombic,$Cmc21$) above 210°C. We have synthesized phase pure α-BiPd which has a monoclinic structure with the space group $P2_1$ with b as its unique axis. Due to its low melting point (650°C), we have chosen to make the sample using a modified Bridgeman technique. The sample was made by control heating the individual components (Bi, 99.999 pure and Pd, 99.99 pure) in a high purity Alumina crucible with a pointed bottom which is kept in a quartz tube that is sealed under a vacuum of 10^{-6} mbar. Initially the contents were heated up up to 650 C (melting point of BiPd) in 12 hrs and then kept at 650°C for 12 hrs. Thereafter it was slow cooled to 590°C with a rate of 1 °C/hr and finally the furnace was switched off. We obtained high quality poly and single crystals of few mm size with mass ranging from 10 to 50 mg. A piece of the from the melt was crushed into a fine powder for powder x-ray diffraction measurement using Cu K_α radiation in a commercial diffractometer which confirmed that the crystal formed α-BiPd ($P2_1$) structure. The unit cell of this monoclinic structure is shown in Fig. 1. The structure consists four inequivalent sites for Bi and four inequivalent sites for Pd having 16 atoms in the unit cell. It also has alternate layer of Bi and Pd sheets with short Pd-Pd The values for the lattice constants estimated from the Rietveld fit [14] are a= 5.6284(±0.0004)Å,b= 10.6615(±0.0004)Å, c= 5.6752(±0.0004)Å, α=γ=90 and β=101. The resistivity, susceptibility and heat capacity measurements are performed using commercial (Quantum Design, USA) setups.

3. Results and discussions

Fig. 2(c) shows the temperature dependence of the resistivity (ρ(T)) from 1.5 to 300 K. NMR measurements have been done with Professor hang's group (Okayama, Japan) and penetrastion depth studies are done in collaboration with Professo H. Yuan's group (Hang hou, China). Fermi surface measurements are performed with Dr. Edward Yelland (St. Andrews, Scotland) and μSR measurements were done in collaboration with Professor Ted Frogan's group (Birmingham, UK). The high quality of the sample is clearly evident from the large residual resistivity ratio (ρ(300K)/ρ(4K)) of 170. The inset 2(a) shows the superconducting transition below 3.8 K with a width of less than 2 mK. The inset 2(b) shows the low temperature resistivity from 4 to 30 K. The solid line is a fit to the equation,

$$\rho(T) = \rho_0 + A \times T^n \tag{1}$$

The fit yields a value of 0.3 $\mu\Omega$ cm for ρ_0, 9.2 � 10^{-5} $\mu\Omega$ cm/K^3 for A and n=3. This T^3 dependence of the resistivity is interpreted by the Wilsons theory [15] which takes into account

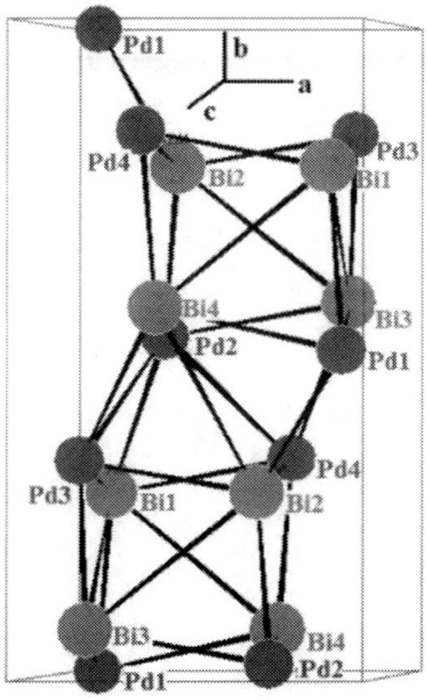

Figure 1. α-BiPd has Monoclinic($P2_1$) structure with 16 atoms in a unit cell (8 formula unit). It contains four inequivalent Bi sites and four inequivalent Pd sites.[27]

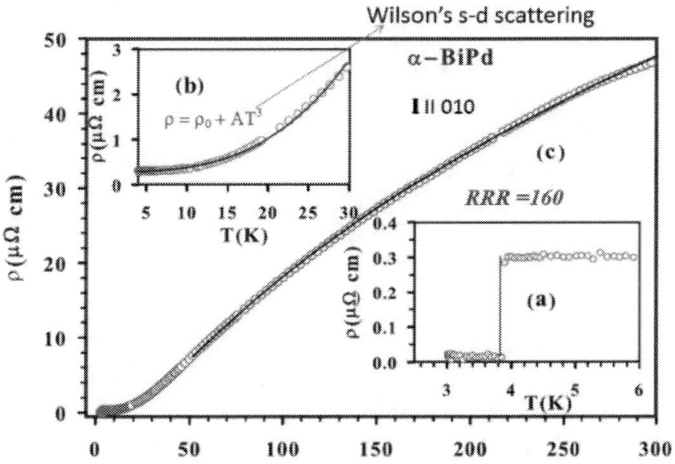

Figure 2. Temperature dependence of the electrical resistivity of the monoclinic ($P2_1$) α-BiPd. The insets a and b are described in the text. The solid lines are fits to the equations described in the text.[27]

the inter band s-d phonon induced scattering within the Debye approximation. However, the same theory predicts a linear temperature dependence for the resistivity at higher temperatures (T>θ_D, where θ_D is the Debye temperature) which is not observed in α-BiPd (see Fig. 2(c)). The Wilsons theory [15] neither considers the actual structure of the density of states of the electrons at the Fermi level nor the effect of unharmonicity of the phonon mode. But the contribution from these effects in the case of α-BiPd are small. There is yet another mechanism (parallel resistor model) suggested by Fisk and Webb [16] and later by Wisemann et al [17] which accounts for the significant deviation of the resistivity from the linear temperature dependence at high temperatures (100 K<T<300 K). This has been seen in many other compounds where the ρ value is rather high (>100 $\mu\Omega$ cm). The strong deviation from linearity and possible tendency towards saturation occur because the mean free path becomes short, of the order of few atomic spacings. When that happens, the scattering cross section will no longer be linear in the scattering perturbation. Since the dominant temperature-dependent scattering mechanism is electron-phonon interaction here, the ρ will no longer be proportional to the mean square atomic displacement, which is proportional to T for a harmonic potential. Instead, the resistance will rise less rapidly than linearly in T and will show negative curvature ($d^2\rho/dT^2 < 0$). This behavior is also seen in our previous studies on silicides and germanides [18, 19, 20]. Wisemann et al [17] describe the ρ(T) of these compounds (which is known as the parallel resistor model) where the expression of ρ(T) is given by,

$$\frac{1}{\rho(T)} = \frac{1}{\rho_1(T)} + \frac{1}{\rho_{max}}, \tag{2}$$

where ρ_{max} is the saturation resistivity which is independent of temperature and ρ_1(T) is the ideal temperature-dependent resistivity. Further, the ideal resistivity is given by the following expression,

$$\rho_1(T) = \rho_0 + C_1(\frac{T}{\theta_D})^3 \int_0^{\theta_D/T} \frac{x^3 dx}{(1 - exp(-x))(exp(x) - 1)}, \tag{3}$$

where ρ_0 is the residual resistivity and the second term is due to phonon-assisted electron scattering similar to the s-d scattering in transition metal compounds. θ_D is the Debye temperature and C_1 is a numerical constant. Eqn.(2) can be derived if we assume that the electron mean free path l is replaced by $l + a$ (a being an average interatomic spacing). Such an assumption is reasonable, since infinitely strong scattering can only reduce the electron mean free path to a. Chakraborty and Allen [21] have made a detailed investigation of the effect of strong electron-phonon scattering within the framework of the Boltzmann transport equation. They find that the interband scattering opens up new *nonclassical channels* which account for the parallel resistor model. The high temperature resistivity fit (50 K<T<310 K) shown in the Fig. 3(c) yields a value of 164 $\mu\Omega$ cm for ρ_{max}, 0.33 $\mu\Omega$ cm for ρ_0, 76.3 $\mu\Omega$ cm/K^3 for C_p and 168.6 K for θ_D. This value to θ_D is close to the value obtained from the heat capacity data (described later) which suggests that the parallel resistor model can successfully explain the high temperature dependence of *ρ(T)* of α-BiPd.

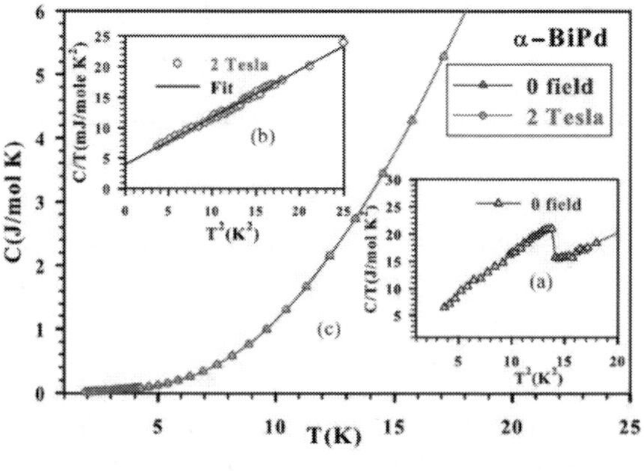

Figure 3. Temperature dependence of the heat capacity of the monoclinic ($P2_1$) α-BiPd from 2 to 35 K. The insets are explained in the text. Solid line is a fit to the equation(3) described in the text.[27]

The temperature dependence of the heat-capacity (C_p) from 2 to 30 K of α-BiPd is shown in Fig. 3. The inset shows the low temperature C_p vs T data. The jump in C_p at 3.7 K (ΔC= 30

Journal of Physics: Conference Series **683** (2016) 012028 doi:10.1088/1742-6596/683/1/012028

mJ/mol K) clearly shows bulk superconducting ordering in this sample below this temperature. The temperature dependence of C_p is fitted to the expression,

$$C_p = \gamma T + \beta T^3 , \tag{3}$$

where γ is due to the electronic contribution and β is due to the lattice contribution. The value of the ratio $\Delta C_p/\gamma T_c$ is 0.2 which is significantly reduced from the BCS value of 1.43. Low values of $\Delta C_p/\gamma T_c$ have been observed before in the heat-capacity study of several superconducting compounds [22, 23]. According to these studies the reduced jump across T_c could arise from extrinsic effect (such as inhomogeneity in the sample or magnetic impurities) or from intrinsic effect (such as the existence of regions which do not participate in superconductivity). In our sample of α-BiPd, we estimate the impurity content to be less than 0.1   by volume and the sharpness of the superconducting transition also suggests good homogeneity.It is possible that the reduced jump could arise from two-band superconductivity where one band remains normal. However, detailed Fermi surface measurements are required before we can analyze the data in terms of this model. The fit to the heat capacity data using the eqn. 3 in the temperature range from 5 to 20 K yielded 41 mJ/mol K^2 and 0.9 mJ/mol K^4 for γ and β, respectively. The γ value has been obtained by matching the entropy of normal and superconducting state at T_c as suggested by Stewart *et al* [24]. From the β value of 0.9 mJ/mol K^4, we estimate the θ_D to be 162 K using the relation,

$$\theta_D = \left(\frac{12\,\pi^4\,N_A\,n\,k_B}{5\beta}\right)^{1/3} , \tag{3}$$

where N_A is the Avogadro's number, n is the number of atoms per formula unit, and k_B is the Boltzmann's constant.

3.1. Point Contact spectroscopy measurements

From the resistivity and specific heat measurements on a similar crystal, we estimate the electronic mean free path, l$\sim$2.4 m at low temperatures. The quality of the crystal was also confirmed by observing de Haas-van Alphen (dHvA) oscillation. Before doing point contact measurement, the crystal surface was polished to a mirror finish. To make ballistic point contact, a mechanically cut fine tip made from 0.25mm diameter Ag wire was brought in contact with the (010) (I$\|$b) and (001) (I$\perp$b) crystal faces using a differential screw arrangement in a conventional sample-in-liquid ^{3}He cryostat. I-V characteristics of the junction formed between the tip and the sample were measured at different temperature down to T=0.32 K using conventional 4-probe technique [28]. The dI/dV vs. V spectra was obtained by numerically differentiating the I-V curves. For all spectra reported here, the contact resistance (R_C) in the normal state varied in the range $R_C\sim1\Omega$ to 30Ω. The corresponding contact diameter estimated using the formula, d = $\frac{4\rho l}{3\pi {R_C}^{1/2}}\sim$ 100-500 Å, was much smaller than l. Therefore, all our point contact spectra are taken in ballistic limit. To further understand the nature of superconductivity, we have measured the upper critical field (Hc_2) and its anisotropy along two crystallographic axes (H$\|$b and H$\perp$b) by measuring ac susceptibility as function of magnetic field at different temperatures. We first concentrate on the PCAR spectra at the lowest temperature. From large statistics, we observe two kinds of PCAR spectra, corresponding to I$\|$b and I$\perp$b respectively. Fig. 4(a) and (b) show representative evolution with contact resistance for (dI/dV) versus V spectra for I$\|$b. Figure 4(d) and (e) shows similar spectra I$\perp$b. In both directions, the striking feature is the observation of a pronounced zero bias conductance peak (BCP) which coexists with more conventional gap-like features in the low R_C contacts ((Fig. 4(a) and 4(d))). In addition, for I$\|$b, clear coherence peaks associated with superconducting gaps are observed around 0.1 meV and 0.4 meV respectively. For I$\perp$b, the corresponding structures are observed at 0.4 meV and 0.8 meV respectively. Both the   BCP and gap features disappear at the bulk Tc confirming their superconducting origin. As the contact resistance increased by gradually withdrawing the tip in both directions the BCP slowly vanishes and we recover spectra with only gap-like features. To quantitatively obtain the values of the superconducting energy gaps, we fit the spectra using a two-band Blonder-Thinkham-Klapwijk (BTK) model [25, 26] generalized to take into account broadening effects. Analyzing more than 50 spectra along I||b and I⊥b (Fig. 4(c) and 4(f)),

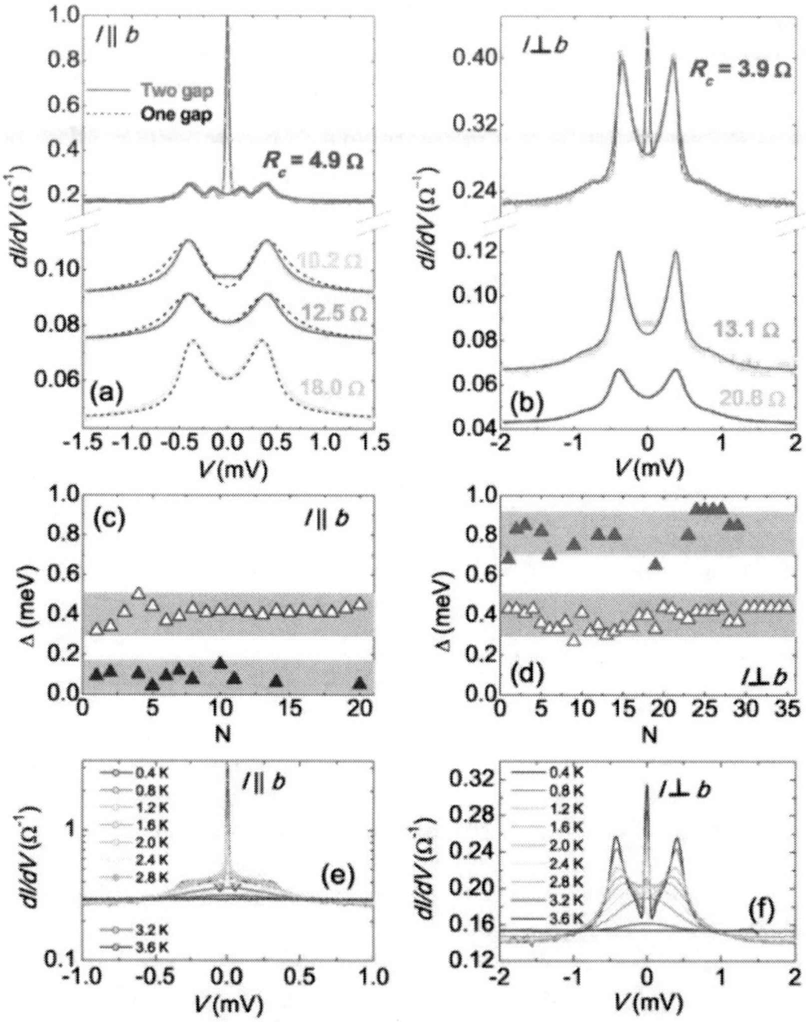

Figure 4. PCAR spectra for different contact resistance at T 0.35K: (a) and (b) I||b and (d) and (e) I⊥b. Solid lines (red) are fits to the modified two gap BTK model. The corresponding fits with a single gap model (black dash lines) are also shown for comparison for some of the spectra. (c), (f) Scatter plot of superconducting energy gap obtained by fitting the modified BTK model to the experimental PCAR spectra for I||b and I⊥b plotted as a function of the serial number of the spectra.The bands are guides to the eye. (e) and (f) Temperature dependence of the PCAR spectra for two low Rc contacts for I||b and I⊥b, respectively.[28]

we observed that the dominant feature is a gap, Δ_1=0.40.1 meV present along both directions. For I||b, in a about 50 of the spectra we can clearly resolve a smaller gap, Δ_2=0.1±0.05 meV with ω~0.2-0.6. On the other hand in 50 of the spectra along I⊥b, we can clearly resolve a larger gap Δ_3=0.80.15 meV with ω~0.1-0.35. We did not obtain any spectra showing the three gaps simultaneously in the same spectra. The large variation in ω and the dispersion in gap values arise from surface roughness which limits our inability to precisely inject current along the desired direction. The temperature variation of the superconducting energy gaps are obtained

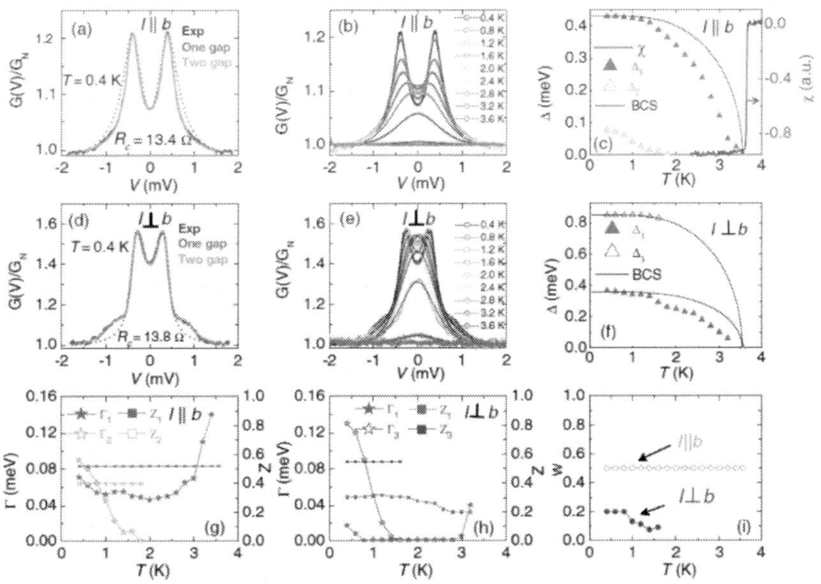

Figure 5. Comparison between one-gap and two-gap fits for PCAR spectra recorded at the lowest temperature for (a) I||b and (d) I⊥b. Temperature variation in the PCAR spectra for (b) I||b and (e) I⊥b. (c) and (f) show the temperature dependence of the superconducting energy gaps extracted from (b) and (e), respectively; the solid black lines are the expected BCS variation in Δ. (g) and (h) show the temperature variation in the broadening parameters and the barrier parameters obtained from the best fits of the spectra in (b) and (e), respectively. (i) Temperature variation in the relative weight factor (ω) for the two gaps used in the best fit of the spectra in (b) and (e). The blue solid line in panel (c) shows the real part of the ac susceptibility (χ) as a function of temperature showing the superconducting transition of BiPd.[28]

by analyzing the temperature dependence of two point contacts along the two directions with large Rc (Fig. 5(a) and 5(c)), where the BCP is suppressed. The temperature dependence of Δ_1, Δ_2 and Δ_3 obtained from a fit of these spectra with the generalized BTK model are shown in Figures 5(b) and 5(d). One can see that Δ_1 has similar temperature variation for both I||b and I⊥b and closes at T_C, suggesting that this gap originates from an isotropic s-wave order parameter. The small gap Δ_2 observed for I||b on the hand decreases rapidly at low temperatures and forms a tail towards T_C as expected for a multiband superconductor. We now focus on the origin of the BCP in the low resistance spectra. Since BCP can arise from several origins it is important to analyze the observed BCP in BiPd critically. First, we look for extrinsic origins of the BCP that are not associated with genuine spectroscopic features. It has been shown that in the case where the point contact is not purely in the ballistic limit, BCP can arise from the current reaching the critical current (Ic) of the point contact. However, in our case such a possibility can be trivially ruled out for two reasons. First, as we have shown before our contact is well in the ballistic limit even after considering error associated with our determination of contact diameter from R_S. More importantly, the conductance spectra at currents larger than I_C

cannot contain any spectroscopic information. In our case however, we observe clear signatures of the superconducting energy gap at bias voltages much larger than voltage range where the BCP appears. Other origins of BCP include (i) magnetic scattering (ii) proximity induced pair tunneling (PIPT) and (iii) Andreev bound state (ABS) when the superconductor has an unconventional symmetry. Magnetic scattering and PIPT can be ruled out since in the former BCP should split under the application of magnetic field and in the latter the BCP should get suppressed at small fields of the order of 0.1 T. None of these is observed in our measurements. We therefore conclude the BCP-s observed here are manifestations of ABS originating from an unconventional component of the order parameter in this material. Further confirmation of the ABS origin of the BCP comes from its evolution with contact size. Since the mean size of the ABS is of the order of the dirty limit coherence length (ξ_0), the BCP originating from ABS gradually disappears as the contact diameter becomes smaller than ξ_0. We can see that for both directions of injection currents the BCP gradually disappears as $d \leq \xi_0$. We therefore conclude that the BCP in BiPd originates from the ABS resulting from an unconventional pairing for which the phase varies on the Fermi surface. While in principle Δ_1, Δ_2 and Δ_3 could also arise from a multiband scenario containing three different bands, this is a very unlikely possibility for the following reasons. First, a simple multiband scenario consisting of multiple s-wave gap functions on different Fermi sheets cannot explain the existence of the pronounced BCP that we observe in our data. Secondly, we do not observe Δ_2 and Δ_3 simultaneously in any of our spectra despite the surface roughness that produces a significant scatter in their individual for both directions of injection current. It is therefore unlikely that these two gaps arise from two different gap functions on different Fermi sheets. Therefore, Δ_2 and $\Delta_3 3$ are likely to be associated with a strongly anisotropic gap function (Δ) for which the observed gap values are different for the two different directions of current injection.

3.2. NMR/NQR measurements

For NMR/NQR measurements [29], the samples were crushed into powders. The T_C at zero and a finite magnetic field H was determined by measuring the ac susceptibility using the in situ NMR/NQR coil.A standard phase-coherent pulsed NMR spectrometer was used to collect data. Measurements below 1.4K were carried out in a ^{3}He/^{4}He dilution refrigerator. Figure 6 shows the ^{209}Bi(I=9/2) NQR spectra of BiPd measured at T = 4.2 K. Four transition lines centered at 12.4, 14.3, 21.4, and 28.8MHz are observed, which correspond to the transitions between the adjacent levels ($\pm 1/2 \Leftrightarrow \pm 3/2$),($\pm 3/2 \Leftrightarrow \pm 5/2$), ($\pm 5/2 \Leftrightarrow \pm 7/2$) and ($\pm 7/2 \Leftrightarrow \pm 9/2$), respectively. The NQR frequency ν_Q=7.31MHz and the asymmetry parameter η=0.35. Figure 7 shows the temperature dependence of $1/T_1$ of ^{209}Bi, which was measured at the $4\nu_Q$ ($\pm 7/2 \Leftrightarrow \pm 9/2$) transition. In the normal state above T_C, $1/T_1$ varies in proportion to T, as seen in conventional superconductors. The $1/T_1$ is enhanced just below Tc over its normal-state value, forming a suppressed, broad coherence peak. The result also indicates an isotropic superconducting gap dominating in α-BiPd. The solid curve below Tc shown in Fig. 7 is a calculation with $2\Delta = 2.7k_BT_C$ and r=7. The parameter 2Δ is slightly smaller than the BCS value of 3.5 k_BT_C. However, a second superconducting gap was reported to appear below 2 K from other measurements (point contact, penetration depth and Muon spin resonance) is not present here. We believe this could due to the disorder effects present in powdered sample. In BiPd, although all atoms break the inversion symmetry along all direction, Pd probably makes the main contribution to the bands near the Fermi level, so that the ASOC is small. Indeed, the band splitting at the Fermi level was calculated to be around 80 meV,which is smaller than the value of Li_2Pt_3B. Nonetheless, we found a clear correlation between the band splitting caused by the asymmetric spin orbit coupling (ASOC) and the height of the coherence peak. Figure 7 shows the $1/T_1$ normalized by its value at T_C against the reduced temperature for LaBiPt and α-BiPd. Generally, the height of the coherence peak can be affected by many factors such as the anisotropy of superconducting

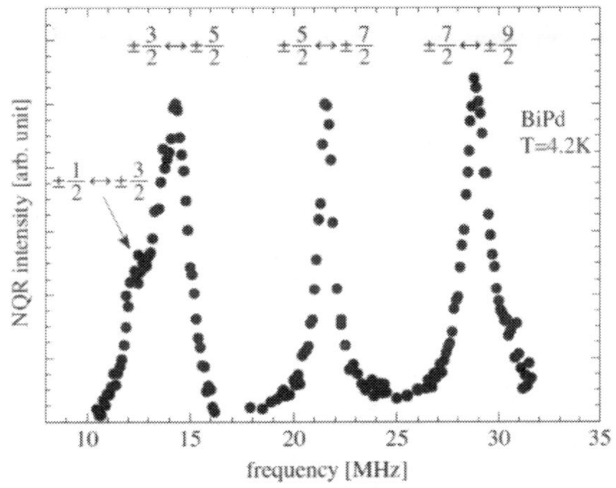

Figure 6. ^{209}Bi NQR spectra of powdered crystal of α-BiPd measured at 4.2 K.[29]

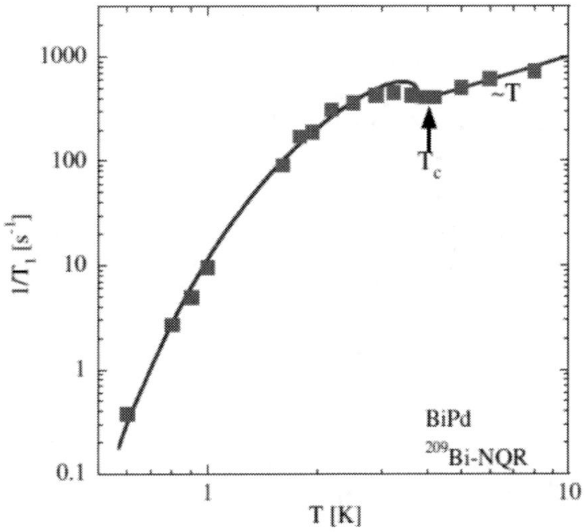

Figure 7. Normalized $1/T_1$ of powdered crystal of α-BiPd (^{209}Bi) against the reduced temperature. The straight line above Tc indicates the $TT_1T=$ const relation.[29]

gap, impurity scattering, and phonon scattering. However, it is possible when a spin-triplet component is induced, the corresponding gap component will have nodes. As a result, the observed height of the coherence peak will be suppressed. Knight shift measurements could suggest whether there exists substantial component of the spin-triplet state in α-BiPd.

3.3. Penetration depth measurements
Temperature dependence of the penetration depth was precisely measured [30] by utilizing a tunnel diode oscillator mounted on a ^{3}He cryostat or a ^{3}He/^{4}He dilution refrigerator. The operating frequency of this oscillator is 7 MHz with a frequency resolution as low as 0.05 Hz, which corresponds to a resolution of penetration depth of 0.1 Å. The penetration depth change is proportional to the shift of the TDO frequency, i.e.,$\Delta\lambda(T)=G\Delta f(T)$, where the G factor is solely determined by the sample geometry. Figure 8(c) shows the changes of the in-plane [$\lambda_{ac}(T)$]

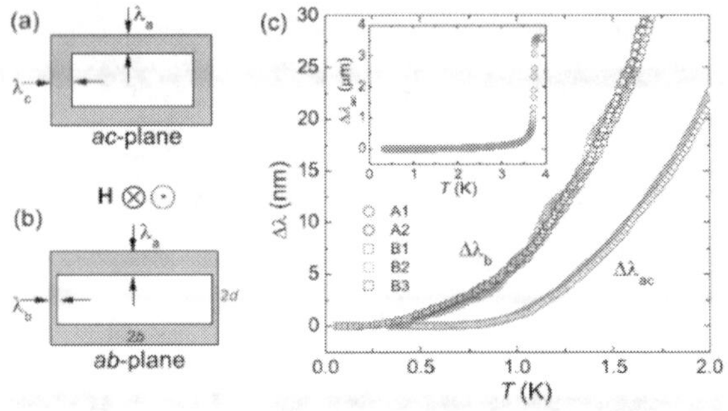

Figure 8. Schematic drawing of magnetic penetration for (a) the isotropic ac plane and (b) the anisotropic ab plane, respectively. A small ac magnetic field H is generated perpendicular to the sample planes. The shade denotes the field-penetrating area. (c) The in-plane ($\Delta\lambda_{ac}$ and out-of-plane ($\Delta\lambda_b$) penetration depth at low temperatures for several BiPd crystals. The G factors are 3.0 Å/Hz and 5.8 Å/Hz for samples A1 and A2, respectively. The inset shows $\lambda_a c$(T) of sample A1 over a wide temperature range.[30]

and out-of-plane [λ_b(T)] penetration depth for BiPd. For each field orientation, several samples were measured and the data are highly reproducible.We note that the samples were cut either along or perpendicular to the b axis.Within the ac plane, the samples are randomly aligned and the good reproducibility of λ_{ac}(T) for samples A1 and A2 indeed suggests an isotropic behavior of the in-plane penetration depth. However, the penetration depth shows distinctly anisotropic behavior for H//b and H⊥b. The in-plane penetration depth λ_{ac}(T) is flattened for T <1 K, showing exponential-type temperature dependence below 1.75 K. On the other hand, the out-of-plane penetration depth λ_b(T) grows much faster with temperature. The inset of Fig. 8(c) plots λ_{ac}(T) of sample A1 over a broad temperature region, where the sharp drop marks a superconducting transition at Tc = 3.7 K, which is a value that is close to that estmited from the electrical resistivity and magnetization. In order to analyze the gap symmetry, we take samples A1 (H‖b) and B1 (H⊥b) as examples and fit their low-temperature penetration depth with various models. Figure 9 shows the temperature dependence of the penetration depth λ_{ac}(T) and λ_b(T). The penetration depth at zero temperature, $\lambda(0)$, can be estimated

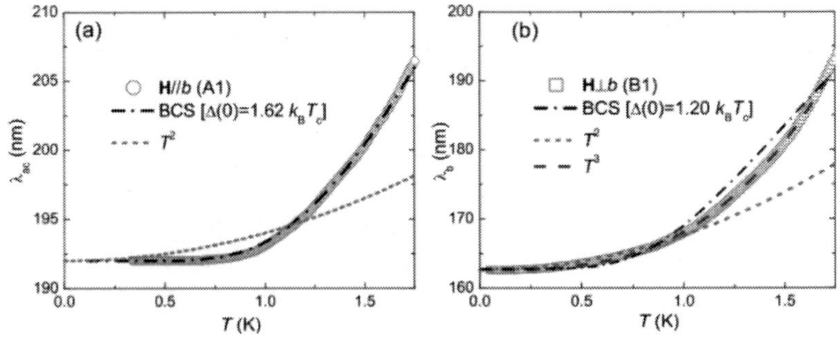

Figure 9. Temperature dependence of the (a) in-plane and (b) out-of-plane penetration depth for BiPd (symbols). The lines represent the fits of experimental data to various models.[30]

by $\lambda(0)$E1.06*$10^{10}/\xi\gamma^{1/2}$T$_C$, where ξ and γ represent the coherence length and the specific-heat Sommerfeld coefficient, respectively. By taking the values of $\xi^{\perp}$= 32 nm,$\xi^{\parallel}$ = 23 nm, and γ = 4 mJ/mol K^2 from the earlier study [27], we obtain $\lambda_b(0)$ =163 nm and $\lambda_{ac}(0)$ E ration depth λ(T) to the BCS model as well as the power-law behaviors. According to the isotropic BCS model in the local limit, the penetration depth can be approximated by the expression at T<<T_C: $\Delta\lambda(T)/\lambda_0 = \sqrt{\pi}\ \Delta(0)/2k_BT \times \exp(-\Delta_0/k_BT)$ where Δ_0 is the superconducting energy gap at T=0 K. The BCS model can describe λ_{ac}(T) with $\Delta(0) = 1.62k_BT_C$ (0.52 meV) [see Fig.9(a)], but gives a poor fit to λ_b(T) in the same temperature range [see Fig. 9(b)]. It is noted that, in the low temperature limit, λ_b(T) can be reasonably fitted by the BCS model with a small gap of 1.2 k_BT_C (0.38 meV). Furthermore, the power-law behavior of $\lambda(T)\sim T^n$ with n = 1 and 2 fails to illustrate the experimental data too, excluding nodal superconductivity for BiPd. Instead, λ_b(T) can be reasonably fitted by $\lambda_b(T)\sim T^3$ at low temperatures. These experimental facts indicate a complex gap structure for BiPd, showing a possible scenario of multiband superconductivity with anisotropic gaps. In the following, we further elucidate this feature by analyzing the superfluid density of α-BiPd. From the temperature dependence of the London penetration depth in two orthogonal field orientations for α-BiPd, anisotropic superconductivity is observed in the penetration depth and its corresponding superfluid density. For T<T_C, the in-plane penetration depth $\lambda_a c$(T) shows BCS-type exponential behavior, while the out-of-plane penetration depth λ_b(T) follows power-law-like temperature dependence. Detailed analysis of the superfluid density ρ_s(T) suggests anisotropic two-band superconductivity for α-BiPd. As a possible scenario, these experimental results can be interpreted in terms of the mixed pairing states in NCS SCs, shedding light on superconductivity without inversion symmetry.

3.4. Muon spin rotation and small angle neutron scattering measurements
The outcome of the TF-μSR experiments are really interesting and suggesting BiPd to be a multiband superconductor, consisting of at least two superconducting components with very different energy gaps and hence critical field. We are also finding indication for low-T_C type-II intermediate mixed state at low fields and temperatures, but conventional type II high-$T_C$superconductivity at high fields in NCS BiPd. Fig. 10 shows the measurements of the TF-μSR depolarization rates at 200 G, a field well below Hc_2 showed a crossover from short (λ=2000 Å) to long (λ=11,500Å) relaxation times between 2 and 3 K, well below T_C (3.8 K) along with a small non-precessing component. This indicates the presence of at least two components with very different critical fields. Both TF-μSR and SANS measurements have shown the presence of an intermediate mixed state below 220 G, due to long range attraction and short range repulsion of vortices. This results in a phase separated mixture of vortex lattice and Meissner regions. Clearly TF-μSR suggest multiple superconductivity gaps. Efforts are being made to observe spontaneius magnetization signal which will confirm the presence of unconventional pairing of Cooper pairs in α-BiPd.

3.5. Fermi surface studies
Fermi surface and quasiparticle masses have been determined using quantum oscillations observed both in the electrical resistivity(the Shubnikov-de Haas (SdH) effect) and in the [32] magnetic torque (the de Haas-van Alphen (dHvA) effect). dHvA effect in BiPd has been studied using a piezoresistive cantilever to measure magnetic torque. Electrical resistivity of a second crystal from the same grain as the dHvA crystal has been measured and found to have a residual resistance ratio R(300 K)/R(1 K)=170; which indicates high sample quality. Torque and transport measurements were made at temperatures 20mK <T<10K in a dilution fridge with a 15/17 T magnet. For the rotation study the angle θ between the crystal b-axis and the applied field B was varied. The rotation axis was parallel to (101) so that $\theta = 90$ corresponds to B// (101). A clear signal contains quantum oscillations even at 10 K and by 1 K many frequencies are observed in α-BiPd. Similar oscillations have also been obseved in magnetic torque oscillation. Due to the high quality single crystals of BiPd (RRR=170) clear quantum

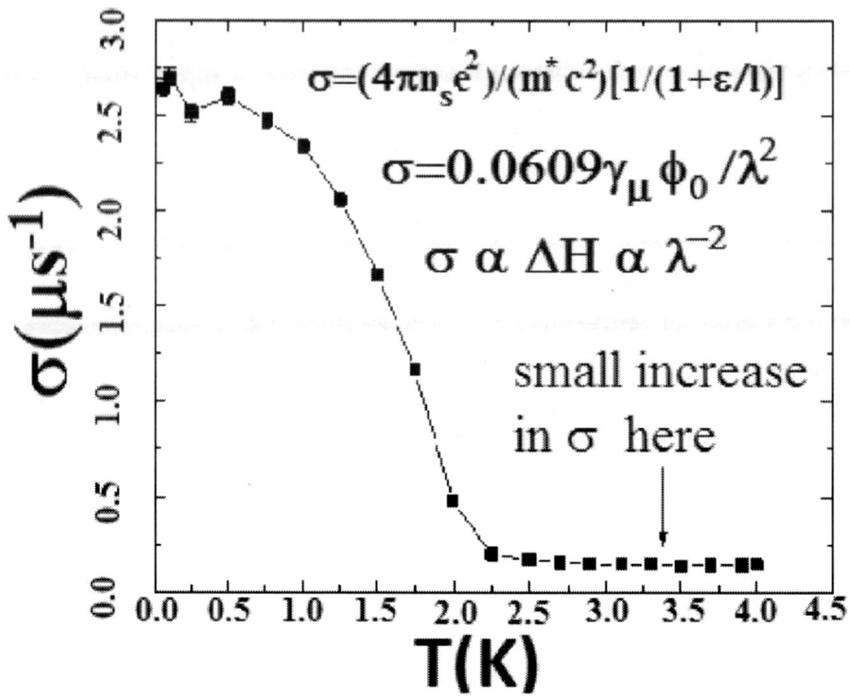

Figure 10. Gaussian TF μSR depolarisation rates in single crystal BiPd versus temperature at applied fields of 200 G and 500 G. right) Gaussian TF μSR depolarisation rates in polycrystal BiPd versus temperature.[31]

oscillation in the magnetic torque and the electrical resistivity has been observed even up to 10 K, which allowed successful construction of the Fermi surface of the noncentrosymmetric superconductor BiPd. These measurements revealed that α-BiPd is a high carrier density, three dimensional metal with a complex Fermi surface of many separate sheets and six spin-orbit split pair bands are crossing the Fermi level. Calculation of density functional theory to determine Fermi Surface is also in good agreement with the measured one and comparison with the specific heat measured at low temperature indicates a fairly low but appreciable renormalization $\sim$2 which is consistent with light quasiparticle masses in the range $\sim$0.8$-$1.5 electron masses deduced from the quantum oscillation measurements. The crystal structure is densely packed with nearest neighbour distances: Pd-Bi 2.74 Å, Pd-Pd 2.84 Å, Bi-Bi 3.31 Å and a density larger than either elemental Bi or Pd. The spatial distribution of states at EF is: 45 (22) inside the Bi (Pd) atomic spheres, and a significant remainder, 33 in the interstitial volume reflecting the dense crystal structure and large overlaps. The symmetry of states at EF within the atomic spheres is predominantly Bi-p and Pd-d with small contributions$\tilde{2}$0 from Bi-s and Pd-p. The calculated DOS N(EF) = 1.0 states $fu^{-1}eV^{-1}$ gives a bare value for the low temperature T-linear term of the specific heat $\gamma_0 = 2.3$ mJ/ mol K^2 compared to the measured value = 4 mJ/ mol K^2, implying a renormalization of 2. This is consistent with typical values of the mass enhancementm m^*/m_b deduced for individual extremal FS orbits. The Fermi surface calculated without spin-orbit results from 6 bands crossing E_F and is compensated, enclosing equal volumes of electrons and holes (1.21 of each). With spin-orbit included, 6 spin-orbit split pairs of bands are involved. The lowest lying 2 pairs (bands 241-244) have similar dispersion throughout the B and give rise to 4 concentric zone-corner-centered ellipsoidal Fermi surfaces enclosing holes (Fig. 10c). Pair 245-

doi:10.1088/1742-6596/683/1/012028

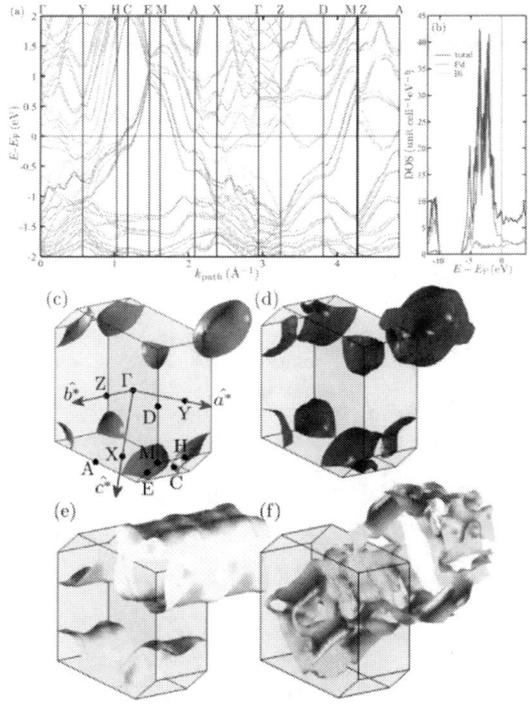

Figure 11. (a) Band structure of BiPd calculated with spin-orbit interaction (dotted and dashed lines) and without (solid lines) (b) Calculated density of states (c-f) Fermi surface sheets of BiPd (a). 4 representative sheets out of 12 are shown: (c) band 241, (d) band 245, (e) band 247, (f) band 252.[32]

246 is similar but the ellipsoids are connected along the b-axis by a neck along E-C (Fig. 10d). Pair 247-248 by contrast is 2-dimensional, approximating a weakly warped cylinder with its axis parallel to b (Fig. 10e), also enclosing holes. The remaining two pairs (bands 249-252) resemble zone-corner centered cylinders but further crossings of EF make a more complex FS topology with necks connecting the cylinders and additional small disconnected sheets (Fig. 10f). Note that the low symmetry of the crystal structure means that the only point symmetries of the Fermi surface are a 2-fold rotation about (010) and inversion, the latter being a consequence of time-reversal symmetry even in a NCS material.

4. Conclusion

We have established multiple superconducting energy gaps in high quality single crystal of α-BiPd using various techniques. It appears that even a small presence of disorder seems to wipe out unconventional pairing and multiple energy gaps [33]. Moreover, the upper critical field estimated from transport measurements seems to be an order of magnitude (0.7 T along b axis) larger as compared to the one estimated from dc magnetization and heat capacity studies ($\sim$ 0.06T along b). The reason for this behavior could either due to large reversible region (pin free) in the vortex state or due to twin boundaries as suggested by a recent work [34]. More puzzling is the absence of multiple gaps in the STM measurements on Bi deficient BiPd crystals [34, 35]. The STM measurements also show Dirac surface states in the normal state of α-BiPd which is confirmed by ARPES data [36]. Clearly both normal and superconducting state properties of α-BiPd are unusual and warrant more investigations.

References

[1] Bauer E, Hilscher G, Michor H, Paul Ch, Scheidt E W, Gribanov A., Seropegin Yu, Noel H, Sigrist M and Rogl P 2004 *Phys. Rev. Lett.* **92** 027003
[2] Gorkov L P and Rasbha E I 2001 *Phys. Rev. Lett.* **87** 037004
[3] Akazawa T, Hidaka H, Fujiwara T, Kobayashi T C, Yamamoto E, Haga Y, Settai R and Onuki Y 2004 *J. Phys.: Condens. Matter* **16** L29 and references cited therein
[4] Kimura N, Ito K, Saitoh K, Umeda Y, Aoki H and Terashima T 2005 *Phys. Rev. Lett.* **95**, 247004
[5] Sugitani I, Okuda Y, Shishido H, Yamada T, Thanizhavel A, Yamamoto E, Matsuda T D, Haga Y, Takeuchi T, Settai R and Onuki Y 2006 *J. Phys. Soc. Jpn.* **75** 043703
[6] Giorgi A L, Szklarz E G, Krikorian N H and Krupka M C 1970 *J. Less-Common Met.* **22** 131
[7] Hanawa M, Muraoka Y, Tayama T, Sakakibara T, Yamaura J and Hiroi Z 2001 *Phys. Rev. Lett.* **87** 187001
[8] Badica P, Kondo T, and Togano K 2005 *J. Phys. Soc. Jpn.* **74** 1014
[9] Yuan H Q, Agterberg D F, Hayashi N, Badica P, Vandervelde D, Togano K, Sigrist M and Salamon M B 2006 *Phys. Rev. Lett.* **97** 017006
[10] Bauer E, Kaldarar H, Lackner R, Michor H, Steiner W, Scheidt E-W, Galatanu A, Marabelli F, Wazumi T, Kumagai K and Feuerbacher M 2007 *Phys. Rev.* B **76** 014528
[11] Bauer E, Khan R T, Michor H, Royanian E, Grytsiv A, Melnychenko-Koblyuk N, Rogl P, Reith D, Podloucky R, Scheidt E -W, Wolf W and Marsman M 2009 *Phys. Rev.* B. **80** 064504
[12] Xu R, de Groot R A and van der Lugt W 1992 *J. Phys.: Condens. Matter.* **4** 2389
[13] Bhatt Y C and Schubert K 1979 *J. Less-Common Met.* **64(2)** 17
[14] Rodriguez-Carvajal J 1993 *Physica* B **192** 55. FULL PROF Package for X-ray powder diffraction program available at http://www.ill.eu/sites/fullprof/
[15] Wilson A H 1958 *Theory of Metals* (Cambridge University Press, Cambridge, England)
[16] Fisk Z and Webb G W 1976 *Phys. Rev. Lett.* **36** 1084
[17] Wiesmann H, Gurvitch M, Lutz H, Ghosh A K, Schwarz B, Strongin M, Allen P B and Halley J W 1977 *Phys. Rev. Lett.* **38** 782
[18] Ramakrishnan S, Ghosh K and Girish Chandra 1992 *Phys. Rev.* B **45** 10769
[19] Ramakrishnan S, Ghosh K and Girish Chandra 1992 *Phys. Rev.* B **46** 2958
[20] Ghosh K, Ramakrishnan S and Girish Chandra 1993 *Phys. Rev.* B **48** 4152
[21] Chakraborty B and Allen P B 1979 *Phys. Rev. Lett.* **42** 736
[22] Ramakrishnan S, Ghosh L, Chinchure Arvind D., Marathe V R and Girish Chandra 1995 *Phys. Rev.* B **52** 6284 and references cited therein
[23] Vining C B, Shelton R N, Braun H F and Pelizzone M 1983 *Phys. Rev.* B **27** 2800
[24] Stewart G R, Meisner G P and Ku H C in *Proceedings of the IV^{t}h International Conference in d- and f- band superconductivity* edited by Buckel W and Weber W (Kernforschungszentrum, Karlsruhe,1982), p. 81.
[25] Blonder G E, Tinkham M and Klapwijk T M 1982 *Phys. Rev.* B **25** 4515
[26] Gonnelli R S, Daghero D, Ummarino G A, Stepanov V A, Jun J, Kazakov S M and Karpinski J 2002 *Phys. Rev. Lett.* **89** 247004
[27] Bhanu Pratap, Thamizhavel A and Ramakrishnan R 2011 *Phys. Rev.* B **84** 064518
[28] Mintu Mondal 2012 *Phys. Rev.* B **86** 094520
[29] Matano K, Maeda S, Sawaoka H, Muro Y, Takabatake T, Joshi B, Ramakrishnan S, Kawashima K, Akimitsu J and Zheng Guo-qing 2013 *J. Phys. Soc. Jpn.* **82** 084711
[30] Jiao L, Zhang J L, Chen Y, Weng Z F, Shao Y M, Feng J Y, Lu X, Joshi B. Tamizhavel A and Ramakrishnan S 2014 *Phys. Rev.* B **89** 060507(R)
[31] Forgan T *et al*, to be published
[32] Yelland E *et al*, to be published
[33] Unusual features (multiple gaps etc.) seem to vanish with samples which have an RRR of 70 (to be published)
[34] Darren C Peets 2014 *J. Phys.: Conference Series* **568** 022037
[35] Sun Z, Enayat M, Maldonado A, Lithgow C, Yelland E, Peets D. C, Yaresko A, Schnyder A P and Wahl P 2015 *Nature Communications* |6:6633|DOI: 10.1038/ncomms7633|, 30 March
[36] Madhab Neupane *et al*, arXiv:1505.03466v1 [cond-mat.] 13 May 2015

Journal of Physics: Conference Series **683** (2016) 012029 doi:10.1088/1742-6596/683/1/012029

Search for spontaneous magnetization in noncentrosymmetric superconductors

A Sumiyama[1], D Kawakatsu[1], J Gouchi[1], I Kawasaki[1], A Yamaguchi[1], G Motoyama[2], Y Hirose[3], R Settai[3] and Y Ōnuki[4]

[1] Graduate School of Material Science, University of Hyogo, Kamigori, Hyogo 678-1297, Japan
[2] Department of Material Science, Shimane University, Matsue, Shimane 690-8504, Japan
[3] Department of Physics, Niigata University, Nishi-ku, Niigata 950-2181, Japan
[4] Department of Physics and Earth Sciences, University of the Ryukyus, Nishihara, Okinawa 903-0213, Japan

E-mail: `sumiyama@sci.u-hyogo.ac.jp`

Abstract. We have measured the magnetization of two noncentrosymmetric superconductors: Ir_2Ga_9 and $CePt_3Si$ using a SQUID magnetometer, neither of which has been tested for the time-reversal symmetry in the superconducting state. For Ir_2Ga_9, the magnetization change ΔM below the superconducting transition temperature $T_c \sim 2.2$ K was smaller than 10^{-5} G in zero magnetic field, suggesting that ΔM originates in the Meissner effect induced by the possible residual field. When $CePt_3Si$ was cooled in zero magnetic field, ΔM along the c axis first decreased below $T_c^+ \sim 0.75$ K (high-T_c phase) and then increased below $T_c^- \sim 0.45$ K (bulk superconducting phase). Since this behavior cannot be explained by the Meissner effect, it may be a sign of a spontaneous magnetization.

1. Introduction

Since the discovery of noncentrosymmetric superconductor $CePt_3Si$[1], various kind of superconductors, of which crystal structures lack a center of inversion symmetry, have been investigated, because such structure allows the coexistence of spin-singlet and spin-triplet states and may induce superconductivity other than the conventional BCS (s-wave, spin-singlet pairing) state. Recently, $LaNiC_2$ has been reported to be in the broken time-reversal symmetry (BTRS) state, based on the observation of a spontaneous magnetic field below the superconducting transition temperature T_c by μSR measurements[2], although NQR measurements indicated conventional BCS-type superconductivity[3]. Very recently, we have investigated the magnetization of $LaNiC_2$ and found that a spontaneous magnetization in the c-axis direction appears below T_c and is closely related to the crystal structure lacking inversion symmetry along the c axis[4]. It will be interesting to investigate other noncentrosymmetric superconductors and clarify whether a spontaneous magnetization in the superconducting state is a common property or peculiar to $LaNiC_2$. In this paper, we have investigated two noncentrosymmetric superconductors Ir_2Ga_9 and $CePt_3Si$, a spontaneous magnetic field of which has not been verified in the superconducting state by μSR measurements thus far.

The noncentrosymmetric superconductor Ir_2Ga_9 has the monoclinic (distorted Co_2Al_9-type) structure with the space group Pc. It was reported to be a type-II BCS superconductor first[5], and then classified as type-I through the study of a single crystal[6]. The de Haas-van Alphen experiments have revealed that splitting energies between the two Fermi surfaces caused by the antisymmetric spin-orbit interaction are 290 and 130K for α and β branches, respectively[7]. No experimental results that suggest unconventional superconductivity have been reported yet.

The heavy-fermion superconductor $CePt_3Si$ has the crystal structure (tetragonal, space group $P4mm$) and lacks the inversion symmetry along the c axis. A large number of studies have been done on its unconventional superconductivity[8], and yet the spontaneous magnetic field in the superconducting state remains unknown; a large internal field (~160 G) by the coexisting antiferromagnetism has been observed below $T_N \sim 2.2$ K by μSR measurements, while a slight change of the internal field below T_c, which is at the limit of the measurement accuracy, is ascribed to a coupling of the superconducting and magnetic order parameters and/or to the decrease of the hyperfine contact contribution acting on the muon[9].

2. Experimental

A single crystal of Ir_2Ga_9 was grown by the Ga-self flux method. The details of the growth conditions were given in the previous paper[7]. The critical temperature determined in dc resistivity measurements was 2.2 K. A piece of sample with flat (001) and (100) planes and about the size of 2 mm was made out of the ingot.

A single crystal of $CePt_3Si$, which has the tetragonal structure (space group $P4mm$) was grown by the Bridgman method. The details of the sample preparation were given in previous papers[10, 11]. A piece of crystal was cut from the ingot into a cubic shape with edges of about 3 mm along the [100] (a-axis), [010] and [001] (c-axis) directions. It was already used for the measurements of magnetic susceptibility[12] and the Josephson effect[13]. Although it shows a single peak at 0.4 K in the temperature dependence of specific heat[12], the Meissner effect appears below 0.75 K, suggesting the inclusion of a trace of high-T_c phase ($T_c^+ \sim$0.75 K) besides the bulk superconducting phase ($T_c^- \sim$0.45 K).

The magnetization was measured using a SQUID magnetometer linked to an astatic pair of pick-up coils. The sample and a reference superconductor In was inside one of the pick-up coils, and another reference Ta was inside the other. To eliminate the Meissner effect by the residual field, the earth's magnetic field was reduced by a double-layered μ-metal shield and a Cryoperm 10® shield outside the vacuum can. Outside the pick-up coils, a solenoid coil was wound to apply a small magnetic field. The residual field at In or Ta position was determined with an accuracy of 50 μG by applying a magnetic field to compensate the Meissner effect induced by the residual field. The details of the measurements are described in our previous paper[14].

3. Results and discussion

Figure 1 shows the temperature dependence of dc magnetic susceptibility around the superconducting transition. The sample was first cooled below T_c in zero field and then H_{dc}=15 mOe was applied. The change in susceptibility indicates the diamagnetism caused by a shielding current. The sample was then warmed to temperatures above T_c and the diamagnetic susceptibility (zero-field cooled susceptibility: χ_{ZFC}) was measured. In the same field, the sample was cooled again to measure field cooled susceptibility (χ_{FC}), which indicates the Meissner effect. Both samples show full diamagnetism $4\pi\chi_{ZFC} = -1$, which is confirmed in comparison with the full diamagnetism of In.

For Ir_2Ga_9, a sharp transition is observed at 2.2 K, and χ_{ZFC} deviates from zero at a higher temperature ~2.7 K. For $CePt_3Si$, notwithstanding that the sample showed a single peak at 0.4 K in specific heat measurements, both χ_{ZFC} and χ_{FC} show transition below 0.75K, suggesting that a high-T_c phase ($T_c^+ \sim$0.75 K) is still contained. The Meissner fraction χ_{FC}/χ_{ZFC} is 0.37 for Ir_2Ga_9($H \parallel c$) and 0.12, 0.06 and 0.1 for $CePt_3Si$ ($H \parallel c$, $H \parallel a_1$ and $H \parallel a_2$), respectively, where a_1 and a_2 indicate two a-axis directions of the rectangular parallelepiped sample. The larger Meissner fraction as well as the narrower transition width suggests that the quality of Ir_2Ga_9

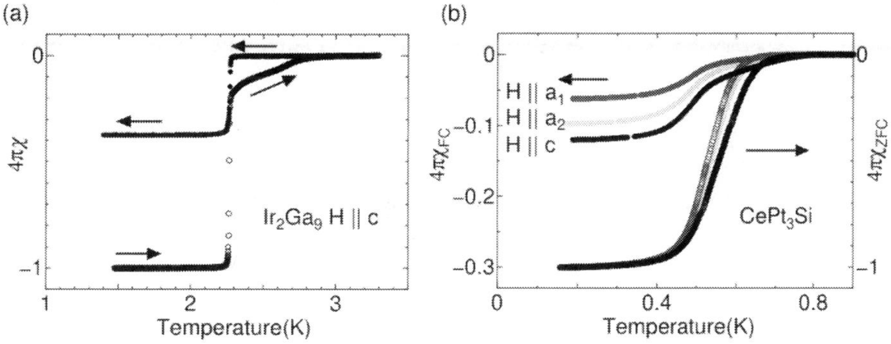

Figure 1. Temperature dependence of magnetic susceptibility χ for (a) Ir_2Ga_9 and (b) $CePt_3Si$, as obtained by the field-cooled and the zero-field cooled methods. The applied field is 15 mOe.

crystal is better than that of $CePt_3Si$; lattice defects and/or impurities that act as pinning sites are less included in Ir_2Ga_9.

The temperature dependences of magnetization change $\Delta M = M(T) - M(3\mathrm{K})$ for Ir_2Ga_9 and $\Delta M = M(T) - M(1\mathrm{K})$ for $CePt_3Si$ were measured in the cooling process in various magnetic fields, as shown in Fig. 2. The magnetization was calculated on the assumption that the magnetization change due to the full diamagnetism in Fig. 1 corresponds to $-1/4\pi \times 15$ mOe neglecting the demagnetizing effect. For Ir_2Ga_9, a sharp transition is observed within the range of applied fields (< 75 mOe). The magnetization change of $CePt_3Si$, on the other hand, clearly shows a superconducting transition below $T_c^+ \sim 0.75$ K even in the largest field. This result suggests that the relatively large magnetization change in the high-T_c phase in spite of its small amount, which is below the limit of detection in specific heat measurements, cannot be explained by some mechanism that is susceptible to magnetic field, such as Josephson network between the high-T_c inclusions.

A test of a spontaneous magnetization in Ir_2Ga_9 is shown in Fig. 3. As shown in the inset, the residual field was different between Ta and In positions. Hereafter, we use the residual field at the In position as the residual field H_{res} at the sample position, and regard $H = H_a + H_{res}$ instead of the applied field H_a as the magnetic field at the sample position. The magnetization

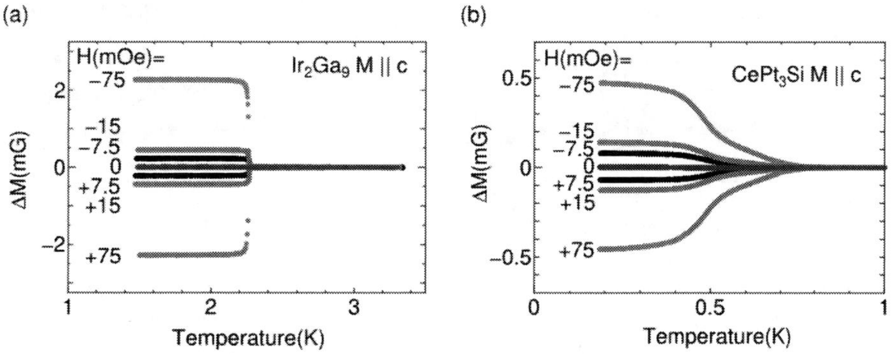

Figure 2. Temperature dependence of magnetization change ΔM in various magnetic fields, where (a) $\Delta M = M(T) - M(3\mathrm{K})$ for Ir_2Ga_9 and (b) $\Delta M = M(T) - M(1\mathrm{K})$ for $CePt_3Si$.

Journal of Physics: Conference Series **683** (2016) 012029 doi:10.1088/1742-6596/683/1/012029

change $\Delta M = M(T) - M(3\text{K})$ in $H = 0$ increases slightly at $T_c \sim 2.2$ K. However, this increase can be compensated by applying H =0.05 mOe, which is much smaller than the typical residual field. Even if we compensate the residual field using the value at the In position, the residual field inevitably exists at the sample position by its spatial variation. Thus, the observed magnetization change can be explained by the Meissner effect induced by the possible residual field and cannot be regarded as a spontaneous magnetization within our experimental accuracy.

As shown in Fig. 3(b), the change in ΔM in $H = 0$ was reproduced in three cooling processes (1, 3, 4) and also in the warming process (2). This behavior indicates that the motion of the magnetic flux is reversible in Ir_2Ga_9 and contrasts with that of $LaNiC_2$, in which an abrupt fluctuation appeared at T_c in the warming process[4].

Figure 4 shows the magnetization change $\Delta M = M(T) - M(1\text{K})$ along the c axis of $CePt_3Si$. Because of the inclusion of the high-T_c phase ($T_c^+ \sim 0.75$ K), ΔM begins to change below 0.75 K and shows a complicated behavior; in zero magnetic field, ΔM first decreases below 0.75 K and then begins to increase below the bulk $T_c^- \sim 0.45$ K. This behavior is reproducible both in the cooling and warming processes, as shown in Fig. 4(b). When a small magnetic field is applied, ΔM at each temperature increases or decreases according to the direction of the applied magnetic field regardless of whether the temperature is above or below T_c^-. Even in H =0.75 mOe, the abrupt upturn at T_c^- is still observed. These results indicate that the Meissner effect induced by the applied field points in the same direction in the two superconducting phases, and therefore the opposite evolution of ΔM in zero magnetic field cannot be explained by the Meissner effect induced by the possible residual field.

The magnetization change $\Delta M = M(T) - M(1\text{K})$ along the a axis is also complicated, as shown in Fig. 5. Although the two a-axis directions of $CePt_3Si$ should be crystallographically equivalent, the present crystal shows a difference in field-cooled susceptibility between the two a-axis directions, which are denoted as a_1 and a_2 axes, as shown in Fig. 1. In zero magnetic field, ΔM along the a_2 axis changes little below either T_c^+ or T_c^-, suggesting the absence of a spontaneous magnetization. In contrast, ΔM along the a_1 axis increases rapidly below T_c^+ and decreases slowly below T_c^- in H =0. By applying magnetic fields, the Meissner effect is induced below T_c^+ in the a_2-axis direction, while in the a_1-axis direction the Meissner effect between T_c^- and T_c^+ is small as compared with the increase in ΔM appeared in zero magnetic field; even in $H = 0.75$ mOe, where the change in ΔM below 1 K is compensated at 0.2 K, the increase in

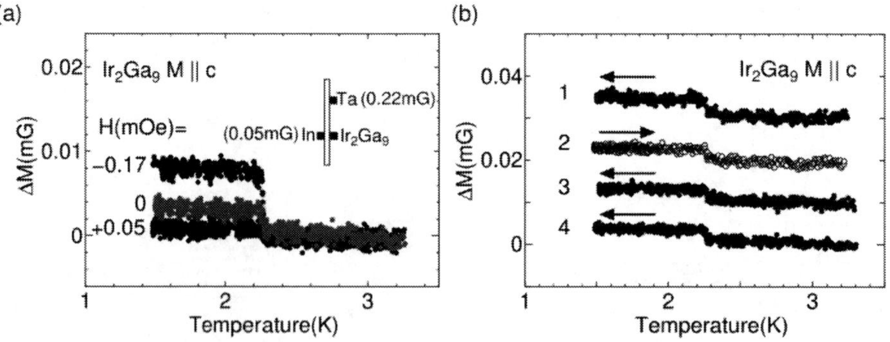

Figure 3. Temperature dependence of magnetization change $\Delta M = M(T) - M(3\text{K})$ of Ir_2Ga_9 in (a) small magnetic fields and (b) zero magnetic field, where the data in (b) were measured in the cooling process (1,3,4) or the warming process (2). The data are shifted vertically by 0.01 mG for clarity. The inset in (a) indicates the residual field at the position of the standard superconductors (In, Ta).

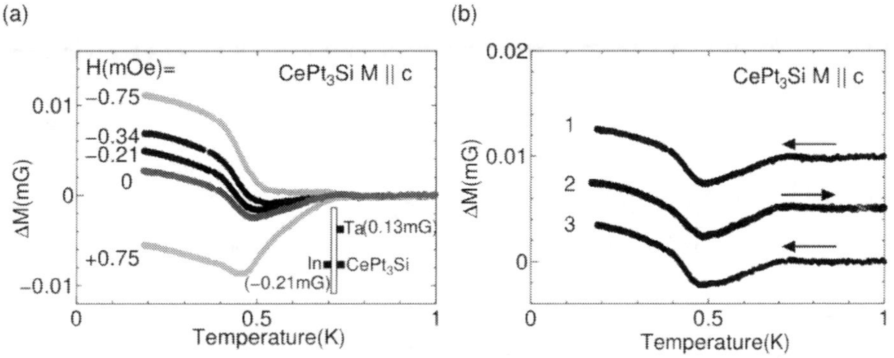

Figure 4. Temperature dependence of magnetization change $\Delta M = M(T) - M(1\mathrm{K})$ along the c axis of $CePt_3Si$ in (a) small magnetic fields and (b) zero magnetic field, where the data in (b) were measured in the cooling process (1,3) or the warming process (2). The data are shifted vertically by 0.005 mG for clarity. The inset in (a) indicates the residual field at the position of the standard superconductors (In, Ta).

ΔM below T_c^+ is still seen.

Since this crystal shows a single peak at 0.4 K in specific heat measurements, it should contain only a small amount of high-T_c (T_c^+) phase. If we focus on the bulk T_c^- phase in H =0, ΔM changes a little (< 0.002 mG) below T_c^- in both a_1- and a_2-axis directions, while the increase in ΔM along the c axis is larger ($\sim$ 0.05 mG). This result may suggest that a spontaneous magnetization of the bulk T_c^- phase appears along the c axis. The high-T_c phase, on the other hand, also shows a change in ΔM below T_c^+, which may also be ascribed to a spontaneous magnetization.

Although it is still unclear how the high-T_c phase exists in the sample, the result that the magnetization changes remarkably above T_c^- in spite of its small amount may suggest that the demagnetizing factor of the shape of the high-T_c inclusions enlarges the apparent change in ΔM. Then the anisotropy observed above T_c^- depends on the state of the shape

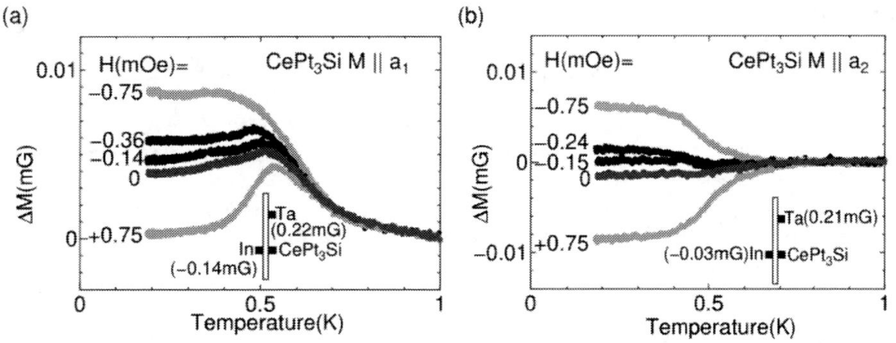

Figure 5. Temperature dependence of magnetization change $\Delta M = M(T) - M(1\mathrm{K})$ along the two a axes of $CePt_3Si$: (a) the a_1 axis and (b) a_2 axis, which correspond to those in Fig. 1. The insets indicate the residual field at the position of the standard superconductors (In, Ta) for each measurement.

and the configuration of the high-T_c inclusions rather than the crystallographic directions. One possible explanation for the origin of such high-T_c inclusions has been proposed in our previous papers[15, 16]: extended atomic bonds around crystallographic defects, dislocations, and stacking faults induce a lower pressure than the surroundings and lift the local T_c. This explanation also allows the appearance of a spontaneous magnetization below T_c^+, if the bulk T_c^- phase shows a spontaneous magnetization below T_c^-. This explanation, however, contradicts the NMR measurements; the bulk T_c^- phase and the high-T_c phase indicate unconventional and conventional BCS superconductivity, respectively[17]. To ensure the spontaneous magnetization of $CePt_3Si$, further work on single crystals of different quality is in progress.

4. Conclusion

We have measured the temperature dependence of magnetization of Ir_2Ga_9 and $CePt_3Si$, and found no spontaneous magnetization within the experimental accuracy for Ir_2Ga_9. The magnetization change caused by the high-T_c inclusions and the bulk superconducting phase of $CePt_3Si$, on the other hand, cannot be explained by the Meissner effect induced by the residual field and suggests the appearance of a spontaneous magnetization.

Acknowledgments

This work was supported by the Topological Quantum Phenomena (No. 25103719) Grant-in Aid for Scientific Research on Innovative Areas from the Ministry of Education, Culture, Sports, Science and Technology (ME T) of Japan.

References

[1] Bauer E, Hilscher G, Michor H, Paul Ch, Scheidt E W, Gribanov A, Seropegin Yu, Noel H, Sigrist M and Rogl P 2004 *Phys. Rev. Lett.* **92** 027003
[2] Hillier A D, Quintanilla J and Cywinski R 2009 *Phys. Rev. Lett.* **102** 117007
[3] Iwamoto Y, Iwasaki Y, Ueda K and Kohara T 1998 *Phys. Lett. A* **250** 439
[4] Sumiyama A, Kawakatsu D, Gouchi J, Yamaguchi A, Motoyama G, Hirose Y, Settai R and Ōnuki Y 2015 *J. Phys. Soc. Jpn.* **84** 013702
[5] Shibayama T, Nohara M, Aruga Katori H, Okamoto Y, Hiroi Z and Takagi H 2007 *J. Phys. Soc. Jpn.* **76** 073708
[6] Wakui K, Akutagawa S, Kase N, Kawashima K, Muranaka T, Iwahori Y, Abe J and Akimitsu J 2009 *J. Phys. Soc. Jpn.* **78** 034710
[7] Takeda M, Teruya A, Nakamura A, Harima H, Hedo M, Nakama T and Ōnuki Y 2015 *J. Phys. Soc. Jpn.* **84** 024701
[8] Bauer E, Kaldarar H, Prokofiev A, Royanian E, Amato A, Sereni J, Brämer-Escamilla W and Bonalde I 2007 *J. Phys. Soc. Jpn.* **76** 051009
[9] Amato A, Bauer E and Baines C 2005 *Phys. Rev. B* **71** 092501
[10] Yasuda T, Shishido H, Ueda T, Hashimoto S, Settai R, Takeuchi T, Matsuda T,D, Haga Y and Ōnuki Y 2004 *J. Phys. Soc. Jpn.* **73** 1657
[11] Takeuchi T, Hashimoto S, Yasuda T, Shishido H, Ueda T, Yamada M, Obiraki Y, Shiimoto M, Kohara H, Yamamoto T, Sugiyama K, Kindo K, Matsuda T D, Haga Y, Aoki Y, Sato H, Settai R and Ōnuki Y 2004 *J. Phys.: Condens. Matter* **16** L333
[12] Nakatsuji K, Sumiyama A, Oda Y, Yasuda T, Settai R and Ōnuki Y 2006 *J. Phys. Soc. Jpn.* **75** 084717
[13] Sumiyama A, Nakatsuji K, Tsuji Y, Oda Y, Yasuda T, Settai R and Ōnuki Y 2005 *J. Phys. Soc. Jpn.* **74** 3041
[14] Sumiyama A, Kawakatsu D, Gouchi J, Yamaguchi A, Motoyama G, Hirose Y, Settai R and Ōnuki Y 2014 *JPS Conf. Proc.* **3** 015017
[15] Aoki Y, Sumiyama A, Motoyama G, Oda Y, Yasuda T, Settai R and Ōnuki Y 2007 *J. Phys. Soc. Jpn.* **76** 114708
[16] Aoki Y, Sumiyama A, Motoyama G, Yamaguchi A, Oda Y, Yasuda T, Settai R and Ōnuki Y 2011 *J. Phys. Soc. Jpn.* **80** 065001
[17] Mukuda H, Nishide S, Harada A, Iwasaki K, Yogi M, Yashima M, Kitaoka Y, Tsujino M, Takeuchi T, Settai R, Ōnuki Y, Bauer E, Itoh K M and Haller E E 2009 *J. Phys. Soc. Jpn.* **78** 014705

Journal of Physics: Conference Series **683** (2016) 012030 doi:10.1088/1742-6596/683/1/012030

^{75}As-NQR study of the hybridization gap semiconductor $CeOs_4As_{12}$

M Yogi[1], N Higa[2], H Niki[1], T Kawata[3], C Sekine[3]

[1]Department of Physics and Earth Sciences, Faculty of Science, University of the Ryukyus, Nishihara, Okinawa 903-0213, Japan
[2]Graduate School of Science, University of the Ryukyus, Nishihara, Okinawa 903-0213, Japan
[3]Graduate School of Engineering, Muroran Institute of Technology, Muroran, Hokkaido 050-8585, Japan

E-mail: myogi@sci.u-ryukyu.ac.jp

Abstract. We performed an ^{75}As nuclear quadrupole resonance (NQR) measurement on $CeOs_4As_{12}$. The ^{75}As-NQR spectrum shape demonstrates that the Ce-site filling fraction of our high-pressure synthesized sample is close to unity. A presence of the $c-f$ hybridization gap is confirmed from the temperature dependence of the nuclear spin-lattice relaxation rate $1/T_1$. An increase of $1/T_1$ below $\sim$3 K indicates a development of the spin fluctuations. The $1/T_1$ for $CeOs_4As_{12}$ shows similar behavior as that for $CeOs_4Sb_{12}$ with different magnitude of the $c-f$ hybridization gap. An absence of phase transition in $CeOs_4As_{12}$ may be caused by the increase of the $c-f$ hybridization, which increases the gap magnitude and reduces the residual density of state inside the gap.

1. Introduction

The filled skutterudite compounds with the chemical formula RT_4X_{12} (R = rare earth; T = Fe, Ru or Os; X = P, As or Sb) have unique body-centered cubic structure (space group: $Im\bar{3}$, T_h^5 No.204); a rare-earth ion is surrounded by the cage consists of twelve pnictogen atoms. As a consequence, strong hybridization occurs between conduction and f-electrons which gives rise to various physical phenomena, such as metal-insulator transition, exotic multiband superconductivity, unconventional heavy fermion state, and so on [1, 2, 3].

Most Ce-based filled skutterudite compounds show insulating or semiconducting behavior, on the basis of which they are called hybridization-gap semiconductors or Kondo semiconductors. The magnitude of the $c-f$ hybridization gap is related to a lattice constant of the crystal and collapses for Sb-based compounds [4, 5]. Among them, $CeOs_4Sb_{12}$ is one of unique compound in Ce-base filled skutterudite which indicates an apparent spin fluctuations below $\sim$25 K and antiferromagnetic (AFM) order below $\sim$1 K [6]. The AFM order is suppressed by applying magnetic field $\mu_0 H \approx 1$ T [7]. However, in the magnetic field above $\sim$1 T, another ordered phase with anomalous $H-T$ phase diagram were reported [4, 8, 9].

Recently, Yan *et al.* reported similar band alignments for $CeOs_4Sb_{12}$ and $CeOs_4As_{12}$ from density-functional-theory calculations [10]. From their calculations, $CeOs_4Sb_{12}$ and $CeOs_4As_{12}$ are predicted to become topological Kondo insulators at low temperatures. Macroscopic experiment also observed similar behavior between $CeOs_4Sb_{12}$ and $CeOs_4As_{12}$ [4, 11]. To clarify

Journal of Physics: Conference Series **683** (2016) 012030 doi:10.1088/1742-6596/683/1/012030

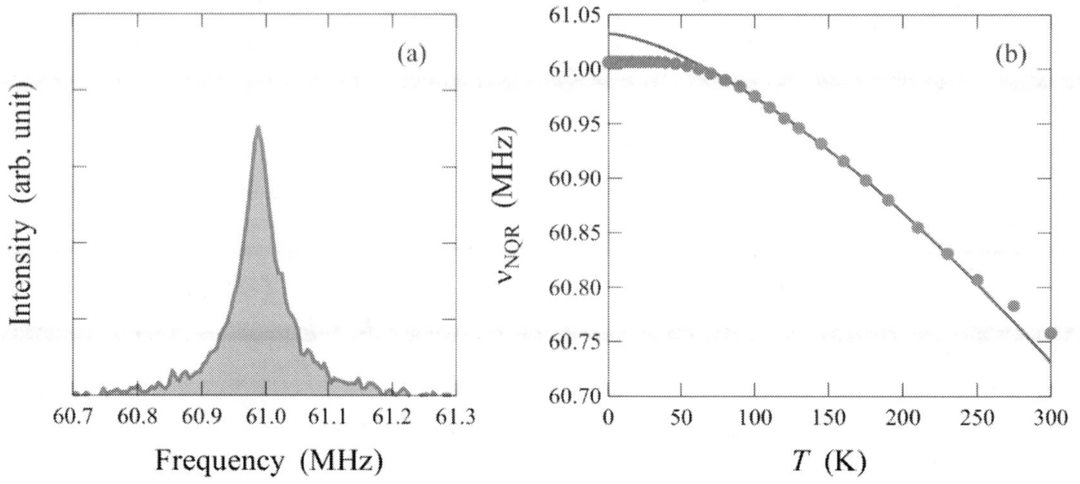

Figure 1. (a) ^{75}As-NQR spectrum of $CeOs_4As_{12}$ at $T = 70$ K. (b) Temperature dependence of the resonance frequency $\nu_{\rm NQR}$. Solid curve indicates an empirical relation $\nu_{\rm NQR} = \nu_q(1-\alpha T^{3/2})$ (see text).

its electric state from a microscopic viewpoint, we have carried out ^{75}As-NQR measurement on $CeOs_4As_{12}$.

2. Experimental

Polycrystalline of $CeOs_4As_{12}$ was prepared at high-temperature and high-pressure using a wedge-type cubic anvil high-pressure apparatus [5]. The compound was prepared by reaction of stoichiometric amounts of each metal and arsenic powder at 4 GPa. The reaction temperature and time were 800 °C and 90 min, respectively. The sample was characterized by powder -ray diffraction using CuKα_1 radiation and silicon as a standard. For the NQR measurement, the crystal was powdered to facilitate applied rf-field penetration. The pulse NQR measurement was performed on ^{75}As nucleus ($I = 3/2$) by a conventional spin-echo method using a conventional phase-coherent spectrometer in the temperature range 0.2-300 K. The NQR spectrum was obtained by a summation of the FFT spectrum as a function of frequency. The nuclear spin-lattice relaxation time T_1 was measured by a saturation-recovery method.

3. Results and Discussion

The NQR Hamiltonian is described as

$$\mathcal{H}_Q = \frac{h\nu_Q}{6}[3I_z^2 - I^2 + \frac{\eta}{2}(I_+^2 + I_-^2)], \tag{1}$$

with $\nu_Q \equiv \frac{3eQV_{zz}}{2I(2I-1)h}$ and $\eta \equiv \frac{V_{xx}-V_{yy}}{V_{zz}}$, where ν_Q is the nuclear quadrupole frequency, and η is the asymmetry parameter. Here, Q is the nuclear quadrupole moment, and $V_{\alpha\alpha} \equiv \partial^2 V/\partial\alpha^2$ ($\alpha = x, y, z$) are three components of the electric field gradient (EFG) tensor, where V is the electrostatic potential at the nuclear position. In the case of $I = 3/2$, one resonance line appears at frequency,

$$\nu_{\rm NQR} \equiv \nu_Q\sqrt{1 + \frac{\eta^2}{3}}. \tag{2}$$

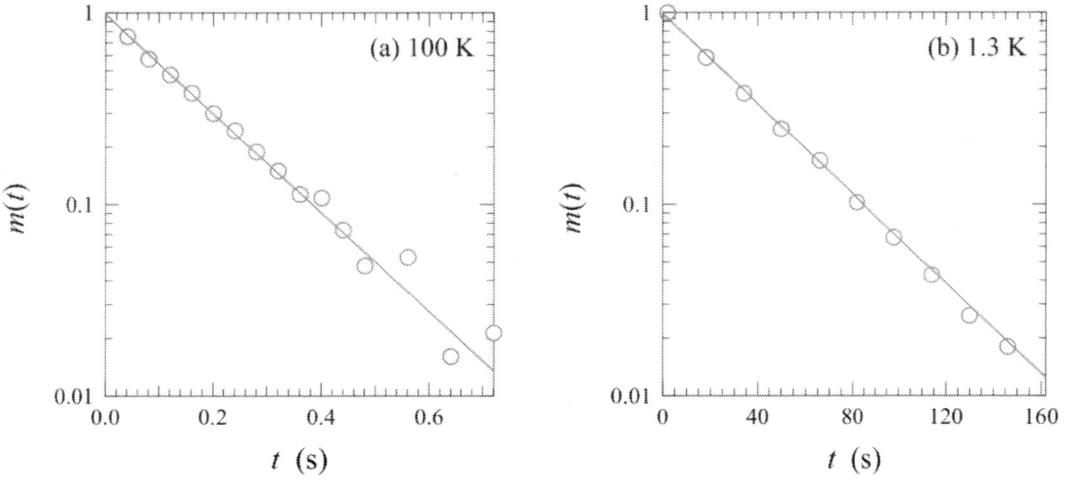

Figure 2. Recovery curves $m(t)$ of the nuclear magnetization $M(t)$ at $T = 100$ and 1.3 K.

Figure 1(a) shows ^{75}As-NQR spectrum at 70 K. The spectrum has almost symmetric shape without additional peaks indicating the filling fraction of Ce atom is close to unity.

The temperature dependence of the NQR frequency $\nu_{\rm NQR}$, which is determined from the peak position of the spectrum, is shown in Fig. 1(b). The $\nu_{\rm NQR}$ increases with decreasing temperature and becomes almost constant below $\sim$ 50 K. Usually, thermal lattice expansion is one of the origin of the temperature dependence of $\nu_{\rm NQR}$ and an empirical relation

$$\nu_{\rm NQR} \equiv \nu_q(1 - \alpha T^{3/2}) \tag{3}$$

was often applied in simple metals such as Cd, In, and Sb [12]. Here, α and ν_q are fitting parameters. This relation can be applied above $\sim$ 80 K as shown in the solid curve in Fig. 1(b). However, a deviation from the fit is observed above 250 K, which might be related to different temperature dependence of ν_Q and η.

The nuclear spin-lattice relaxation time T_1 of ^{75}As-NQR was measured at the peak position of the spectrum. Figure 2 shows the recovery curves of the nuclear magnetization $m(t) \equiv 1 - M(t)/M_0$ as a function of t, where M_0 and $M(t)$ are the respective nuclear magnetizations for the thermal equilibrium condition and at time t after the saturation pulse at 100 and 1.3 K. T_1 was obtained by fitting the recovery data with a theoretical recovery curve for $I = 3/2$, which is expressed in a simple exponential function as

$$m(t) = \exp(-3t/T_1). \tag{4}$$

The solid lines shown in Fig. 2 indicate fits of the recovery data using the fitting function.

Figure 3(a) shows temperature dependence of $1/T_1$ for $\rm CeOs_4As_{12}$. An evident decrease of $1/T_1$ with temperature indicates an existence of the $c - f$ hybridization gap, similar to other Ce-based filled skutterudite compounds. To estimate a value of the gap, Arrhenius plot of $1/T_1$ is displayed in Fig. 3(b). It is apparent that $1/T_1$ does not obey simple exponential behavior as $1/T_1 \propto \exp(-\Delta/k_{\rm B}T)$ in the wide temperature range. This may be caused by the complicated structure of the electric density of states (DOS) and the gap. Indeed, the electrical resistivity was not described by a simple Arrhenius function expected for a semiconductor with a single energy gap [11]. Instead, a phenomenological expression which has three exponential terms with

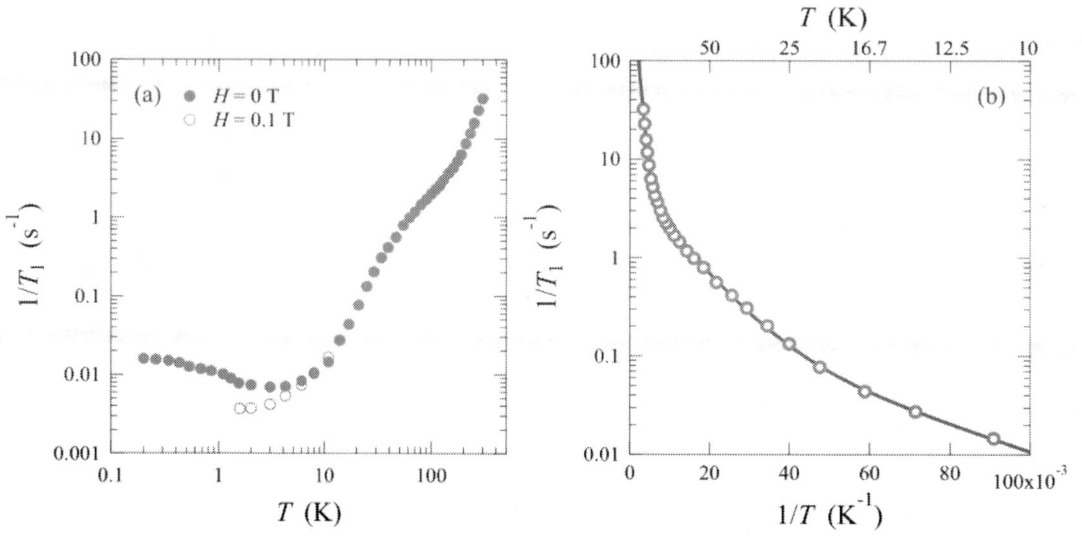

Figure 3. (a) Temperature dependence of $1/T_1$ at $\mu_0 H = 0$ T (closed circles) and 0.1 T (open circles), respectively. (b) Arrhenius plot of $1/T_1$ at $\mu_0 H = 0$ T. Solid curve is a fit using the phenomenological expression. (see text)

Table 1. Evaluated values of the hybridization gap for $CeOs_4Pn_{12}$ (Pn = P, As and Sb). The gap for $CeOs_4P_{12}$ is cited from Ref. 13. The lattice constants are also listed for comparison.

material	Δ_1/k_B (K)	Δ_2/k_B (K)	Δ_3/k_B (K)	lattice constant (Å)
$CeOs_4P_{12}$	-	500	-	8.0626 [14]
$CeOs_4As_{12}$	1059	114	31	8.5296 [5]
$CeOs_4Sb_{12}$	283	32	8	9.3011 [15]

different gap were introduced to explain its temperature dependence. Hence, we try to fit $1/T_1$ using the similar phenomenological expression as follows:

$$\frac{1}{T_1} = \sum_{i=1}^{3} A_i \exp(-\Delta_i/k_B T). \quad (5)$$

Solid curve in Fig. 3(b) shows a fit of the data above 10 K using the expression. Estimated values of the gap are listed in Table 1. As for $CeOs_4Sb_{12}$, an evident spin fluctuations below ~25 K is suppressed by applying high magnetic field ~15 T [8]. Therefore, we applied the same procedure for the $1/T_1$ of $CeOs_4Sb_{12}$ at $\mu_0 H = 15$ T and estimated the gap values. In addition, Magishi *et al.* have reported that the $1/T_1$ of $CeOs_4P_{12}$ shows activated-type temperature dependence above 160 K as $1/T_1 \propto \exp(-\Delta/k_B T)$ with $\Delta/k_B \sim 500$ K [13]. These gap values and the lattice constants are summarized in Table 1. This result reveals that the gap magnitude becomes smaller with increasing lattice constant, although the multi-gap analysis was not applied in $CeOs_4P_{12}$.

Next we focus on low temperature region of $1/T_1$. If we consider a simple semiconductor or insulator, $1/T_1$ decreases exponentially with decreasing temperature. On the other hand, several Kondo semiconductors or insulators show $T_1T =$ const. behavior at low temperature due to an existence of the residual density of states inside the $c - f$ hybridization gap [16, 17]. In comparison with these compounds, $CeOs_4As_{12}$ shows different behavior at low temperature; the

$1/T_1$ starts to increase below ~ 3 K which reveals a development of spin fluctuations similar to $CeOs_4Sb_{12}$. In $CeOs_4Sb_{12}$, the spin fluctuations are suppressed by applying magnetic field. Therefore, to investigate a similarity for both compounds, we have measured $1/T_1$ under small magnetic field $\mu_0 H = 0.1$ T. $1/T_1$ shows magnetic field dependence below $\sim$8 K as shown in Fig. 3(a). Such magnetic field sensitive behavior was also observed in the electrical resistivity [11]. To clarify the electric state of $CeOs_4As_{12}$ more properly, it is necessary to measure $1/T_1$ under various magnetic fields.

4. Summary

In summary, we have carried out ^{75}As-NQR measurement on $CeOs_4As_{12}$. The ^{75}As-NQR spectrum indicates that the Ce-site filling fraction is close to unity for our high-pressure synthesized sample. The temperature dependence of $1/T_1$ shows an evident decrease with decreasing temperature at high temperature region. An increase of $1/T_1$ below ~ 3 K indicates a development of spin fluctuations. The temperature dependence of $1/T_1$ for $CeOs_4As_{12}$ shows similar behavior as that for $CeOs_4Sb_{12}$ with different magnitude of the $c-f$ hybridization gap. This result is consistent with the same band alignments predicted by theoretical calculations. An absence of phase transition in $CeOs_4As_{12}$ may be caused by an increase of the $c-f$ hybridization, which increases the gap magnitude and reduces the residual density of state inside the gap. More detailed NMR experiment under various magnetic fields at low temperatures are required to clarify the electric state of $CeOs_4As_{12}$.

Acknowledgments

This work was supported by a Grant-in-Aid for Scientific Research on Innovative Areas Heavy Electrons (Nos. 20102004 and 20102007) of The Ministry of Education, Culture, Sports, Science, and Technology, Japan.

References

[1] Sekine C, Uchiumi T, Shirotani I and Yagi T 1997 *Phys. Rev. Lett.* **79** 3218
[2] Yogi M, Nagai T, Imamura Y, Mukuda H, Kitaoka Y, Kikuchi D, Sugawara H, Aoki Y, Sato H and Harima H 2006 *J. Phys. Soc. Jpn.* **75** 124702
[3] Sanada S, Aoki Y, Aoki H, Tsuchiya A, Kikuchi D, Sugawara H and Sato H 2005 *J. Phys. Soc. Jpn.* **74** 246
[4] Sugawara H, Osaki S, Kobayashi M, Namiki T, Saha S R, Aoki Y and Sato H 2005 *Phys. Rev.* B **71** 125127
[5] Sekine C, Abe R, Takeda K, Matsuhira K and Wakeshima M 2008 *Physica* B **403** 856-858
[6] Yogi M, Kotegawa H, Zheng G-q, Kitaoka Y, Ohsaki S, Sugawara H and Sato H 2005 *J. Phys. Soc. Jpn.* **74** 1950
[7] Iwasa K, Itobe S, Yang C, Murakami Y, Kohgi M, Kuwahara K, Sugawara H, Sato H, Aso N, Tayama T and Sakakibara T 2008 *J. Phys. Soc. Jpn.* Suppl. A **77** 318-320
[8] Yogi M, Niki H, Yashima M, Mukuda H, Kitaoka Y, Sugawara H and Sato H 2009 *J. Phys. Soc. Jpn.* **78** 053703
[9] Tayama T, Ohmachi W, Wansawa M, Yutani D, Sakakibara T, Sugawara H and Sato H 2015 *J. Phys. Soc. Jpn.* **84** 104701
[10] Yan B, Müchler L, Qi X-l, Zhang S-c and Felser C 2012 *Phys. Rev.* B **85** 165125
[11] Baumbach R E, Ho P C, Sayles T A, Maple M B, Wawryk R, Cichorek T, Pietraszko A and Henkie Z 2008 *Proc Natl Acad Sci USA* **105** 17307-17311
[12] Christiansen J, Heubes P, Keitel R, Klinger W, Lowffler W, Sandner W and Witthuhn W 1976 *Z. Physik* B **24** 177
[13] Magishi K, Sugawara H, Saito T and Koyama K 2012 *J. Phys.: Conf. Ser.* **391** 012039
[14] Jeitschko W and Braun D 1977 *Acta Cryst.* B **33** 3401
[15] Braun D J and Jeitschko W 1980 *J. Less-Common Met.* **72** 147
[16] Nakamura K, Kitaoka Y, Asayama K, Takabatake T, Tanaka H and Fujii H 1994 *J. Phys. Soc. Jpn.* **63** 433-436
[17] Yogi M, Niki H, Kawata T and Sekine C 2014 *JPS Conf. Proc.* **3** 011046

Journal of Physics: Conference Series **683** (2016) 012031 doi:10.1088/1742-6596/683/1/012031

Non-Fermi Liquid Properties in a Cubic Pr-Based Compound $PrPb_3$ under Magnetic Fields

T Yoshida[1], Y Machida[1], K Izawa[1], T Onimaru[2], and H S Suzuki[3]

[1]Department of Condensed Matter Physics, Tokyo Institute of Technology, Meguro, Tokyo, 152-8551, Japan

[2]Department of Quantum Matter, AdSM, Hiroshima University, Higashi-Hiroshima, Hiroshima, Japan

[3]National Institute for Material Science, Tsukuba, Ibaraki, 305-0047, Japan

E-mail: `yoshida.t.ba@m.titech.ac.jp`

Abstract. We have studied the electrical resistivity ρ of the Pr-based cubic compound $PrPb_3$ with the Γ_3 doublet ground state. The temperature dependence of the resistivity is found to exhibit a non-Fermi liquid behavior with convex curve at high temperatures, as suggestive of the putative realization of the quadrupole Kondo effect. At low temperatures under magnetic fields, we observe anomalies associated with antiferroquadrupole (AFQ) ordering, incommensurate/commensurate transition inside the AFQ phase, and field-induced phase transition. The constructed phase diagram based on these observations well reproduces the one determined by the specific heat and magnetization measurements. Deep inside each phase, the resistivity turns to show the Fermi liquid behavior with the quadratic temperature dependence. The estimated slope of the T^2 term is almost independent of field in contrast to our previous work of $PrIr_2Zn_{20}$ and $PrRh_2Zn_{20}$ in which the slope is largely enhanced at the edge of the AFQ phase.

1. Introduction

Investigations of non-Kramers systems with an even number of f-electrons such as Pr^{3+} are of great significance in the field of condensed matter physics. Especially in the case of Γ_3 doublet lying on the crystal electric field (CEF) ground state, novel electronic states irrelevant to the time-reversal symmetry are expected to emerge at low temperatures in contrast to Kramers systems. In such Γ_3 systems, the local quadrupole moment of f-electron can be scattered by conduction electrons via two equivalent scattering channels, and the channel frustration leads to the imperfect screening of the quadrupole moments, preventing the quasi-particle formation. Therefore, the breakdown of the Landau's Fermi liquid (FL) theory allows us to expect the realization of non-trivial physics. Indeed, the two-channel picture suggested the emergence of a quadrupole Kondo effect with non-Fermi liquid (NFL) properties of $\rho \propto T^{0.5}$ and $C/T \propto -\log T$ [1, 2]. In addition, the imperfect screening results in partial releasing of the entropy by only a half of $R \log 2$ even at 0 K, so that the quadrupole Kondo state cannot be the ground state, and instead another state must be realized. Several possibilities are proposed as a way of releasing the residual entropy, such as the ordering of localized Majorana particles by spontaneous channel symmetry breaking [3] and the NFL-FL crossover arising from the channel anisotropy due to external and/or internal perturbations [4].

Journal of Physics: Conference Series **683** (2016) 012031 doi:10.1088/1742-6596/683/1/012031

However, the experimental confirmations of these proposals have not been done due to lack of suitable materials. One of a few candidates is the Heusler type compounds $PrMg_3$ and $PrAg_2In$. Although these compounds were reported to show $-\log T$ dependence of specific heat and $T^{0.5}$ like dependence of resistivity [5, 6] as expected from the quadrupole Kondo effect, the effect of lattice distortions cannot be neglected because the Heusler structure has a possibility for the random site exchange leading to the local splitting of Γ_3 doublet and partial quenching of the quadrupole degrees of freedom. Thus the intrinsic features arising from the Γ_3 doublet may be masked. Recently the cage compound $PrIr_2$ n_{20} with the Γ_3 doublet ground state was discovered as a new candidate and was reported to show the $T^{0.5}$ like dependence of the resistivity [7]. However, the convex temperature dependence of the resistivity can arise from the anharmonic phonon characteristic of the cage compounds as discussed in the β-pyrochlore oxide [8, 9]. Thus, whether such phenomenon is originated from the quadrupole Kondo effect remains elusive. Another candidate is $La_xPr_{1-x}Pb_3$ with the Γ_3 doublet ground state in which the specific heat was reported to exhibit $-\log T$ dependence, and is expected to be a first example of the realization of quadrupole Kondo effect [10, 11]. However, the arguments are based on the experiments using the diluted sample by La on the Pr sites, so that the local breaking of the cubic CEF symmetry around Pr site similar to the Heusler type compounds cannot be excluded. In contrast, the physical properties of the non-doped $PrPb_3$ have barely been explored except for the quadrupole ordered phase, even though the system is suitable for studying quadrupole Kondo physics arising from Γ_3 doublet without the randomness and the cage structure.

In this paper, we report the electrical resistivity of $PrPb_3$, which has seldom been reported ever, as a part of systematic understanding of Γ_3 systems.

2. $PrPb_3$

$PrPb_3$ is the $AuCu_3$ type cubic compound (space group: $Pm\bar{3}m$) with the Γ_3 doublet ground state. The first excited Γ_4 triplet is separated by about 15 K [12] and thus the Γ_3 doublet is isolated at low temperatures. The ground state carries two types of quadrupole moment $O_2^0 \equiv (3J_z^2 - \boldsymbol{J}^2)/2$ and $O_2^2 \equiv \sqrt{3}(J_x^2 - J_y^2)/2$. The O_2^0 type antiferroquadrupole (AFQ) ordering was observed at 0.4 K by the specific heat and elastic modulus measurements [13, 14, 15]. The constructed phase diagram from these experiments was well reproduced by the simplified two-sublattice mean field calculation [12]. In addition, it was also reported that this AFQ phase has two modulated structures: commensurate (C) and incommensurate (IC) phase [16]. Although nature of the ordered state has been extensively studied, the effect of coupling between the quadrupole moments and conduction electrons has not been fully examined.

The electrical resistivity is one of promising probes to unveil such effect, but there has been no satisfactory report devoted to the resistivity of $PrPb_3$. This is mainly because Pb thin layer tends to form on the surface of the host crystal by exposing the air and/or applying stresses [17, 18]. In addition, since Pb undergoes a superconducting transition at about 7 K, the Pb layer gives a fatal error in the precise estimation of the intrinsic transport properties.

3. Experiments

The electrical resistivity ρ was measured by a standard DC four-probe method under magnetic fields by using a single crystal of $PrPb_3$. The current and field were applied in parallel with the [100] direction. The Au wires with a diameter of $\phi = 25\mu$m were attached to the sample by a spot-welding technique for the electrical contacts. The measurements were performed in the temperature range of $0.04 < T < 10$ K and the field range of $0 < B < 9$ T. The sample geometry is $2.25 \times 1.15 \times 1.25$ mm, where the largest length is parallel to the [100] direction. The residual resistivity ratio is estimated to be ~ 300, indicating a good sample quality.

In order to avoid the Pb layer formation on the sample surfaces, the sample was shaped by cleavage. We also attempted to shorten the exposing time to the air within about one hour during the sample setting.

4. Results and discussion

First, let us discuss the resistivity obtained at relatively high temperatures. From Figure 1(a), one immediately notice that there is a clear discrepancy below T_C of Pb pointed by the black arrow between our data and the previous data taken from Ref. [18] in which the Pb formation on the surface was recognized. In addition, since the field dependence of the resistivity at 4 K shown in Fig. 1(b) displays no anomaly corresponding to the superconducting transition, the Pb contribution is likely to be negligibly small in our sample. Therefore, we can succeed to prepare the setting to uncover the intrinsic electrical resistivity of $PrPb_3$ with negligible contributions of the Pb layer for the first time.

Now, we shall move on to the discussion of the low temperature resistivity. Fig. 2(a) shows the temperature dependence of the resistivity below 2 K under several fields. Interestingly, the resistivity exhibits an unusual behavior with a convex curve in the temperature range above around 0.5 K. This is obviously different from the expectation of the FL theory with $\rho \propto T^2$. As shown in Fig. 2(b), the exponent of the temperature dependence in this region is close to 0.5. A power of the temperature dependence of the resistivity less than unity indicates that the system does not reach the coherent state, implying that the equivalence of two scattering channels for the quadrupole moments of Pr^{3+} strongly enhances the fluctuations and prevents the formation of the quasi-particles. In this sense, ρ is supposed to follow $T^{0.5}$ dependence for the single-impurity quadrupole Kondo model as mentioned Section 1. It is interesting to note that the NFL behavior was also observed in the specific heat measurements as a logarithmic temperature dependence, which is also consistent with the single-impurity quadrupole Kondo effect [10]. From these experimental facts, one may expect that the quadrupole Kondo effect

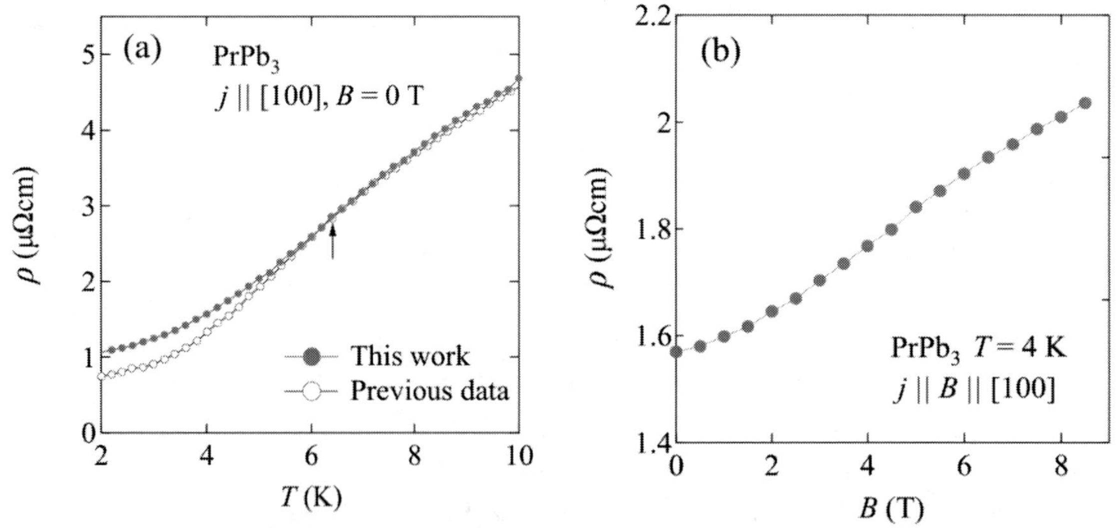

Figure 1. (a) The temperature dependence of the resistivity at 0 T above 2 K together with the previous data, in which Pb layer formation was recognized [18]. The black arrow indicates the superconducting transition temperature of Pb. There is a discrepancy between the two data below T_C of Pb because the resistivity is affected by the superconductivity of Pb in the previous report. (b) The field dependence of the resistivity at 4 K. There is no anomaly accompanied by the superconductivity of Pb.

Journal of Physics: Conference Series **683** (2016) 012031 doi:10.1088/1742-6596/683/1/012031

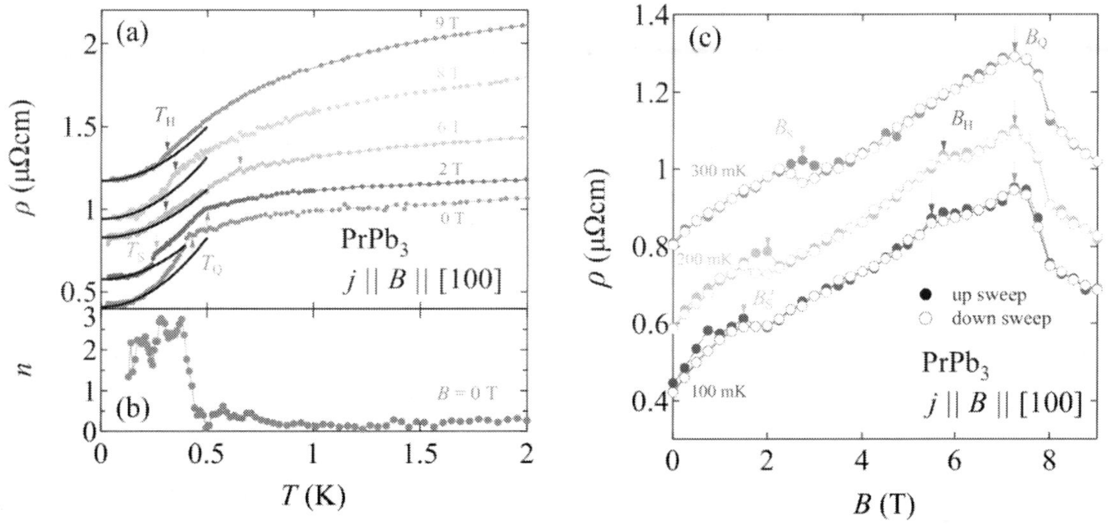

Figure 2. (a) The temperature dependence of the resistivity under several fields. The red, blue and green arrows indicate the transition temperature of the AFQ phase T_Q, high-field phase T_H and incommensurate/commensurate T_S, respectively. The solid lines represent the results of the fitting by a form of $\rho = \rho_0 + AT^2$. (b) The temperature dependence of the exponent n of the resistivity at $B = 0$ T which is estimated from the equation $\rho = \rho_0 + AT^n$. (c) The field dependence of the resistivity below 0.3 K. The closed and open circles represent the up and down sweep measurement, respectively. The anomalies observed at B_Q, B_H and B_S correspond to those at T_Q, T_H and T_S in the temperature dependence (a), respectively. B_S' represents the shoulder like anomaly.

is the case in $PrPb_3$. However such expectation is not conclusive because the single-impurity model is obviously not appropriate to the present material. Therefore, further discussions are required in order to specify the origin of the NFL behavior in $PrPb_3$.

On further cooling, the resistivity shows kinks or anomalies under each field as indicated by arrows in Fig. 2(a). The red, green and blue arrows correspond to the AFQ ordering T_Q, incommensurate/commensurate (IC-C) transition T_S, and transition to the high-field (HF) phase T_H. The corresponding anomalies are also found at B_Q, B_S and B_H in the field dependence of the resistivity as shown in Fig. 2(c) and these characteristic temperatures and fields are summarized in Fig. 3(a). We note that the obtained phase diagram is consistent with the one determined from the specific heat and magnetization measurements [13]. Our resistivity measurement first detects the anomalies associated with HF phase transition. However, the nature of this phase including the order parameter is unclear at this stage. The clear hysteresis accompanied by the IC-C transition is observed in the field dependence of the resistivity (Fig. 2(c)), as already reported by the magnetization measurements and the neutron diffraction experiments [19, 16]. The hysteresis becomes ambiguous below 0.1 K and changes to a shoulder like anomaly at B_S' because the phase boundary reaches the critical end point around 0.1 K. Curiously enough, the shoulder like anomalies corresponding to the crossover temperatures seem to intersect with $T = 0$ line at finite field, although the entropy of the IC phase is larger than the C phase. The IC phase is forbidden to exist at $T = 0$ and must change to another state in order to release the residual entropy. This unknown state is suggested to exist below 0.2 K by the specific heat and magnetization measurements [13, 20], but we cannot detect any anomalies corresponding to this state from the resistivity measurements within our experimental resolution.

Well below the anomalies, the resistivity exhibits the quadratic temperature dependence

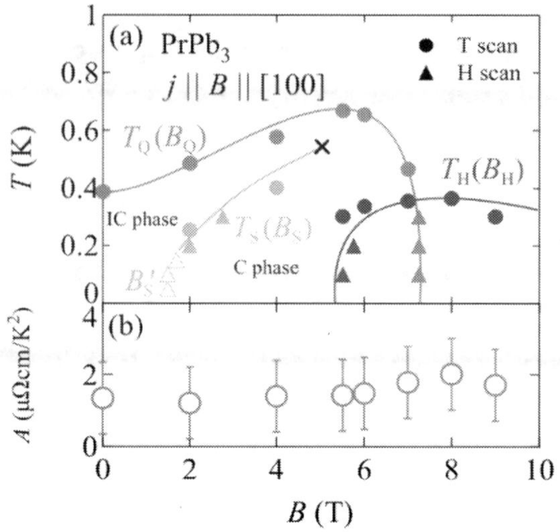

Figure 3. (a) The phase diagram of $PrPb_3$ obtained from the resistivity measurements. The red, green and blue marks represent the antiferroquadrupole, incommensurate/commensurate and high-field phase transitions, respectively, The open green triangles indicate the shoulder like anomaly B'_S without accompanying the hysteresis around T_S (B_S). The black cross mark indicates the termination of the phase boundary referred from Ref. [13]. The solid lines are guide to the eyes. (b) The field dependence of the A coefficient determined by fitting the resistivity data to the form of $\rho = \rho_0 + AT^2$.

$\rho = \rho_0 + AT^2$ as indicated by the solid lines in Fig. 2(a). This is because the FL behavior is attained due to freezing the quadrupole degree of freedom by ordering. Fig. 2(b) clearly shows the attainment of the FL behavior with the power $n \sim 2$ below T_Q. The estimated A coefficient under each field is insensitive to the applied magnetic field within the experimental error as shown in Fig. 3(b). With knowledge of the large specific heat $C/T \sim 6$ J/K^{2}mol at 0 T reported in Ref. [10], the Kadowaki-Woods ratio A/γ^2 is estimated to be $\sim 4 \times 10^{-3} a_0$, which is much less than the universal value a_0. This means that the large C/T is not due to the mass enhancement of itinerant quasi-particles, consistent with the small cyclotron mass $\sim 2m_0$ estimated by the dHvA experiments [21].

We have previously studied the electrical transport coefficients of $PrIr_2$ n_{20} and $PrRh_2$ n_{20} which have a Γ_3 doublet at the CEF ground state as well as $PrPb_3$ [22, 23, 24]. These compounds have a common feature with $PrPb_3$ that the resistivity exhibits the $T^{0.5}$ like NFL behavior above the AFQ ordering and the T^2 like FL behavior below the AFQ ordering. Taking into account this commonality, we can claim that the NFL behavior with convex temperature dependence of the resistivity is widely realized in non-Kramers systems with Γ_3 doublet ground state, likely due to the quadrupole Kondo effect regardless of the crystal structure as long as the cubic symmetry is guaranteed. On the other hand, these compounds have the different feature that the A coefficient is greatly enhanced when the AFQ ordering is collapsed by the magnetic field, while it is nearly field insensitive in $PrPb_3$ irrespective of the ground state. Given these differences, the HF phase in $PrPb_3$ emerged near the AFQ phase boundary would not be linked to the non-trivial heavy Fermion state in $PrIr_2$ n_{20} and $PrRh_2$ n_{20}.

5. Summary

We have succeeded to measure the intrinsic electrical resistivity of $PrPb_3$ with the Γ_3 doublet ground state by avoiding the formation of Pb thin layer on the host crystal. We found the

NFL behavior of the resistivity with a convex curve in the wide temperature and field range, as suggestive of the realization of the quadrupole Kondo effect. Well below the ordering temperatures of the AFQ phase or the field-induced HF phase, the FL behavior with the T^2 dependence is found to be realized. From the field dependence of the resistivity, the transition to the field-induced HF phase is clearly detected for the first time in the resistivity measurements. The constructed phase diagram based on the resistivity measurements is consistent with the previous one determined by the specific heat and magnetization measurements. The A coefficient of the T^2 dependence of the resistivity is almost independent of field.

Acknowledgments
This work is partially supported by Grants-in-Aid for Scientific Research (Nos. 23340099 and 25400361) from the Japan Society for the Promotion of Science.

References

[1] Nozières P and Blandin A 1980 *J. Phys. (Paris)* **41** 193
[2] Cox D L and Zawadowski A 1998 *Adv. Phys.* **47** 599
[3] Hoshino S, Otsuki J and Kuramoto Y 2013 *J. Phys. Soc. Jpn.* **82** 044707
[4] Yotsuhashi S and Maebashi H 2002 *J. Phys. Soc. Jpn.* **71** 1705
[5] Tanida H, Suzuki H, Takagi S, Onodera H and Tanigaki K 2006 *J. Phys. Soc. Jpn.* **75** 073705
[6] Yatskar A, Beyermann W P, Movshovich R and Canfield P C 1996 *Phys. Rev. Lett.* **77** 3637
[7] Onimaru T, Matsumoto K T, Inoue Y F, Umeo K, Sakakibara T, Karaki Y, Kubota M and Takabatake T 2011 *Phys. Rev. Lett.* **106** 177001
[8] Dahm T and Ueda K 2007 *Phys. Rev. Lett.* **99** 187003
[9] Hiroi Z, Yonezawa S, Nagao Y and Yamaura J 2007 *Phys. Rev. B* **76** 014523
[10] Kawae T, Shimogai M, Mito M, Takeda K, Ishii H and Kitai T 2001 *Phys. Rev. B* **65** 012409
[11] Kawae T, Yamamoto T, Yurue K, Tateiwa N, Takeda K and Kitai T 2003 *J. Phys. Soc. Jpn.* **72** 2141
[12] Tayama T, Sakakibara T, Kitami K, Yokoyama M, Tenya K, Amitsuka H, Aoki D, Onuki Y and Kletowski Z 2001 *J. Phys. Soc. Jpn.* **70** 284
[13] Sato Y, Morodomi H, Ienaga K, Inagaki Y, Kawae T, Suzuki H S and Onimaru T 2010 *J. Phys. Soc. Jpn.* **79** 093708
[14] Niksch M, Assmus W, Lüthi B, Ott H R and Kjems J K 1982 *Helv. Phys. Acta.* **55** 688
[15] Onimaru T, Sakakibara T, Harita A, Tayama T, Aoki D and Onuki Y 2004 *J. Phys. Soc. Jpn.* **73** 2377
[16] Onimaru T, Sakakibara T, Aso N, Yoshizawa H, Suzuki H S and Takeuchi T 2005 *Phys. Rev. Lett.* **94** 197201
[17] Kletowski Z, Sławiński P and Cichorek T 1996 *J. Magn. Magn. Mater.* **162** 277–279
[18] Morie T 2007 *Ph.D. Thesis* The University of Tokyo
[19] Sakakibara T, Tayama T, Onimaru T, Aoki D, Onuki Y, Sugawara H, Aoki Y and Sato H 2003 *J. Phys.: Condens. Matter* **15** S2055
[20] Sato Y, Makiyama S, Kawae T, Onimaru T, Suzuki H S, Jackson M, Paulsen C, Amara M and Galéra R 2014 *JPS Conf. Proc.* **3** 011052
[21] Aoki D, Katayama Y, Settai R, YInada, Onuki Y, Harima H and Kletowski Z 1997 *J. Phys. Soc. Jpn.* **66** 3988
[22] Ikeura T, Matsubara T, Machida Y, Izawa K, Nagasawa N, Matsumoto K T, Onimaru T and Takabatake T 2014 *JPS Conf. Proc.* **3** 011091
[23] Machida Y, Yoshida T, Ikeura T, Izawa K, Nakama A, Higashinaka R, Aoki Y, Sato H, Sakai A, Nakatsuji S, Nagasawa N, Matsumoto K, Onimaru T and Takabatake T 2015 *J. Phys: Conf. Ser.* **592** 012025
[24] Yoshida T, Machida Y, Izawa K, Shimada Y, Nagasawa N, Onimaru T and Takabatake T 2015 *Physics Proceedia* accepted

Journal of Physics: Conference Series **683** (2016) 012032 doi:10.1088/1742-6596/683/1/012032

High Magnetic Field Study of Elastic Constants of the Cage-structure Compound $SmBe_{13}$

S Mombetsu[1], T Murazumi[1], K Hiura[1], S Yamazaki[1] Y Shimizu[2], H Hidaka[1], T Yanagisawa[1], H Amitsuka[1], S Yasin[3,4], S Zherlitsyn[3], and J Wosnitza[3]

[1] Department of Physics, Hokkaido Universsity, Sapporo 060-0810, Japan
[2] Institute for Solid State Physics, University of Tokyo, Kashiwa 277-8581, Japan
[3] Hochfeld-Magnetlabor Dresden (HLD-EMFL), Helmholtz-Zentrum Dresden-Rossendorf, D-01314 Dresden, Germany

E-mail: monmon@phys.sci.hokudai.ac.jp

Abstract. Ultrasonic measurements were performed on the cage-structure compound $SmBe_{13}$. We have investigated the magnetic field–temperature phase diagram of this material by using pulsed magnetic fields. We found that the low-temperature magnetic order is suppressed by a magnetic field of 43 T for $H \parallel [001]$, which is smaller than the estimated value from mean-field approximation assuming the Γ_8 quartet crystal-electric-field ground state and simple antiferromagnetic order. We found that the elastic constant C_{44} shows softening below the ordering temperature and has a local minimum below 7 T. These facts suggest that the low-temperature state is not a simple antiferromagnetically ordered state. In addition, no elastic anomaly due to rattling modes was found in the present measurements.

1. Introduction

Sm-based compounds have been studied extensively and a rich variety of physical properties has been found, such as metal-insulator transitions under pressure in SmX (X = Te, Se, S) [1, 2], dense Kondo behavior and unusual magnetic order in $SmSn_3$ and $SmIn_3$ [3], as well as possible topological in-gap surface states in the Kondo insulator SmB_6 [4, 5]. Among them, cage-structure compounds have recently been attracting much attention because of their novel physical properties, for instance, possible octupole order in $SmRu_4P_{12}$ [6], magnetically robust heavy-fermion behavior in $SmOs_4Sb_{12}$ [7], and magnetic-field insensitive magnetic order in $SmTr_2Al_{20}$ (Tr = Ti, V, Cr) [8, 9]. In these materials, strong c-f hybridization and multipole degrees of freedom, which are enhanced by the highly symmetrical cage structure, are considered to play important roles in the physical properties. Furthermore, in cage-structure compounds, large anharmonic vibrations of the guest ion, rattling, have been observed by several measurements, such as an Einstein-phonon contribution in specific heat, an ultrasonic dispersion, and in Raman scattering [10, 11, 12, 13]. The possible contribution of these rattling modes to the magnetically robust heavy-fermion state has also been discussed for $SmOs_4Sb_{12}$ [14].

[4] Present address: American University of the Middle East (AUM), College of Engineering and Technology, Eqaila, Kuwait

Journal of Physics: Conference Series **683** (2016) 012032 doi:10.1088/1742-6596/683/1/012032

$SmBe_{13}$ has the Na n_{13}-type crystal structure (Fm$\bar{3}$c), which is the same as for the heavy-fermion superconductor UBe_{13}. The Sm ion is surround by a cage structure which consists of 24 Be atoms. This material shows a phase transition at $T_M \approx 8.8$ K, which was revealed by specific-heat measurements using a polycrystalline sample [15].

Recently, our group has succeeded to grow single crystals of $SmBe_{13}$ and performed measurements of magnetization, specific heat, -ray diffraction (RD), muon-spin relaxation (μSR) and electrical resistivity. The magnetic susceptibility shows a Curie-Weiss behavior above T_M with positive Weiss temperature, and exhibits an antiferromagnetism (AFM)-like cusp anomaly at T_M [16]. Our recent μSR study confirmed that the low-temperature phase is a magnetically ordered state [17]. In the magnetic-field and temperature dependencies of the magnetization, several anomalies were found below T_M, which suggest existence of subphases below T_M [16]. These facts imply that this material has a complex magnetically ordered state. It should be noted that many $M\mathrm{Be}_{13}$ (M = Gd–Er, and Np) compounds show helical magnetic structures at low temperatures and have positive Weiss temperatures; further, $HoBe_{13}$ has a multi-magnetic phase diagram [18, 19, 20]. Consequently, helical magnetic order is one of the candidates for the low-temperature state of $SmBe_{13}$ [16].

In order to understand this low-temperature state in $SmBe_{13}$, it is necessary to clarify the degrees of freedom of the $4f$ electrons. It is suggested that the valence of the Sm ions in this material is nearly trivalent, and that the $4f$ electrons are well localized even at low temperatures [16]. Thus, it is important to determine the crystalline-electric-field (CEF) level scheme of these localized $4f$ electrons of Sm^{3+}. In a cubic CEF, the sextet ($J = 5/2$) of Sm^{3+} splits into a Γ_7 doublet and a Γ_8 quartet. Thus far, the Γ_7 doublet was assumed to be the ground state with a gap of 12.5 and 30 K to the first excited quartet, estimated by use of specific-heat and magnetization measurements, respectively [15, 21]. However, as another possibility, the Γ_8 ground state with a gap of 90 K is proposed from our recent analysis, where the Einstein-phonon contribution with Einstein energy of about 170 K in the specific heat is taken into account properly [16]. Thus, it is desirable to confirm the CEF level scheme by other measurements.

In the present study, we have investigated the phase boundary of the low-temperature phase using pulsed magnetic fields. Since the Sm ion has a relatively small Landé g-factor, high-field measurements are necessary to study the low-temperature state. Indeed, in these measurements, we could clearly observed an elastic anomaly at 43 T and 1.5 K. By ultrasonic measurements in static magnetic fields, we also investigated the CEF level scheme and magnetic-field dependence of the transition temperature, T_M.

2. Experimental Details

A single-crystalline $SmBe_{13}$ sample grown by an Al-flux method was used in the present study. The sample was annealed at 700°C for 2 weeks. Ultrasonic measurements were carried out by means of the conventional pulse-echo method. Ultrasonic waves were excited by $LiNbO_3$ transducers of 100 μm thickness, which were attached on the polished surfaces of the sample. The ultrasonic frequencies of longitudinal C_{11} and transverse C_{44} modes were 108 MHz (3rd harmonics) and 20 MHz (fundamental frequency), respectively. The measurements in static magnetic fields were carried out by using a superconducting magnet. Pulsed magnetic fields up to 61.3 T were generated by a magnet at the Dresden High Magnetic Field Laboratory (Helmholtz- entrum Dresden-Rossendorf).

3. Results

In Figs. 1 and 2, we show the temperature dependence of the elastic constants, $C_{11}(T)$ and $C_{44}(T)$, in zero magnetic field. For both modes, no clear anomaly is observed from 300 K down to T_M. Although an Einstein mode has been observed in specific heat [16], this material does not show any ultrasonic dispersion. One of the possible explanations for the absence of

Journal of Physics: Conference Series **683** (2016) 012032 doi:10.1088/1742-6596/683/1/012032

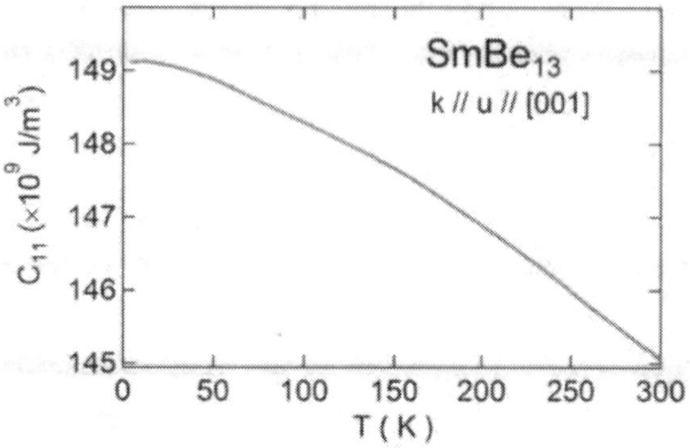

Figure 1. Temperature dependence of C_{11} in zero magnetic field. The wave vector, $\boldsymbol{k}$, and the polarization, $\boldsymbol{u}$, of the sound wave are aligned along [001].

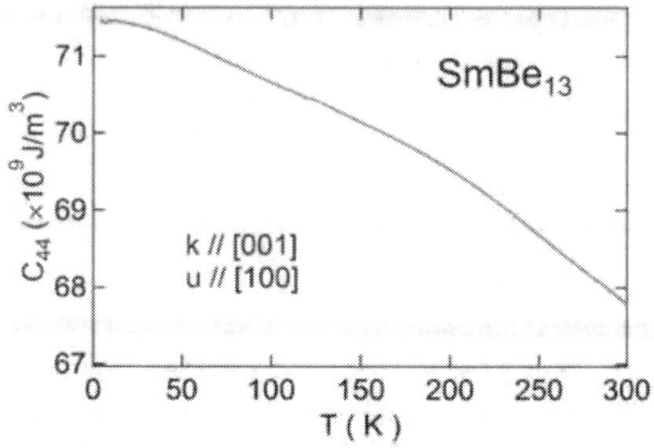

Figure 2. Temperature dependence of C_{44} in zero magnetic field.

ultrasonic dispersions is that the Einstein mode does not couple to the acoustic phonon. Another possibility is that such a dispersion exists only above the present temperature range, *i.e.*, above room temperature. Actually, the Einstein energy of about 170 K is higher than that of filled-skutterudites, which show some ultrasonic dispersions due to rattling [11]. Further ultrasonic measurements at higher temperatures would be needed.

Figure 3 shows $C_{11}(T)$ in various magnetic fields for $H \parallel [001]$. In zero magnetic field, $C_{11}(T)$ shows a softening below 14 K. In low magnetic field, elastic anomaly at T_{M} is not very clear. Below 2 T, we define the T_{M} as the temperature at which $\mathrm{d}C_{11}/\mathrm{d}T$ takes a local maximum, and the error bar is defined as the full width at half maximum. On the other hand, the elastic anomaly at T_{M} is clear above 3 T. Moreover, a difference between zero-field cooling (FC) and field cooling (FC) is observed below T_{Hys} in magnetic fields between 2 and 9 T. These facts suggest that there is a subphase below T_{M} above 2 T. The difference between FC and FC processes has also been observed in our recent magnetization measurements [16].

Here, we consider the origin of the softening of $C_{11}(T)$ below 14 K. This softening could be explained by strain-quadrupole coupling, since the 4f electrons of Sm^{3+} can have quadrupole degrees of freedom [22]. Since the Γ_8 quartet has quadrupole degrees of freedom, a softening toward low temperatures is expected in the case of a Γ_8 quartet CEF ground state. Thus, a Γ_8 ground state is one of the possible CEF models. It is also possible that the elastic constant show a softening in the case of a Γ_7 doublet ground state with a small gap of about 12.5 K to the Γ_8 quartet excited state. On the other hand, Γ_7(0 K)–Γ_8(30 K) is inconsistent with our result, because in this case the elastic constant is expected to have a local minimum at around 15 K and should not show a softening below 15 K.

In Fig. 4, $C_{44}(T)$ in various fields is shown. Elastic anomaly in $C_{44}(T)$ at T_{M} is not so clear compared with that of $C_{11}(T)$, and we cannot determine T_{M} from $C_{44}(T)$. The arrows in the figure represent the T_{M} determined from $C_{11}(T)$. $C_{44}(T)$ also shows softening at low temperatures. Compared to the softening of $C_{11}(T)$, the softening starts rapidly near T_{M}, which suggests that this softening is caused by the phase transition and/or precursory phenomenon of the phase transition, *i.e.*, short-range ordering. This softening of $C_{44}(T)$ is enhanced in a magnetic field of about 5 T, where the difference between FC and FC in $C_{11}(T)$ is largest, and suppressed at higher magnetic fields. This significant change in the behavior of $C_{44}(T)$ also implies a modulation of the low-temperature ordered state by the magnetic field.

Journal of Physics: Conference Series **683** (2016) 012032 doi:10.1088/1742-6596/683/1/012032

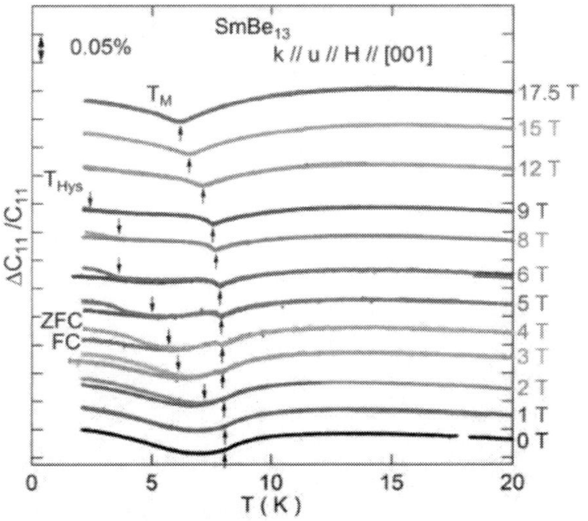

Figure 3. Temperature dependence of C_{11} in various magnetic fields.

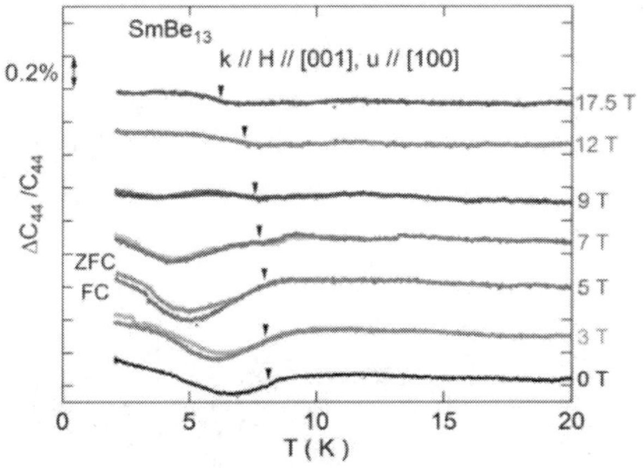

Figure 4. Temperature dependence of C_{44} in various magnetic fields. The arrows indicate T_{M} determined from C_{11}.

In Fig. 5, we show the relative change of $C_{11}(H)$ observed in pulsed magnetic fields. Note that the absolute value of the elastic constant is always the same before and after pulsed magnetic fields. Thus, eddy-current heating is negligible. At 1.5 K, $C_{11}(H)$ exhibits a minimum at $H_{\mathrm{M}} \approx 43$ T as indicated by a dashed line. A broad minimum is also seen at around 32 T in the data obtained at 4.2 K. Since the temperature dependence of the elastic constant, $C_{11}(T)$, has a local minimum at T_{M} in high magnetic fields, this minimum in the pulsed-field data could be due to the phase transition to the paramagnetic state.

Figure 6 shows the H-T phase diagram obtained from our C_{11} data in static and pulsed magnetic fields. As can be seen, T_{M} is clearly reduced with applied magnetic fields. The extrapolated phase boundary from the low-field region nicely merge with the magnetic field, H_{M}.

Next, we show a model calculation of the phase-transition fields for AFM order by means of a mean-field approximation. Since Sm ions locate at 8a sites, which are the lattice points of a simple cubic structure when only Sm ions are considered, we assume G-type AFM state of simple cubic structure, *i.e.*, the directions of neighboring magnetic moments are opposite with each other. In our calculation, we consider that the $4f$ electrons are well localized at Sm^{3+}, which is confirmed by our recent physical property measurements [16]. Here, we take into account only the ground-state multiplet of $J = 5/2$, and the Hamiltonian is

$$H = H_{\mathrm{CEF}} - K\boldsymbol{J}\cdot\langle\boldsymbol{J}\rangle + g_J\mu_B\boldsymbol{J}\cdot\boldsymbol{H}. \tag{1}$$

The inter-site coupling constant, K, is assumed to be -4 and -3.07 K for free Sm-ions and the Γ_8(0 K)–Γ_7(90 K) CEF-state case, respectively, which cause a transition at about 8 K in each case. By computing the self-consistent equation in the case of applying magnetic fields along [001], we obtained following results. In the case of free Sm^{3+} ions, *i.e.*, $H_{\mathrm{CEF}} = 0$, AFM moments are aligned perpendicular to external magnetic fields, and a magnetic field of 75 T is estimated to suppress the AFM at 1.5 K. In the case of the Γ_8(0 K)–Γ_7(90 K) CEF state, magnetic moments are aligned along the easy axis ⟨100⟩, and a magnetic field of 57 T would be needed to break the AFM order. These estimated values of magnetic field regarding to the upper

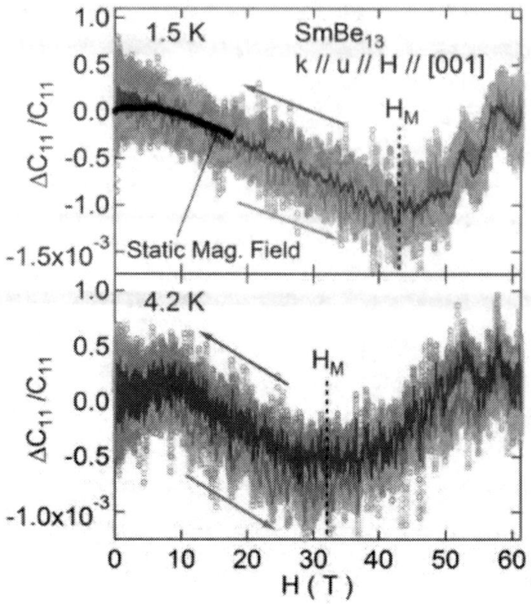

Figure 5. Relative change of C_{11} in pulsed magnetic field. Light-colored circles represent raw data, and deep-colored lines are smoothed curves.

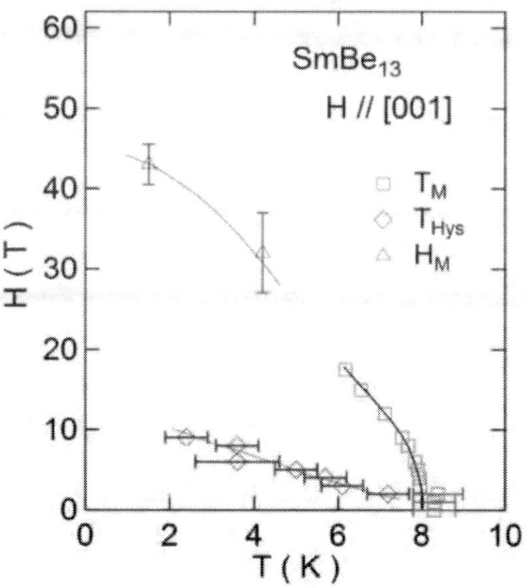

Figure 6. *H*-*T* phase diagram obtained from the result of our C_{11} measurements. The lines are guide to the eye.

phase boundary of G-type AFM phase are larger than the present experimental result of 43 T. This mismatch between the experimentally and theoretically obtained upper-phase boundary implies that the actual ordered state of $SmBe_{13}$ in high magnetic fields for $H \parallel [001]$ is not a simple G-type AFM state. It is also possible that multipole degrees of freedom play also a role at high magnetic fields, since the $4f$-electron system has multipole degrees of freedom in the case of Γ_8 CEF ground state. For further discussion, measurements in pulsed magnetic fields aligned along other axes would be necessary.

4. Summary

We have performed ultrasonic measurements for a single-crystalline $SmBe_{13}$ in static and pulsed magnetic fields in order to shed more light on the low-temperature magnetic order of this compound. We have observed an elastic anomaly at 43 T at 1.5 K, which could be an upper phase boundary of the magnetically ordered state. The relative weakness of the low-temperature phase against magnetic fields, compared to results using mean-field calculations, strongly suggests that the ordered state is not simple AFM. By measurements in static fields, some indication for the existence of a subphase at lower temperatures was found, which has also been suggested by magnetization measurements.

5. Acknowledgment

We acknowledge the support of the HLD at H DR, member of the European Magnetic Field Laboratory (EMFL). Research at HLD was supported by a Grant-in-Aid-for Scientific Research (Grant No.26400342) and The Strategic Young Reseacher Overseas Visits Program for Accelerating Brain Circulation from the Japan Society for Promotional Science.

References

[1] Jayaraman A, Narayanamurti V, Bucher E and Maines R G 1970 *Phys. Rev. Lett.* **25** 368.
[2] Jayaraman A, Narayanamurti V, Bucher E and Maines R G 1970 *Phys. Rev. Lett.* **25** 1430.
[3] Kasaya M, Liu B, Sera M, Kasuya T, Endoh D, Goto T and Fujimura T 1985 *J. Magn. Magn. Mater.* **52** 289.
[4] A. Menth, E. Buehler, and T. H. Geballe 1969 *Phys. Rev. Lett.* **22** 295.
[5] Dzero M, Sun K, Galitski V and Coleman P 2010 *Phys. Rev. Lett.* **104** 106408.
[6] Aoki Y, Sanada S, Kikuchi D, Sugawara H and Sato H 2008 *Physica* B **403** 1574.
[7] Sanada S, Aoki Y, Aoki H, Tsuchiya A, Kikuchi D, Sugawara H and Sato H 2005 *J. Phys. Soc. Jpn.* **74** 246.
[8] Higashinaka R, Maruyama T, Nakama A, Miyazaki R, Aoki Y and Sato H 2011 *J. Phys. Soc. Jpn.* **80** 093703.
[9] Sakai A and Nakatsuji S 2011 *Phys. Rev.* B **84** 201106(R).
[10] Goto T, Nemoto Y, Sakai K, Yamaguchi T, Akatsu M, Yanagisawa T, Hazama H, Onuki K, Sugawara H and Sato H 2004 *Phys. Rev.* B **69** 180511(R).
[11] Yanagisawa T, Saito H, Hidaka H, Amitsuka H, Araki K, Akatsu M, Nemoto Y, Goto T, Ho P-C, Baumbach R E and Maple M B 2008 *J. Phys. Soc. Jpn.* **77** 074607.
[12] Ogita N, Kondo T, Hasegawa T, Takasu Y, Udagawa M, Takeda N, Ishikawa K, Sugawara H, Kikuchi D, Sato H, Sekine C and Shirotani I 2006 *Physica* B **383** 128.
[13] Matsuhira K, Sekine C, Wakeshima M, Hinatsu Y, Namiki T, Takeda K, Shirotani I, Sugawara H, Kikuchi D and Sato H 2009 *J. Phys. Soc. Jpn.* **78** 124601.
[14] Hattori K, Hirayama Y and Miyake K 2005 *J. Phys. Soc. Jpn.* **74** 3306.
[15] Bucher E, Maita J P, Hull G W, Fulton R C and Cooper A S 1975 *Phys. Rev.* B **11** 440.
[16] Hidaka H to be submitted.
[17] in preparation.
[18] Vigneron F 1991 *Phys. Scr.* **44** 27.
[19] Hiess A, Bonnet M, Burlet P, Ressouche E, Sanchez J -P, Waeren-borgh J C, Zwirner S, Wastin F, Rebizant J, Lander G H and Smith J L 1996 *Phys. Rev. Lett.* **77** 3917.
[20] Dervenagas P, Burlet P, Bonnet M, Bourdarot F, Hiess A, Bud'ko S L, Caneld P C, Lander G H, Kim J S and Stewart G R 2000 *Phys. Rev.* B **61** 405.
[21] Besnus M J, Panissod P, Kappler J P, Heinrich G and Meyer A 1983 *J. Magn. Magn. Mater.* **31–34** 227.
[22] Goto T and Lüthi B 2003 *Adv. Phys.* **52** 67.

Transport properties of single-crystalline Ising magnet $SmPt_2Si_2$

K. Fushiya, T. D. Matsuda, R. Higashinaka, and Y. Aoki
Department of Physics, Tokyo Metropolitan University, Hachioji, Tokyo 192-0397, Japan
E-mail: fushiya-kengo@ed.tmu.ac.jp, aoki@tmu.ac.jp

Abstract. The electronic transport properties of a $SmPt_2Si_2$ single crystal are measured, in which magnetically disordered (paramagnetic) Sm ions are expected to remain partially in the antiferromagnetically (AFM) ordered state occurring below $T_I = 5.1$ K. In the paramagnetic state, the resistivity exhibits a shallow minimum at $\sim$ 11 K and a pronounced negative magnetoresistance, suggesting the occurrence of the Kondo effect and/or AFM short-range ordering. Below T_I, the resistivity increases sharply, indicating a decrease in the carrier density caused by a superzone gap formation associated with the AFM transition. As regards the Hall effect, the extraordinary component is negligibly smaller than the normal component. The positive sign of the normal Hall coefficient indicates that the hole Fermi surfaces dominate the electrical transport properties. The difference in the transport properties of the two ordered phases is discussed.

1. Introduction

Over the past decade, there has been increased interest in the strongly correlated electron behaviors appearing in Sm-based intermetallic compounds. Typical examples are the unusual magnetic-field-insensitive heavy-fermion behavior in $SmOs_4Sb_{12}$ [1, 2, 3], the magnetic-field-insensitive phase transitions and significantly enhanced Sommerfeld coefficients in SmT_2Al_{20} [4, 5, 6], the metal-insulator (MI) transition and magnetic-field-induced charge ordering in $SmRu_4P_{12}$ [7, 8, 9], and the MI transition under pressure in $SmX (X = S$, Se, and Te) [10, 11, 12].

RT_2X_2 (R: rare earth; T: transition metal; X: Si and Ge) materials exhibit a variety of exotic strongly correlated electron behaviors. For R = Sm, however, there are only a few reports concerning systematic physical property investigations using single crystals [13]. Recently, we succeeded in growing $SmPt_2Si_2$ single crystals, and determined that this material is a strong Ising magnet [14]. The magnetic field H vs temperature T phase diagram determined for $H \parallel c$ is shown in figure 1. Under zero-field conditions, an antiferromagnetic (AFM) ordering (named phase I) appears at $T_I = 5.1$ K. In phase I, a Curie term remains in the magnetic susceptibility $\chi(T)$, indicating the partial existence of magnetically disordered (paramagnetic) Sm ions [14]. The large electronic specific heat coefficient $\gamma = 350$ mJ/K^{2}mol in phase I indicates the existence of heavy quasiparticles, which may be formed on the sublattice of partially magnetically disordered Sm ions. A metamagnetic anomaly appears at the boundary between phases I and II. The magnetic structure of phase II is expected to be $\uparrow\uparrow\downarrow$ from the magnetization M value. In this paper, we investigate the electronic transport properties of $SmPt_2Si_2$ using a single crystal.

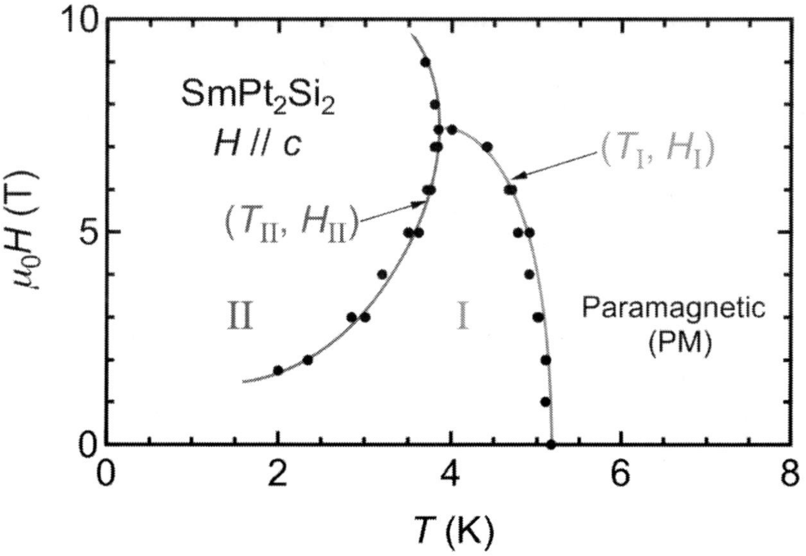

Figure 1. H-vs-T phase diagram of $SmPt_2Si_2$ for $H \parallel c$ (data below 7 T taken from Ref. [14] and data above 7 T determined from the present measurements).

2. Experimental

$SmPt_2Si_2$ single crystals were grown using the Sn-flux method and raw materials of 3N (99.9 pure) Sm, 4N Pt, 4N Si, and 5N Sn. These materials were inserted in an alumina crucible with an off-stoichiometric composition of Sm:Pt:Si:Sn = 1:3:1:40, and sealed in a quartz tube. The quartz tube was heated to 1150 °C, maintained at this temperature for 1 day, and then cooled to 650 °C at a rate of -2 °C/h, which required approximately 10 days in total. Single crystals were obtained by spinning the ampoule in a centrifuge in order to remove excess Sn flux. The transport properties were measured using a standard AC four-probe technique in a commercial Physical Property Measurement System (Quantum Design Inc.).

3. Results and discussion

The temperature dependence of the electrical resistivity $\rho(T)$ of $SmPt_2Si_2$ measured for the current $j \parallel a$ under zero-field conditions is shown in figure 2 (a). The residual resistivity ratio RRR ($\equiv \rho(300\text{ K})/\rho(2\text{ K})$) is 2.3. The overall temperature dependence of ρ exhibits a metallic behavior with a slight downward curvature. There is no anomaly associated with a charge-density-wave transition, such as that observed in $LaPt_2Si_2$ at 112 K [15].

The low-temperature $\rho(T)$ data measured under various magnetic fields is shown in figure 2 (b). Under zero-field conditions, ρ exhibits a steep increase below T_{I}, indicating a carrier-density decrease caused by a superzone gap formation associated with the AFM transition. The magnitude of the increase in ρ is only approximately 1 . According to band-structure calculations [16, 17], $SmPt_2Si_2$ has five Fermi surfaces with strong two-dimensional characteristics. Considering the present $j \parallel a$ configuration, the observed small increase in ρ may indicate that the magnetic ordering wave vector is parallel to the c axis.

Under applied magnetic fields of $\mu_0 H > 2$ T, phase II appears in the measured temperature range. At the I → II phase transition, ρ exhibits a decrease. This change in ρ is more clearly shown in the field dependence of ρ (see figure 2 (c)). For $\mu_0 H > 7.5$ T, phase I does not exist.

At the paramagnetic (PM) $\rightarrow$ II trandition, ρ exhibits a sharp increase, indicating the first-order character of the transition.

Under zero-field conditions, $\rho(T)$ exhibits a shallow minimum at 11 K ($\equiv T_{\rm min}$). This feature indicates the existence of a contribution with $d\rho/dT<0$, which can be attributed to the Kondo effect and/or a short-range ordering possibly present above $T_{\rm I}$. This interpretation is supported by pronounced negative magnetoresistance (MR) appearing at $T>T_{\rm I}$. Figure 2 (d) shows $\Delta\rho \equiv \rho(H)-\rho(0)$ vs T for various applied fields. It is apparent from this figure that the negative MR develops with decreasing temperature. As reported in Ref. [14], the $4f$-electron contribution to the entropy S_{4f} at $T_{\rm I}$ is 4.6 J/mol·K, which is 80 of $R\ln 2$. This decrease in S_{4f} can be attributed to the Kondo effect and/or the short-range ordering. Comparison of $S_{4f}(T_{\rm I}) = 4.6$ J/mol·K with a single-impurity Kondo model calculation [18] provides the upper bound of the Kondo temperature $T_{\rm K}$, which is 2.3 K.

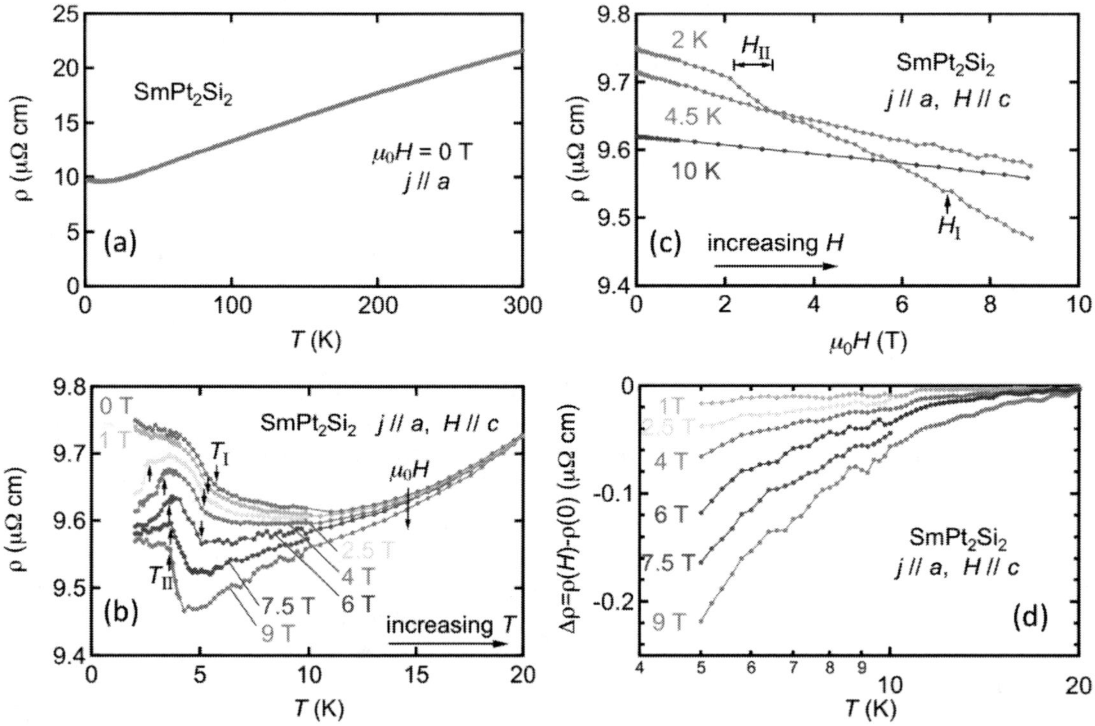

Figure 2. Temperature dependence of resistivity ρ for $j \parallel a$ (a), low-temperature part of ρ for various magnetic fields (b), H dependence of ρ at 2, 4.5, and 10 K (c), and transverse magnetoresistance $\Delta\rho \equiv \rho(H)-\rho(0)$ vs T (d).

The temperature dependence of the Hall coefficient $R_{\rm H} \equiv \rho_{\rm H}/H$, where $\rho_{\rm H}$ represents the Hall resistivity, and that of the Hall mobility $\mu_{\rm H} = R_{\rm H}/\rho$ are shown in figure 3 (a). Usually, the $\rho_{\rm H}$ of magnetic materials is expressed as $\rho_{\rm H} = R_0 H + R_{\rm S} M$, i.e., the sum of the normal component originating from the cyclotron motion of the conduction electrons and the extraordinary component caused by the magnetic scattering of the conduction electrons from the magnetic ions [19]. Comparison of $R_{\rm H}(T)$ with $\chi(T)$, as shown in figure 3 (b), indicates that the extraordinary component is extremely weak in $SmPt_2Si_2$. $R_{\rm H}(T)$ gradually increases with decreasing temperature and begins to saturate below $\sim$ 50 K, while $\chi(T)$ exhibits a strong temperature dependence below $\sim$ 50 K, corresponding to the Curie-Weiss behavior; note that the two components in ρ_H cannot be distinguished based on the field dependence, because both ρ_H and M exhibit almost linear H dependence (not shown). Therefore, we assume $R_H \simeq R_0$ in the following analysis.

$SmPt_2Si_2$ is a compensated metal, i.e., it has equal numbers of electrons and holes. Therefore, we use a two-carrier model. (Note that this is a simplified model, as $SmPt_2Si_2$ has five Fermi surfaces [16, 17].) In the measured temperature range, R_H is always positive. This feature indicates that holes dominate the transport properties; the hole mobility is higher than that of the electrons (in the two-carrier model, $\mu_H = \mu_h - \mu_e > 0$, where μ_h and μ_e represent the hole and electron mobilities, respectively). At room temperature, $R_H = 1.22 \times 10^{-10}$ m^3/C (this value corresponds to a carrier density of $\sim$ 8.87/f.u. in a single-carrier model). R_H increases with decreasing temperature and almost saturates at a value of 2.3×10^{-10} m^3/C below $\sim$ 20 K in the PM state. As shown in figure 3 (c), R_H exhibits an increase below T_I and, at T_{II}, it exhibits a further increase. This feature is also apparent in the ρ_H-vs-H curve measured at 2 K shown in figure 3 (d). This observation indicates that the dominance of the hole mobility increases in a stepwise manner at the PM→I→II phase transitions.

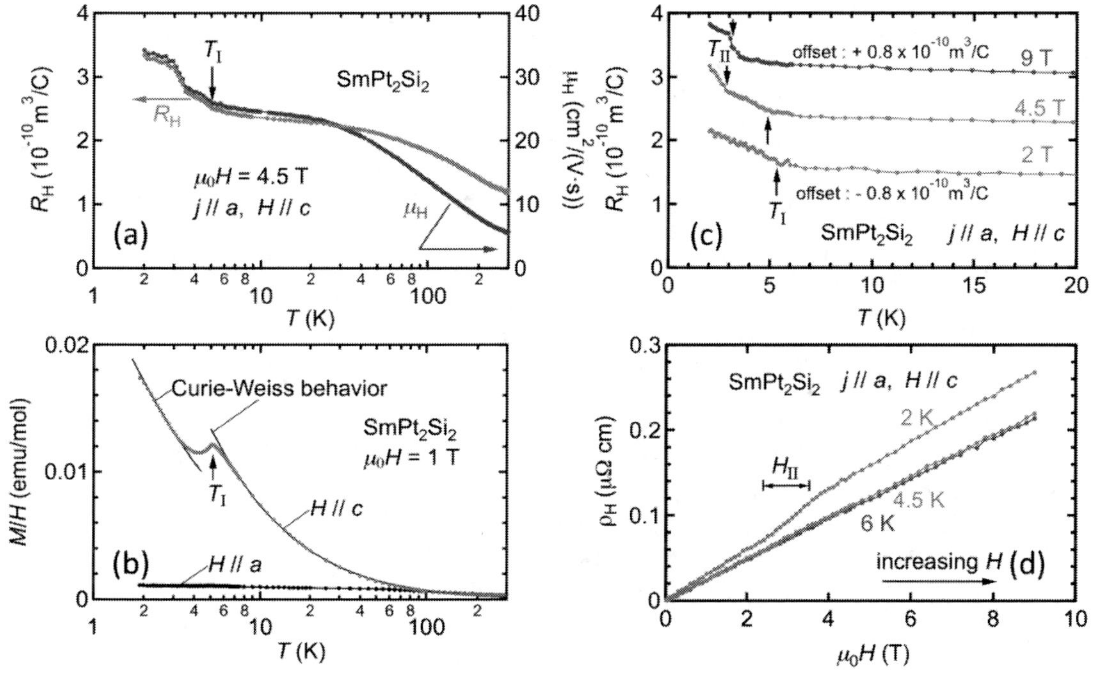

Figure 3. Temperature dependences of Hall coefficient R_H and Hall mobility $\mu_H = R_H/\rho$ of $SmPt_2Si_2$ measured at 4.5 T (a), χ for $H \parallel a$ and $H \parallel c$ (data taken from Ref. [14]) (b), low-temperature part of $R_H(T)$ at 2, 4.5, and 9 T (c), and H dependence of ρ_H at 2, 4.5, and 6 K (d).

4. Summary

The electronic transport properties of a $SmPt_2Si_2$ single crystal have been measured. In the paramagnetic state, the resistivity ρ exhibits a shallow minimum at $\sim$ 11 K and a pronounced negative magnetoresistance, suggesting the occurrence of the Kondo effect and/or AFM short-range ordering. Below T_1, ρ increases sharply, indicating a decrease in the carrier density caused by a superzone gap formation associated with the AFM transition. As regards the Hall effect, the extraordinary component is negligibly smaller than the normal component. The positive sign of the normal component indicates that the hole Fermi surfaces dominate the electrical transport properties. This feature is further enhanced in the two magnetically ordered states.

Acknowledgments

We thank Prof. H. Sato for valuable discussions and S. Iwami for his technical support. This work was supported by a Grant-in-Aid for Scientific Research on Innovative Areas Heavy Electrons (No. 20102007) from the Ministry of Education, Culture, Sports, Science, and Technology, by Grants-in-Aid for Scientific Research C (No. 23540421) and for Young Scientists B (No. 21740261) from the Japan Society for the Promotion of Science and by JSPS KAKENHI grant numbers: 15J07736, 15H03693, and 15H05884 (J-Physics).

References

[1] Sanada S, Aoki Y, Aoki H, Tsuchiya A, Kikuchi D, Sugawara H and Sato H 2005 *J. Phys. Soc. Jpn.* **74** 246
[2] Yuhasz M W, Frederick A N, Ho -C P, Butch P N, Taylor J B, Sayles A T, Maple B M, Betts B J, Lacerda H A, Rogl P and Giester G 2005 *Phys. Rev.* B **71** 104402
[3] Fushiya K, Miyazaki R, Higashinaka R, Yamada A, Mizumaki M, Tsutsui S, Nita K, Uruga T, Suemitsu B, Sato H and Aoki Y 2015 *Phys. Rev.* B **92** 075118
[4] Higashinaka R, Maruyama T, Nakama A, Miyazaki R, Aoki Y and Sato H 2011 *J. Phys. Soc. Jpn.* **80** 093703
[5] Sakai A and Nakatsuji S 2011 *Phys. Rev.* B **84** 201106(R)
[6] Yamada A, Higashinaka R, Miyazaki R, Fushiya K, Matsuda D T, Aoki Y, Fujita W, Harima H and Sato H 2013 *J. Phys. Soc. Jpn.* **82** 123710
[7] Sekine C, Uchiumi T, Shirotani I and Yagi T in *Science and Technology of High Pressure* ed. Manghnani H M, Nellis J W and Nicol F M (Universities Press Hyderabad, 2000) p. 826
[8] Matsuhira K, Hinatsu Y, Sekine C, Togashi T, Maki H, Shirotani I, Kitazawa H, Takamasu T and Kido G 2002 *J. Phys. Soc. Jpn.* **71** Suppl., p. 237
[9] Matsumura T, Michimura S, Inami T, Hayashi Y, Fushiya K, Matsuda D T, Higashinaka R, Aoki Y and Sugawara H 2014 *Phys. Rev.* B **89** 161116(R)
[10] Lapierre F, Hass C K and Ehrenreich H 1981 *Solid State Commun.* **40** 347
[11] Elmiger W M and Wachter P 1987 *J. Magn. Magn. Mater.* **63** 612
[12] Jayaraman A, Narayanamurti V, Bucher E and Maines G R 1970 *Phys. Rev. Lett.* **25** 368
[13] Hiebl K and Rogl P 1985 *J. Magn. Magn. Mater.* **50** 39
[14] Fushiya K, Matsuda D T, Higashinaka R, Akiyama K and Aoki Y 2014 *J. Phys. Soc. Jpn.* **83** 113708
[15] Nagano Y, Araki N, Mitsuda A, Yayama H, Wada H, Ichihara M, Isobe M and Ueda Y 2013 *J. Phys. Soc. Jpn.* **82** 064715
[16] Hase I and Yanagisawa T 2013 *Physica.* C **484** 59
[17] Kim S, Kim K and Min I B 2015 *Sci. Rep.* **5** 15052
[18] Tsvelick M A and Wiegmann B P 1983 *Advances in Physics* **32** 4
[19] Hurd M C in *The Hall Effect and Its Applications* ed. Chien L C and Westgate R C (Plenum, New York, 1980) p.1.

Journal of Physics: Conference Series **683** (2016) 012034 doi:10.1088/1742-6596/683/1/012034

Low temperature superconducting phase of the ternary pnictide $SrPd_2Sb_2$ with $CaBe_2Ge_2$-type structure

N Kase[1], H Suzuki[1], T Tsukamoto[2], T Nakano[1], and N Takeda[2]

[1]Graduate school of science and technology, Niigata University, Niigata 950-2102, Japan
[2]Department of materials science and technology, Niigata University, Niigata 950-2102, Japan

E-mail: n-kase@eng.niigata-u.ac.jp

Abstract. We find that low temperature superconducting phase (LSP) of $SrPd_2Sb_2$ with the $CaBe_2Ge_2$-type structure at T_c = 1.4 K. Superconducting transition is found at T_c = 1.4 K in electrical resistivity $\rho(T)$. Specific heat $C(T)$ shows a clear anomaly at T_c = 1.2 K, which shows that superconductivity is of bulk nature. From $\rho(T)$ in several magnetic fields, the upper critical field $\mu_0 H_{c2}(0)$ is estimated to be 36 mT, which is strongly suppressed, and is 16.7 times smaller than that of 1.9 K phase.

1. Introduction

Recently, new superconductor $SrPt_2As_2$ with the $CaBe_2Ge_2$-type structure is reported at superconducting transition temperature T_c = 5 K, and the superconductivity coexists with charge-density wave[1]. The T_c of $SrPt_2As_2$ is relatively high among other non-ironbased 122 arsenides.

It is known that $SrPd_2Sb_2$ possesses two polymorphs[2], namely, low-temperature phase (LT-$SrPd_2Sb_2$) with $CaBe_2Ge_2$-type structure and high-temperature phase (HT-$SrPd_2Sb_2$) with the $ThCr_2Si_2$-type structure. We reported both polymorphs of $SrPd_2Sb_2$ show superconductivity at T_c = 1.9 K (LT-$SrPd_2Sb_2$) and 0.6 K (HT-$SrPd_2Sb_2$)[3]. T_c of LT-$SrPd_2Sb_2$ is 3 times higher than that of HT-$SrPd_2Sb_2$.

Furthermore, we newly find that LT-$SrPd_2Sb_2$ includes two superconducting phase, namely 1.4 and 1.9 K phase. We report here new superconducting phase of LT-$SrPd_2Sb_2$ with 1.4 K phase and its superconducting state by performing electrical resistivity $\rho(T)$, and specific heat $C(T)$ measurements. In this article, low (high) temperature superconducting phase of $SrPd_2Sb_2$ is represented by LSP(HSP)-$SrPd_2Sb_2$; LT-$SrPd_2Sb_2$ includes both LSP-$SrPd_2Sb_2$ (1.4 K) and HSP-$SrPd_2Sb_2$ (1.9 K) samples. $SrPd_2Sb_2$ has two polymorphs and three superconducting phases, HT-$SrPd_2Sb_2$ (T_c = 0.6 K, $ThCr_2Si_2$-type structure), LSP-$SrPd_2Sb_2$ (T_c = 1.4 K, $CaBe_2Ge_2$-type structure), and HSP-$SrPd_2Sb_2$ (T_c = 1.9 K, $CaBe_2Ge_2$-type structure).

2. Experimental Details

Polycrystalline samples of LSP- and HSP-$SrPd_2Sb_2$ were synthesized from a solid-state reaction. The samples were treated in Ar-filled grove box, to avoid degradation of Sr element due to reaction with air and/or moisture. The stoichiometric mixture of Sr (ingot, 99.9%), Pd (powder,

99.9%), and Sb (shot, 99.99%) was sealed in evacuated quartz tube. The ampule was heated at 900 ℃ maintained for 20 h, and then at 720 ℃ for 72 h, followed by natural cooling to room temperature. To synthesize the sintered pellets, samples were ground, pelletized, and heated at 900 ℃ in an evacuated quartz tube, and then treated as before. Appearance of the poly-crystalline LSP- and HSP-$SrPd_2Sb_2$ sample is a brass color.

The crystal structure of the synthesized polycrystalline sample was examined by the powder X-ray diffraction (XRD) technique using a conventional X-ray spectrometer equipped with Cu-K radiation and a monochromator (RAD-2X, Rigaku). The intensity data were collected over a 2θ range of 10°–80° with a step width of 0.01° and a counting rate of 4°/minutes.

Electrical resistivity $\rho(T)$ was measured by a dc-four-probe method using delta mode with a current source (Model 6221, Keithley) and a nano voltmeter (Model 2182A, Keithley), down to 0.3 K in zero and various applied magnetic fields. Specific heat $C(T)$ was measured by an adiabatic heat-pulse and relaxation method down to 0.3 K using a commercial ^{3}He refrigerator (Heliox-VL, Oxford Instruments).

3. Experimental Results

3.1. X-ray diffraction pattern

Figure 1 shows powder-XRD patterns of poly-crystalline LSP- and HSP-$SrPd_2Sb_2$. These peaks can be indexed with a tetragonal unit cell with the space group of $P4/nmm$. Minor extra peaks due to impurity phase can be detected from the powder-XRD diffraction of both $SrPd_2Sb_2$ samples. We cannot assign what compound is the minor impurity phase. The lattice constants obtained from all indexes are a = 0.4653 nm, c = 1.0629 nm for LSP-$SrPd_2Sb_2$, and a = 0.4637 nm, c = 1.0668 nm for HSP-$SrPd_2Sb_2$, respectively. The cell volume of LSP-$SrPd_2Sb_2$ (0.2301 nm^3) are larger than that of HSP-$SrPd_2Sb_2$ (0.2294 nm^3).

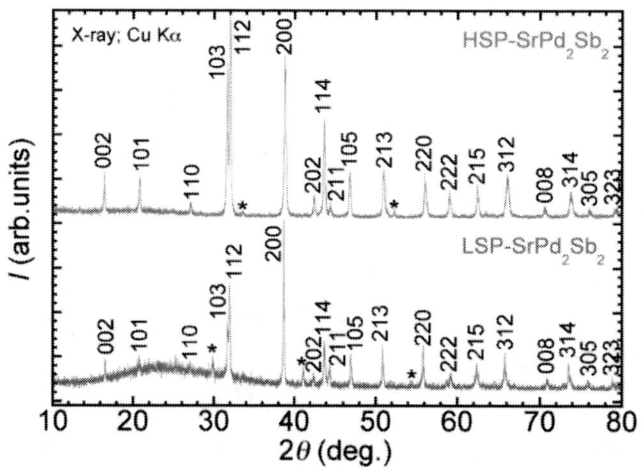

Figure 1. Powder X-ray diffraction patterns of HSP-$SrPd_2Sb_2$(upper) and LSP-$SrPd_2Sb_2$ (down) at room temperature.

3.2. electrical resistivity

Figure 2 shows T-dependence of electrical resistivity $\rho(T)$ of LSP-$SrPd_2Sb_2$ with a current value of I = 0.5 mA and a sample size of 1.5×0.7×0.2 mm^3. The residual resistivity ratio (RRR) is estimated to be 18 and 1.65, as shown in the inset of Fig. 2. The RRR of LSP-$SrPd_2Sb_2$ is higher than that of HSP-$SrPd_2Sb_2$, which indicates that sample quality of LSP-$SrPd_2Sb_2$

Journal of Physics: Conference Series **683** (2016) 012034 doi:10.1088/1742-6596/683/1/012034

is better. From the powder-XRD patterns, minor impurity phase in LSP-$SrPd_2Sb_2$ seems to be larger than that in HSP-$SrPd_2Sb_2$. We consider that the difference in RRR is originated from sample condition in $\rho(T)$ measurements. Sample of LSP-$SrPd_2Sb_2$ was sintered at 900 ℃ after the sample was synthesized by solid state reaction. On the other hand, bulk sample of HSP-$SrPd_2Sb_2$ was not sintered. Thus, condition of sample in $\rho(T)$ is relatively different, and strongly affects the difference in RRR.

Superconductivity is observed at $T_c = 1.4$ K in LSP-$SrPd_2Sb_2$. Zero resistivity was observed at around 1.35 K in $\rho(T)$. The superconducting transition width ΔT is about 0.05 K. The sharp transition suggests good quality of sample.

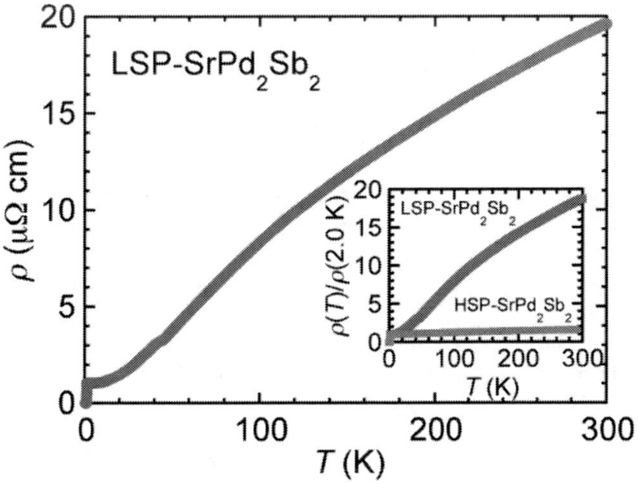

Figure 2. T-dependence of electrical resistivity $\rho(T)$ of LSP-$SrPd_2Sb_2$ with current value of $I = 0.5$ mA. Inset shows $\rho(T)$ normalized by $\rho(2.0\ K)$ of LSP- and HSP-$SrPd_2Sb_2$.

3.3. specific heat
Figure 3 shows T-dependence of specific heat $C(T)$ divided by temperature T of LSP-$SrPd_2Sb_2$ in 0 and 0.05 T. Below 0.6 K, $C(T)$ exhibits an upturn due to a quadrupole splitting of the ^{121}Sb ($I = 5/2$) and ^{123}Sb ($I = 7/2$) nuclear spins. The normal state specific heat can be fitted by the equation $C/T = \gamma + \beta T^2$, where γ is the electronic specific heat coefficient (Sommerfeld constant), and $C_{ph} = \beta T^2$ is the phonon contribution. The parameters were obtained as $\gamma = 13.1\ mJ/(mol{\cdot}K^2)$, $\beta = 7.82\ mJ/(mol{\cdot}K^4)$ for LSP-$SrPd_2Sb_2$. In a simple Debye model of the phonon contribution, the β coefficient is related to the Debye temperature Θ_D according to the equation $\Theta_D = (\frac{12\pi^4}{5\beta}nR)^{1/3}$, where $R = 8.314\ J/(mol{\cdot}K)$ and $n = 5$ for $SrPd_2Sb_2$. From this relationship, we calculated $\Theta_D = 108$ K, for LSP-$SrPd_2Sb_2$.

$C(T)$ shows clear transition at 1.2 K, which suggests that superconductivity of 1.2 K phase is of bulk nature. $\Delta C/\gamma T_c$ is calculated to be 1.0, which is smaller than that of weak-coupling limit.

3.4. upper critical field
The upper critical field $\mu_0 H_{c2}(T)$ is determined by $\rho(T)$ in several magnetic fields. The H-T phase diagram deduced from $\rho(T)$ is presented in Fig. 4. T_c of $\rho(T)$ is defined as the temperature at which the resistivity drops to 50% of ρ_0. $\mu_0 H_{c2}(T)$ linearly increases

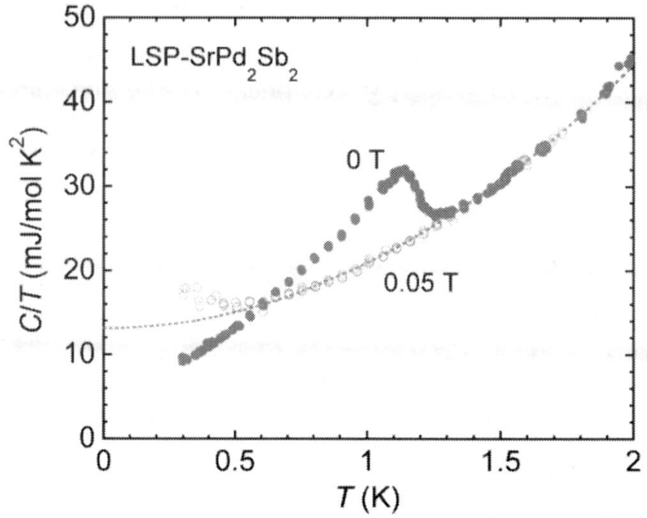

Figure 3. T-dependence of specific heat $C(T)$ of LSP-$SrPd_2Sb_2$ in 0 and 0.05 T. Dotted line shows the fitted result of $C(T)/T = \gamma + \beta T^2$.

with decreasing temperature. This behavior can not be fitted with the Wertharmer-Helfand-Hohenberg (WHH) theory in dirty limit ($h = 0.69$)[4, 5], and is also observed in the $\mu_0 H_{c2}(T)$ of the polycrystalline $SrPt_2As_2$[1]. From the linear extrapolation, $\mu_0 H_{c2}(0)$ is determined to be 36 mT for LSP-$SrPd_2Sb_2$. $\mu_0 H(T)$ of LSP-$SrPd_2Sb_2$ is strongly suppressed from that of HSP-$SrPd_2Sb_2$ ($\mu_0 H_{c2}(0) = 0.6$ T), which is 16.7 times larger[3]. From the relationship $\mu_0 H_{c2}(0) = \Phi_0 / 2\pi\xi(0)^2$, $\xi(0)$ was calculated to be 95.6 nm.

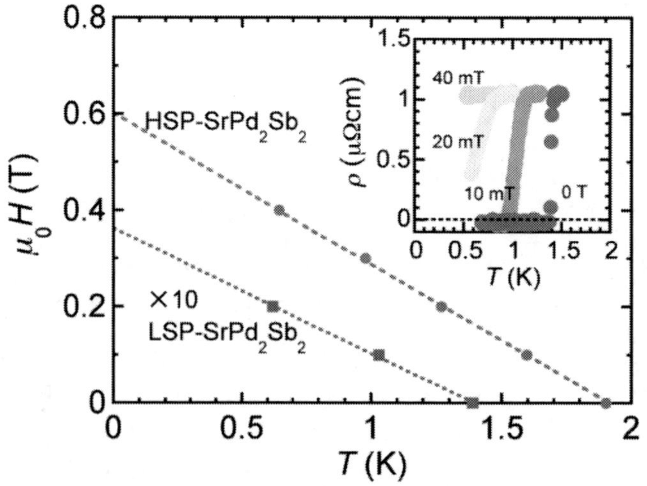

Figure 4. T-dependence of upper critical field $\mu_0 H_{c2}(T)$ of LSP- and HSP-$SrPd_2Sb_2$. Inset shows T-dependence of electrical resistivity $\rho(T)$ of LSP-$SrPd_2Sb_2$ in several magnetic fields.

4. Discussion

We discovery new superconducting phase of $SrPd_2Sb_2$ with $CaBe_2Ge_2$-type structure at $T_c = 1.4$ K. The low temperature superconducting phase (LSP) is clearly different from high temperature superconducting phase (HSP). The upper critical field $\mu_0 H_{c2}(0)$ of LSP-$SrPd_2Sb_2$ is estimated to be 36 mT, which is strongly suppressed from that of HSP-$SrPd_2Sb_2$. Lattice constants of LSP-$SrPd_2Sb_2$ are clearly different, and slightly larger than those of HSP-$SrPd_2Sb_2$. The origin of the difference in T_c of both phases still remains to be clarified.

Acknowledgments

This work was partially supported by the Grant for Promotion of Niigata University Research Projects (No. 524501), the Grant-in-Aid for Young Scientists (B) (No. 26800183) from the Ministry of Education, Culture, Sports, Science, and Technology, Japan, the Sasakawa Scientific Research Grant (26-207) from the Japan Science Society, and the Uchida Energy Science Promotion Foundation (26-1-22).

References

[1] Kudo K, Nishikubo Y, and Nohara M 2010 *J. Phys. Soc. Jpn.* **79** 123710
[2] Mewis A 1986 *Z. Anorg. Allg. Chem.* **536** 7
[3] Kase N, Suzuki H, Tsukamoto T, Nakano T, and Takeda N, submitted to *J. Phys. Soc. Jpn.*
[4] Helfand E and Werthamer N R 1966 *Phys. Rev.* **147** 288
[5] Maki K 1966 *Phys. Rev.* **148** 362

Single crystal growth and heat capacity measurements of triangular lattice $R_2Pt_6Ga_{15}$ (R=rare earth)

Y Matsumoto, T Ueda, and S Ohara

Department of Engineering Physics, Electronics and Mechanics, Graduate school of Engineering, Nagoya Institute of Technology, Nagoya, Aichi 466-8555, Japan

E-mail: `matsumoto.yuji@nitech.ac.jp`

Abstract. We have succeeded in synthesizing the single crystal of $R_2Pt_6Ga_{15}$ (R=La-Nd, Sm-Lu) with hexagonal $Sc_{0.67}Fe_2Si_5$-type structure using Ga self flux method. The crystal structure was confirmed by the powder X-ray method. The unit-cell volume V of $R_2Pt_6Ga_{15}$ follows the lanthanide concentration except R = Ce, Eu and Yb, indicating that the valences of R = La, Pr, Nd, Sm, Gd-Tm, and Lu ion are trivalent, whereas those of R = Ce, Eu and Yb ion are deviate from trivalent. We have measured the specific heat $C(T)$ of $R_2Pt_6Ga_{15}$. It is found that the magnetic order takes place in $R_2Pt_6Ga_{15}$ (R=Pr, Nd, Sm-Tm). Moreover, the multiple phase transitions were observed in $R_2Pt_6Ga_{15}$ (R = Nd, Eu, Gd and Ho).

1. Introduction

In highly correlated f electron system, many anomalous physical properties have been discovered around the quantum critical point. Among the Yb member compounds, $YbRh_2Si_2$ [1], β-$YbAlB_4$ [2] and quasi-crystal $Au_{51}Al_{34}Yb_{15}$ [3] exhibit anomalous quantum critical behaviors which cannot be explained by the conventional Hertz-Millis-Moriya (HMM) theory [4, 5, 6]. In $YbRh_2Si_2$, the magnitude of magnetic frustration effect can be important role to take place these critical behaviors . In rare earth compound, the magnetic ground state is explained by the competition between the Kondo effect and the Ruderman-Kittel-Kasuya-Yosida (RKKY) interaction, so-called Doniach phase diagram [7]. The magnetic frustration effect can also suppress to the magnetic order such like the Kondo effect. The quantum critical behavior due to the magnetic frustration effect in Ce_2Pt_2Pb [8] with Shastry - Sutherland lattice and YbAgGe [9] with distorted kagome lattice have been observed. To search for the compound with the magnetic frustration is possible to discover the novel quantum phenomena.

The R_2Pt_6 $_{15}$ (R = rare earth, = Al, Ga) system with hexagonal $Sc_{0.67}Fe_2Si_5$-type structure (space group $P6_3/mmc$) would be one of the suitable candidate to research the magnetic frustrate system. The R ions coordinate the triangular lattice layer and the nearest R-R distance is about 4.3 Å [10]. While, the distance between triangular-lattice layers is about 8.7 Å along c-axis, which is more than twice as long as the in-plane R ion distance, namely the magnetic R-ions in $R_2Pt_6Ga_{15}$ form a two-dimensional triangular-lattice. In R_2Pt_6 $_{15}$, $R_2Pt_6Al_{15}$ (R=Ce, Gd-Lu) [10, 11, 12, 13, 14, 15] and $R_2Pt_6Ga_{15}$ (R=La, Ce, Pr) [16, 17, 18] have ever been reported. The lattice parameters of a and c of $Ce_2Pt_6Ga_{15}$ are 4.326 and 16.522

Journal of Physics: Conference Series **683** (2016) 012035 doi:10.1088/1742-6596/683/1/012035

Å , respectively, whereas those of $Yb_2Pt_6Al_{15}$ are 4.266 and 16.361 Å , respectively. It is possible to synthesize R_2Pt_6 $_{15}$ (R = heavy rare earth). Therefore, we have searched for new $R_2Pt_6Ga_{15}$ compound and measured the specific heat of $R_2Pt_6Ga_{15}$.

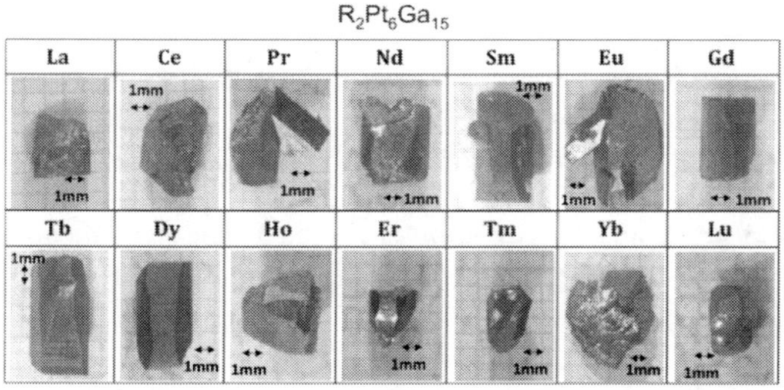

Figure 1. The photographs of single crystal $R_2Pt_6Ga_{15}$.

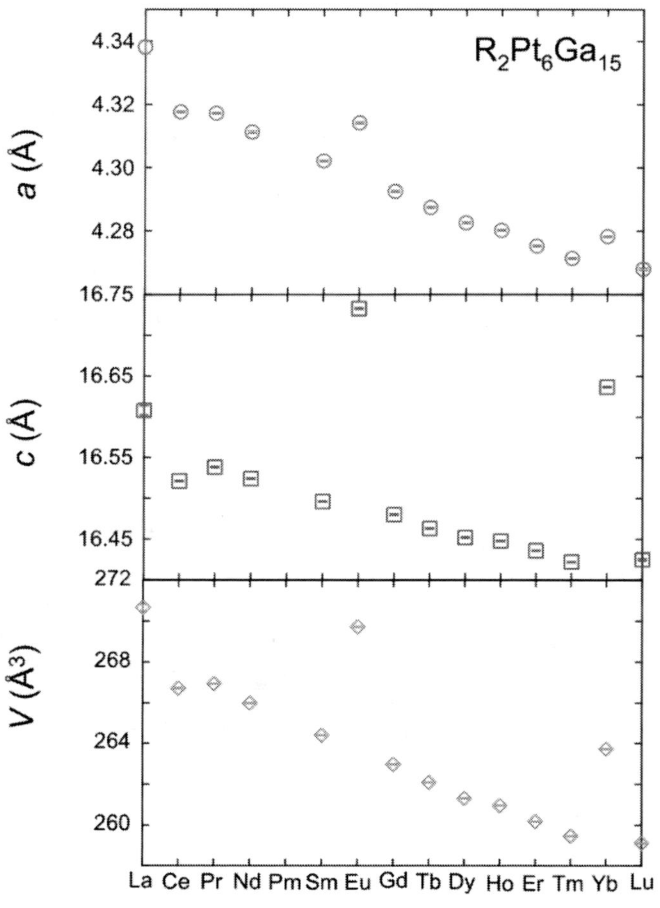

Figure 2. The lattice constants a, c, and unit-cell volume V of hexagonal $R_2Pt_6Ga_{15}$.

2. Experimental

The single crystals of $R_2Pt_6Ga_{15}$ (R=La-Nd, Sm-Lu) were grown in Ga-flux method. Rare earth, Pt and Ga materials were placed in an alumina crucible and sealed in an evacuated quartz tube.

The sealed tubes were heated to 1100 °C, soaked for 5 hours, then cooled down to 300 °C in 160 hours. The excess Ga was spun off in a centrifuge, and then washed out in distilled water. Figure 1 shows the photograph of obtained single crystal $R_2Pt_6Ga_{15}$. Typical sample size is from 2 to 5 mm.

The single phase of the hexagonal $R_2Pt_6Ga_{15}$ (R=La-Nd, Sm-Lu) structure was confirmed by the powder -ray diffraction method. We obtained the lattice parameters of $R_2Pt_6Ga_{15}$ by Rietveld analysis. Figure 2 displays the lattice constants a, c, and unit-cell volume V of $R_2Pt_6Ga_{15}$. V monotonously shrinks from R= La to Lu except R = Ce, Eu and Yb, indicating that the $R_2Pt_6Ga_{15}$ system follows the well-known lanthanide concentration except R = Ce, Eu and Yb. The results indicate that the valences of R = La, Pr, Nd, Sm, Gd-Tm and Lu ion are likely to be trivalent, while those of R = Eu and Yb ion are close to divalent.

The specific heat $C(T)$ measurement from 300 to 2 K was carried out by the relaxation method using a commercial physical property measurement system (PPMS; Quantum Design).

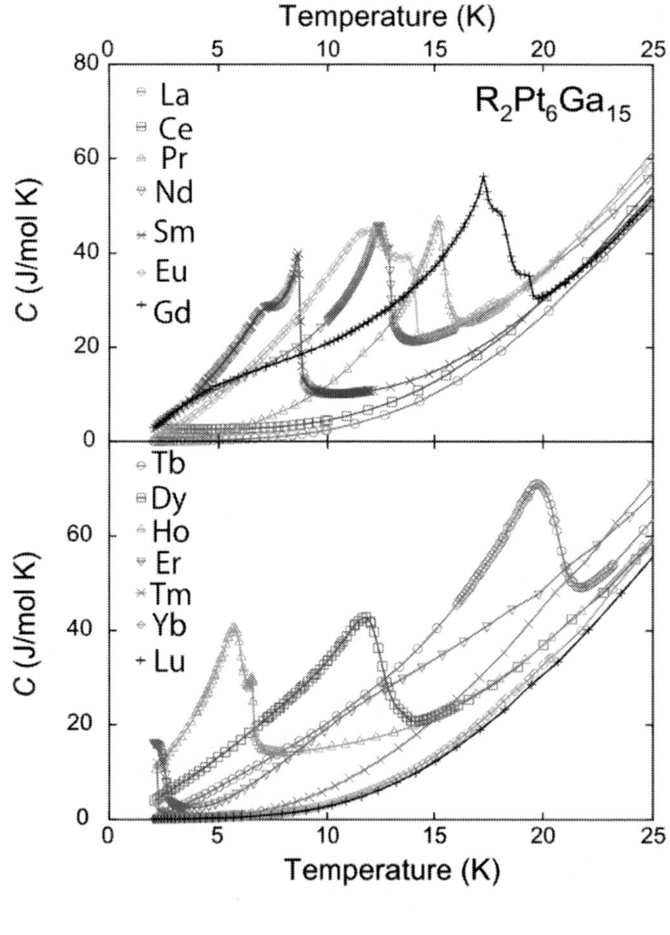

Figure 3. The temperature dependence of heat capacity $C(T)$ of $R_2Pt_6Ga_{15}$. The data of $La_2Pt_6Ga_{15}$ and $Ce_2Pt_6Ga_{15}$ were taken from Yamashita et al [18].

3. Results and discussion

We have measured the $C(T)$ of $R_2Pt_6Ga_{15}$. Figure 3 shows the temperature dependence of $C(T)$ of $R_2Pt_6Ga_{15}$ (R=La-Nd, Sm-Lu) below 25 K. The $C(T)$ of $La_2Pt_6Ga_{15}$ and $Ce_2Pt_6Ga_{15}$ were taken from Yamashita et al [18]. There is no phase transition below 2 K for $R_2Pt_6Ga_{15}$ (R = La, Ce, Yb and Lu). The $C(T)$ of $R_2Pt_6Ga_{15}$ (R = La, Yb and Lu) follow $C(T) = \gamma T + \beta T^3$ expression in low temperature. The γ of $Yb_2Pt_6Ga_{15}$ and $Lu_2Pt_6Ga_{15}$ are 25 and 19 mJ/mol K^2, respectively, and those of β are 2.17 and 1.87 mJ/mol K^4, respectively. The valence of Yb ion is near divalent. The result is consistent to the deviation from the lanthanide concentration for $Yb_2Pt_6Ga_{15}$. Unfortunately, $Yb_2Pt_6Ga_{15}$ does not exhibit quantum critical behavior, although we searched for the novel phenomena around quantum critical point in $Yb_2Pt_6Ga_{15}$.

It is found that there are anomalies of $C(T)$ due to the magnetic order in $R_2Pt_6Ga_{15}$ (R=Pr, Nd, Sm-Tm). Moreover, the multiple phase transitions were observed in $R_2Pt_6Ga_{15}$ (R = Nd, Eu, Gd and Ho). There is an anomaly in the magnetic susceptibility around the phase transition [19], indicating that the phase transitions are magnetic order. The magnetic order temperature T_M corresponding to the middle point between the peak and the onset roughly follows the de Gennes scaling and T_M in $R_2Pt_6Ga_{15}$ with R = light rare earth deviate from the de Gennes scaling, as shown in Fig. 4, where g_J is the Lande g-factor and J is the total angular momentum and we assumed that the valence of Eu ion is divalent. This result and the order of T_M are similar to that of RCu_2Si_2 and $RRhIn_5$ with no geometrical frustration [20, 21]. The suppression of T_M due to the magnetic frustration effect in $R_2Pt_6Ga_{15}$ (R=Pr, Nd, Sm-Tm) would be negligible small. However, whether the magnetic frustration effect is strong or not is dependent to the magnetic structure of $R_2Pt_6Ga_{15}$. If the magnetic structure of $R_2Pt_6Ga_{15}$ is Y-type, the magnetic frustration effect would be observed and the complex magnetic structure is possible. We are measuring the magnetic properties and intend to measure neutron diffraction of $R_2Pt_6Ga_{15}$.

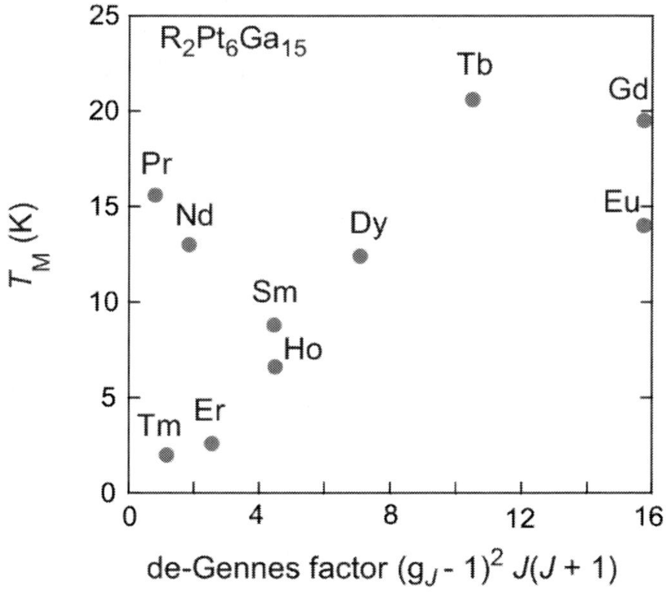

Figure 4. The magnetic oder temperature T_M versus de-Gennes factor of $R_2Pt_6Ga_{15}$.

Acknowledgments

We thank Mr. Taki and Mr. Minamiguchi for technical support. This work was supported by a Grant-in-Aid for Scientific Research Young Scientists B (No 24740248) Grant-in-Aid for Scientific Research C (No 26400333) from the Ministry of Education, Culture, Sports, Science and Technology (ME T) and Japan Society of the Promotion of Science (JSPS).

References

[1] Custers J, Gegenwart P, Wilhelm H, Neumaier K, Tokiwa Y, Trovarelli O, Geibel C, Steglich F, Pepin C, and Coleman P, 2002 *Nature* **424** 524.

[2] Nakatsuji S, Kuga K, Machida Y, Tayama T, Sakakibara T, Karaki Y, Ishimoto H, Yonezawa S, Maeno Y, Pearson E, Lonzarich G G, Balicas L, Lee H, and Fisk Z, 2008 *Nature Phys.* **4** 603.
[3] Deguchi K, Matsukawa S, Sato N K, Hattori T, Ishida K, Takakura H, and Ishimasa T, 2012 *Nature Materials* **11** 1013.
[4] Hertz J A, 1976, *Phys. Rev. B* **14** 1165.
[5] Millis A J, 1993 *Phys. Rev. B* **48** 7183.
[6] Moriya T, and Takimoto T, 1995 *J. Phys. Soc. Jpn* **64** 960.
[7] Doniach S, 1977, *Physica B* **91** 231.
[8] Kim M S, and Aronson M C, 2011 *J. Phys.:Condens. Matter.* **23** 164204.
[9] Sengupta K, Forthaus M K, Kubo H, Katoh K, Umeo K, Takabatake T, and Abd-Elmeguid M M, 2010 *Phys. Rev. B* **81** 125129.
[10] Lutsyshyn Y, Tokaychuk Y, Davydov V, and Gladysevskii R, 2008 *Chem. Met. Alloys* **1** 303.
[11] Niermann J, and Jeitschko W, 2004 *Z. Anorg. Allg. Chem.* **630** 361.
[12] Lutsyshyn Y, Tokaychuk Y, and Gladysevskii R, 2009 *Chem. Met. Alloys* **2** 75.
[13] Deppe M, Hartmann S, Macovei M E, Oeschler N, Nicklas M, and Geibel C, 2008 *New Journal of Phys.* **10** 093017.
[14] Latturner S E, Kanatzidis M G, 2002 *Inorg. Chem.* **41** 5479.
[15] Murashova E V, Tursina A I, Bukhan'ko N G, Gribanov A V, Chernyshev I V, and Seropegin Yu D, 2005 *J. Alloys Comp.* **398** 100.
[16] Lacerda A, Canfield P C, Beyermann W P, Hundley M F, Thompson J D, Sparn G, Fisk Z, Burns C, Barnhart D, Lawson A C, Kwei G H, and Goldstone J, 1992 *J. Alloys Comp.* **181** 191.
[17] Kwei G H, Lawson A C, Larson A C, Morosin B, Larson E M, and Canfield P C 1996 *Acta Cryt. B* **52** 580.
[18] Yamashita T, and Ohara S, 2012 *J. Phys. Conf. series* **400** 042074.
[19] Ueda T, 2015 Graduation thesis, Nagoya Institute of Technology (Japanese)
[20] Dung N D, Ota Y, Sugiyama K, Matsuda T D, Haga Y, Kindo K, Hagiwara M, Takeuchi T, Settai R, Onuki Y, 2009 *J. Phys. Soc. Jpn* **78** 024712.
[21] Hieu N V, Takeuchi T, Shishido H, Tonohiro C, Yamada T, Nakashima H, Sugiyama K, Settai R, Matsuda T D, Haga Y, Hagiwara M, Kindo K, Araki S, Nozue Y, Onuki Y, 2007 *J. Phys. Soc. Jpn* **76** 064702.

Journal of Physics: Conference Series **683** (2016) 012036 doi:10.1088/1742-6596/683/1/012036

Magnetocaloric effect in a cluster-glass system $Ho_5Pd_{2-x}Ni_x$

Saori Toyoizumi[1], Hideaki Kitazawa[2], Kengo Morita[3] and Akira Tamaki[1]

[1]Graduate School of Advanced Science and Technology, Tokyo Denki University, 5 Senju Asahi-cho, Adachi, Tokyo 120-8551, Japan
[2]National Institute for Materials Science, 1-2-1 Sengen, Tsukuba, Ibaraki 305-0047, Japan
[3]School of Engineering, Tokyo Denki University, 5 Senju Asahi-cho, Adachiku, Tokyo 120-8551, Japan

E-mail: 14udb03@ms.dendai.ac.jp

Abstract. In order to investigate the effect of chemical pressure on the large magnetocaloric effect in Ho_5Pd_2, we conducted X-ray diffraction, magnetization, and specific heat measurements on $Ho_5Pd_{2-x}Ni_x (0 \leq x \leq 1.0)$ rare-earth intermetallic compounds. The linear x dependence of the lattice constant a suggests that Ni is replaced with Pd in the case of $Ho_5Pd_{2-x}Ni_x$ $(0 \leq x \leq 0.5)$. The spin-glass transition temperature T_g and paramagnetic Curie temperature θ_p indicate a weak oscillatory x dependence. However, the magnetic entropy change $-\Delta S_m$ and the relative cooling power (RCP) are rapidly suppressed with increasing x. These large reductions in $-\Delta S_m$ and RCP cannot be explained only in terms of normal Ruderman-Kittel-Kasuya-Yoshida (RKKY)-type indirect exchange interactions.

1. Introduction

Magnetic refrigeration of magnetic materials based on the magnetocaloric effect (MCE) has attracted attention owing to its potential application in energy-efficient environment-friendly refrigeration technologies. T. Samanta *et al.* discovered that the rare-earth compound Ho_5Pd_2 exhibits a large relative cooling power (RCP) of 6.32 J/cm^3 at approximately 30 K, and suggested the possibility of antiferromagnetic ordering using magnetization data[1]. Since the temperature with the maximum RCP value is close to the boiling point of hydrogen at 20 K, it is thought that Ho_5Pd_2 is a promising material for use in hydrogen liquefaction in future societies that utilize a hydrogen energy system. Our neutron experiments confirmed that Ho_5Pd_2 belongs to a cluster-glass that is specified by short-range ordering, with an incommensurate wave vector, $\vec{q} = \{0.18, 0.18, 0.18\}$[2]. A. F. Gubkin *et al.* reported on spin-glass behaviors of Ho_5Pd_2 that were obtained using DC and AC magnetization measurements, together with the results of neutron experiments[3]. These results suggest that magnetic ordering below T_g is not antiferromagnetic and that short-range ordering with respect to the incommensurate wave vector must be caused by complex competition between ferromagnetic and antiferromagnetic interactions and partial occupancies at the Ho and Pd sites. Recently, we reported that the magnetic entropy change depends on the number of vacancies at the Ho sites in a $Ho_{5+x}Pd_2$ system $(-0.4 \leq x \leq 0.4)$[4]. Since the ionic radius of Ni is smaller than that of Pd with a similar electronic state, it is thought that the non-magnetic Ni substitution at the non-magnetic Pd sites corresponds to a chemical pressure effect. In order to determine the way in which chemical pressure affects magnetocaloric effect, we have investigated magnetic and thermal properties using measurements of the magnetization and heat capacity of $Ho_5Pd_{2-x}Ni_x(0 \leq x \leq 1.0)$.

2. Experiment

Polycrystalline samples of $Ho_5Pd_{2-x}Ni_x$ were synthesized using an arc-melting method under a pure argon atmosphere. The mixtures of Ho (3N), Pd (3N5), and Ni (3N) with a nominal composition ($x = 0$, 0.1, 0.3, 0.5, and 1.0) were alternately melted and turned upside down five times on a water-cooled copper hearth, in order to ensure their homogenization. The as-cast samples were then annealed at 1123 K for one week in an evacuated quartz tube. The crystal structures of the annealed samples were evaluated using -ray diffraction (RD) with a Rigaku RINT 2100 diffractometer. The -ray diffraction patterns were analyzed using Rietveld analysis, as implemented in the Fullprof Suite software package[5]. The magnetization measurements were conducted using a superconducting quantum interference device (SQUID) magnetometer (MPMS-5S, Quantum Design Ltd.) in the temperature range of 1.9 to 300 K, with DC magnetic field strengths ranging from 0 to 5 T. The specific heat measurements were conducted using the adiabatic heat relaxation method in the temperature range of 1.9 to 300 K, using a physical property measurement system (PPMS, Quantum Design Ltd.).

3. Results and Discussion

Figure 1 shows the RD patterns of $Ho_5Pd_{2-x}Ni_x(0 \leq x \leq 1.0)$ at room temperature. The Rietveld analysis was performed assuming that a normal composition of Ni was substituted for the Pd. This analysis revealed that the main phase of all the samples ($0 \leq x \leq 1.0$) exhibited a cubic Dy_5Pd_2-type structure (space group: $Fd3m$)[6]. In the cases of $x = 0.5$ and 1.0, a small amount of unknown phases (denoted by •) were also clearly detected. The lattice constant a as a function of x is shown in Fig.2 and tabulated in Table 1. The amplitude of a linearly decreased with the replacement of Pd sites with Ni, up to a value of $x = 0.5$, as shown in Fig.2. However, it is of note that the amplitude of a above $x = 0.5$ began to deviate from the linear curve. This implies that the lattice constant a in $Ho_5Pd_{2-x}Ni_x(0 \leq x \leq 0.5)$ follows the well-known Vegard's law and that the majority of the Ni replaced Pd.

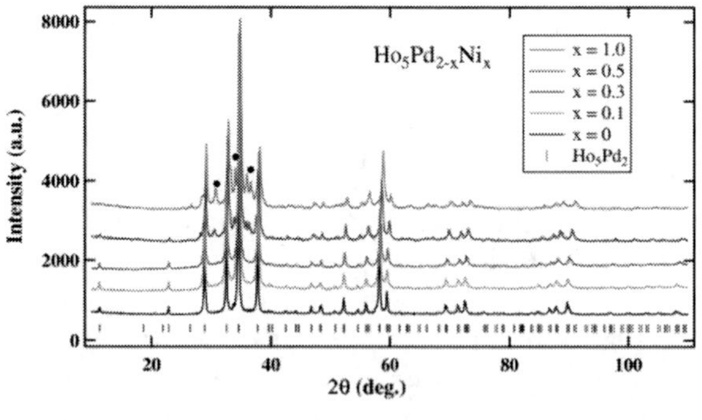

Figure 1. -ray diffraction patterns of $Ho_5Pd_{2-x}Ni_x$ at the room temperature. The RD pattern is shifted along the y-axis direction.

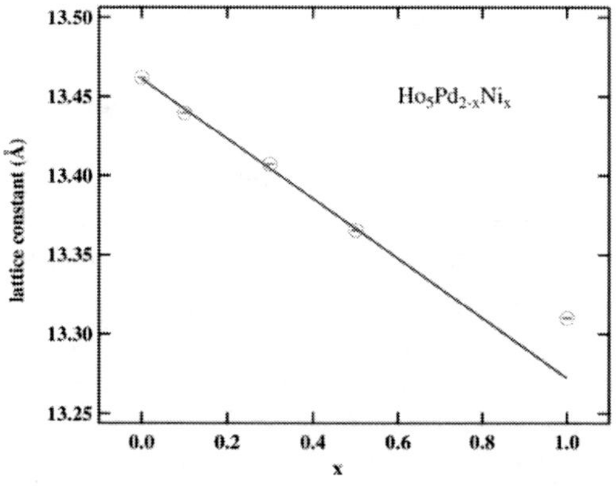

Figure 2. Lattice constant a as a function of x in $Ho_5Pd_{2-x}Ni_x$.The line indicates the least-squares fitting line for $0 \leq x \leq 0.5$.

Journal of Physics: Conference Series **683** (2016) 012036 doi:10.1088/1742-6596/683/1/012036

Table 1. The lattice constant a, the spin-glass transition temperature T_g, the maximum values of the magnetocaloric parameters ($-\Delta S_m^{max}$ and RCP) under 5 T, the effective magnetic moment μ_{eff}, and the paramagnetic Curie temperature θ_p for $Ho_5Pd_{2-x}Ni_x$.

x	lattice constant a (Å)	T_g (K) (from C_p)	$-\Delta S_m^{max}$ (J/kg K)	RCP (J/kg)	μ_{eff} (μ_B/Ho)	θ_p (K)
0.0	13.4620	25.5	15.4	678	11.07	29.7
0.1	13.4397	24.9	12.0	492	10.65	29.2
0.3	13.4071	22.9	11.0	462	10.76	27.5
0.5	13.3655	23.4	9.4	389	10.72	30.9
1.0	13.3099	24.8	8.5	332	10.59	33.8

The temperature dependencies of the heat capacity $C(T)$ and $C(T)/T$ of $Ho_5Pd_{2-x}Ni_x$ under a magnetic field of $B = 0$ are shown in Fig.3 and Fig.4, respectively. The $C(T)$ of Lu_5Pd_2 is also plotted as a reference substance for the lattice part of the specific heat provided in Fig.3. The spin-glass transition temperature T_g is defined as the temperature at the maximum value of $C(T)/T$. A λ-shaped anomaly for $x \geq 0.5$ appears at 45 K, which originated from the phase transitions of unknown phases. When x was changed from 0 to 1.0, the amplitude of T_g initially decreased, with a minimum at $x = 0.3$, and then began to increase again. The increase of $C(T)$ and $C(T)/T$ below 3.5 K suggests either the occurrence of other phase transitions or the nuclear contribution of Ho at lower temperatures[7]. We will confirm this using heat capacity experiments at lower temperatures. Note that the large magnetic contribution of the specific heat remains above T_g.

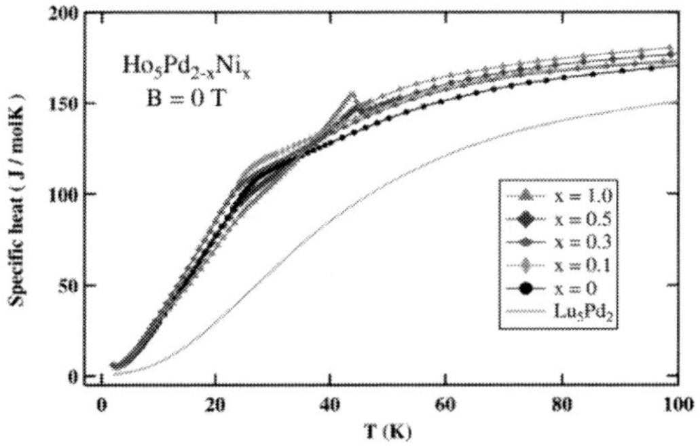

Figure 3. Temperature dependences of the heat capacity $C(T)$ in $Ho_5Pd_{2-x}Ni_x$.

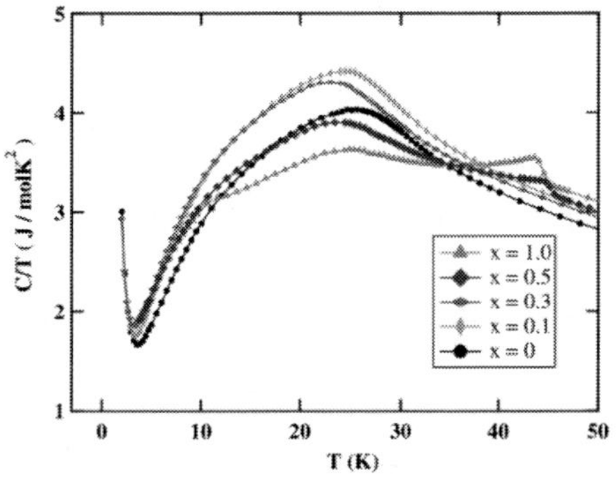

Figure 4. $C(T)/T$ as a function of temperature T in $Ho_5Pd_{2-x}Ni_x$.

The temperature dependence of the magnetic susceptibility $\chi(T)$ of $Ho_5Pd_{2-x}Ni_x$ under a magnetic field of $B = 0.1$ T is shown in Fig.5. The spin-glass transition temperature T_g obtained from the $\chi(T)$ data should be located at approximately the shoulder of the $\chi(T)$ curve. The $\chi(T)$ values obtained in the zero-field cooled (FC) and field cooled (FC) processes began to differ below T_g. However, A. F. Gubkin *et al.* reported that the amplitude of T_g is affected by the magnetic field [3]. Therefore we will adopt the T_g as defined using the $C(T)$ data under zero magnetic field as the spin-glass transition temperature of $Ho_5Pd_{2-x}Ni_x$ in this study. A small hump in the data at 45 K can be seen in the cases of $x = 0.5$ and 1.0. Judging from the anomalies related to $C(T)$ and $\chi(T)$ at 45 K and the existence of unknown phases in the RD patterns, the unknown phases in $x = 0.5$ and 1.0 undergo an amount of magnetic ordering below 45 K. The amplitude of $\chi(T)$ follows the Curie-Weiss law in the temperature range of 60 K to 300 K (not shown in Figure). The Curie-Weiss law is expressed as

$$\chi = \frac{N\mu_{\rm eff}^2}{3k_{\rm B}\left(T - \theta_{\rm p}\right)}, \tag{1}$$

where N, $\mu_{\rm eff}$, $k_{\rm B}$, and $\theta_{\rm p}$ are the number of magnetic atoms per mole, the effective magnetic moment, the Boltzmann constant, and the paramagnetic Curie temperature, respectively. The obtained values of $\mu_{\rm eff}$ and $\theta_{\rm p}$ are listed in Table 1. The positive $\theta_{\rm p}$ of the $Ho_5Pd_{2-x}Ni_x$ indicates that the ferromagnetic interaction dominates.

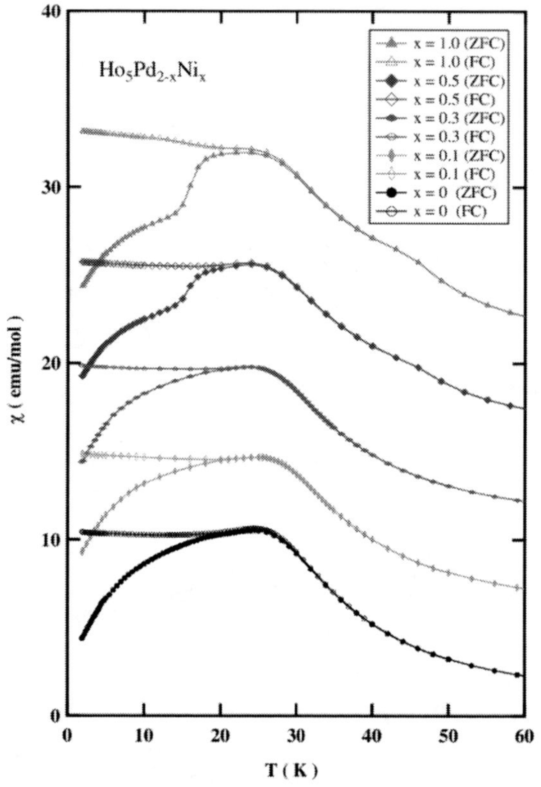

Figure 5. Temperature dependence of the magnetic susceptibility $\chi(T)$ in $Ho_5Pd_{2-x}Ni_x$. Each $\chi(T)$ curve is shifted by +5 emu/mol along the y-axis direction.

In order to clarify the effect of chemical pressure on the MCE, we evaluate two crucial quantities related to MCE. The first crucial quantity is the magnetic entropy change $-\Delta S_{\rm m}$ defined by

$$\Delta S_{\rm m} = \int_0^B \left(\frac{dM}{dT}\right)_B dB, \tag{2}$$

which is calculated using the temperature T and magnetic field B dependence of the magnetization M, according to the Maxwell relation. The second crucial quantity is the RCP[8]. The RCP is an important quality factor for magnetic refrigeration materials, which is related to the amount of heat transfer between the cold and hot reservoirs in an ideal refrigeration cycle.

The value of RCP is evaluated using the product of the maximum value $-\Delta S_{\rm m}^{\rm max}$ and the full width at half maximum $\delta T_{\rm FWHM}$ of the magnetic entropy change :

$$\mathrm{RCP} = \Delta S_{\rm m}^{\rm max} \times \delta T_{\rm FWHM}. \tag{3}$$

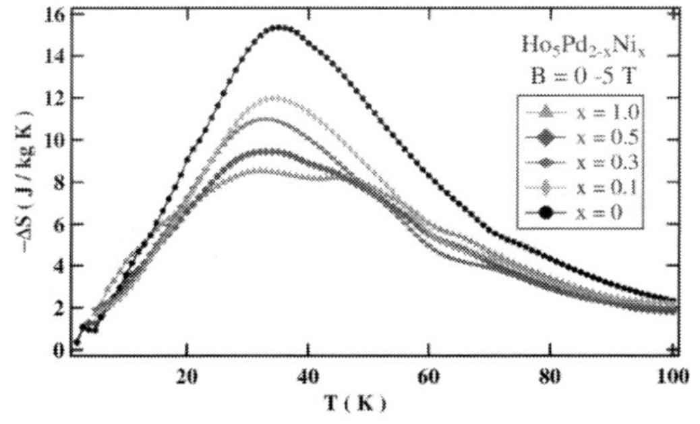

Figure 6. Magnetic entropy changes $-\Delta S_{\rm m}$ as a function of T in $\mathrm{Ho_5Pd_{2-\mathit{x}}Ni_\mathit{x}}$ ($B = 0-5$ T).

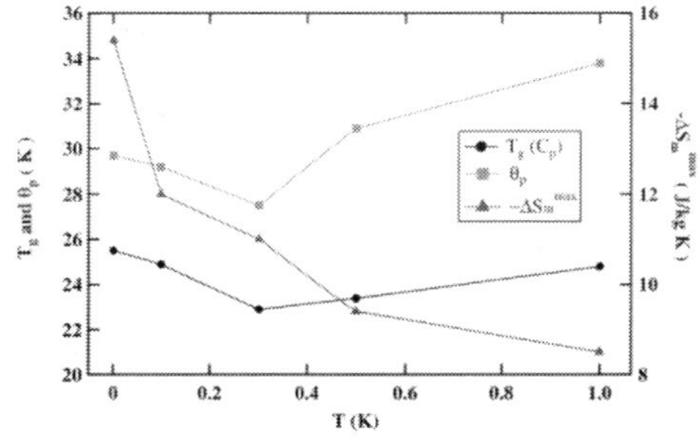

Figure 7. x dependencies of the spin-glass transition temperature $T_{\rm g}$, the paramagnetic Curie temperature $\theta_{\rm p}$, and the maximum magnetic entropy change $-\Delta S_{\rm m}^{\rm max}$ ($B = 0-5$ T).

The magnetic entropy changes $-\Delta S_{\rm m}$ as a function of temperature in $\mathrm{Ho_5Pd_{2-\mathit{x}}Ni_\mathit{x}}$ are shown in Fig.6. The maximum magnetic entropy change $-\Delta S_{\rm m}^{\rm max}$ is located at approximately 35 K, which is above $T_{\rm g}$. The maximum magnetic entropy change $-\Delta S_{\rm m}^{\rm max}$ and the RCP values are listed in Table 1. The magnetic parameters $T_{\rm g}$, $\theta_{\rm p}$, and $-\Delta S_{\rm m}^{\rm max}$ are plotted as a function of x in Fig.7. The maximum $-\Delta S_{\rm m}^{\rm max}$ value of 15.4 J/kgK was obtained for $\mathrm{Ho_5Pd_2}(x = 0)$. The amplitudes of $-\Delta S_{\rm m}^{\rm max}$ and RCP rapidly decreased with increasing x. However, $T_{\rm g}$ and $\theta_{\rm p}$ exhibited different x dependencies. When x increased from 0 to 1.0, both values decreased, with a minimum at approximately $x = 0.3$, before beginning to increase again. The spin-glass behavior represented by $T_{\rm g}$ is well correlated with the strength of magnetic interactions which is characterized by $\theta_{\rm p}$. However, the chemical pressure effect in $\mathrm{Ho_5Pd_{2-\mathit{x}}Ni_\mathit{x}}$ does not monotonically tune the amplitude of the magnetic interactions. Since the magnetic properties in rare-earth intermetallic compounds are mainly governed by Ruderman-Kittel-Kasuya-Yoshida (RKKY)-type indirect exchange interactions, which exhibit an oscillatory dependence on the distance between magnetic ions, the obtained chemical pressure effect on $T_{\rm g}$ and $\theta_{\rm p}$ in $\mathrm{Ho_5Pd_{2-\mathit{x}}Ni_\mathit{x}}$ may be explained using a model based on complex RKKY interactions. However, the large reduction in $-\Delta S_{\rm m}$ in $\mathrm{Ho_5Pd_{2-\mathit{x}}Ni_\mathit{x}}$ cannot be explained using this method. It is necessary to perform neutron scattering experiments on $\mathrm{Ho_5Pd_{2-\mathit{x}}Ni_\mathit{x}}$ in order to collect further microscopic information and elucidate the chemical pressure effect.

4. Conclusion

Since the lattice constant follows Vegard's law in $\mathrm{Ho_5Pd_{2-\mathit{x}}Ni_\mathit{x}} (0 \leq x \leq 0.5)$, a portion of the Pd sites of $\mathrm{Ho_5Pd_2}$ must have been replaced with Ni. The magnetic entropy change and RCP of $\mathrm{Ho_5Pd_{2-\mathit{x}}Ni_\mathit{x}}$ were greatest at $x = 0$, under a magnetic field change of 5 T. The Ni substitution in $\mathrm{Ho_5Pd_2}$ rapidly reduced the amplitudes of the magnetic entropy change and RCP. Since the amplitudes of $T_{\rm g}$ and $\theta_{\rm p}$ exhibited weak oscillatory behavior, the large reduction of magnetic entropy change and RCP cannot be explained by referring only to RKKY interactions.

Journal of Physics: Conference Series **683** (2016) 012036 doi:10.1088/1742-6596/683/1/012036

References

[1] Samanta T, Das I and Banerjee S 2007 *Appl. Phys. Lett.* **91** 082511
[2] Kitazawa H, Kawamura Y, Terada N, Mamiya and H, Suzuki H S 2012 *19th International Conference on Magnetism* (Busan, July 2012) p 106
[3] Gubkin A F, Sherstobitova E A, Terentyev P B, Hoser A and Baranov N V 2013 *J. Phys.: Cond. Mat.* **25** 236003
[4] Toyoizumi S, Kitazawa H, Kawamura Y, Mamiya H, Terada N, Tamura R, Dönni A, Morita K and Tamaki A 2015 *J. Appl. Phys.* **117** 17D101
[5] Rodriguez-Carvajal J 1993 *Physica B* **192** 55
[6] Fornasini M L and Palenzona A 1974 *J.Less Comm.Mat.* **38** 77
[7] Aoki Y, Urakawa J, Sugawara H, Sato H, Markin P E, Bostrem I G and Baranov N V 2000 *Phys. Rev. B* **62** 8935
[8] Gschneidner K A Jr, Pecharsky V K, and Tsokol A O 2005 *Rep. Prog. Phys.* **68** 1479

TMU International Symposium 2015 IOP Publishing
Journal of Physics: Conference Series **683** (2016) 012037 doi:10.1088/1742-6596/683/1/012037

Spin correlation and Majorana spectrum in chiral spin liquids in a decorated-honeycomb Kitaev model

Joji Nasu[1] **and Yukitoshi Motome**[2]

[1]Department of Physics, Tokyo Institute of Technology, Ookayama, 2-12-1, Meguro, Tokyo 152-8551, Japan

[2]Department of Applied Physics, University of Tokyo, Hongo, 7-3-1, Bunkyo, Tokyo 113-8656, Japan

E-mail: nasu@phys.titech.ac.jp

Abstract. Temperature evolution of the spin correlation and excitation spectrum is investigated for the Kitaev model defined on a decorated honeycomb lattice by using the quantum Monte Carlo simulation in the Majorana fermion representation. The ground state of this quantum spin model is given by two kinds of chiral spin liquids: one is topologically trivial with Abelian anyon excitations, and the other is topologically nontrivial with non-Abelian anyon excitations. While lowering temperature, the model exhibits several crossovers in the paramagnetic state, which originate from the fractionalization of quantum spins into Majorana fermions, in addition to a phase transition associated with time reversal symmetry breaking. We show that the spin correlation develops around the crossover temperatures, whereas it shows a slight change at the critical temperature, as in other Kitaev-type models. We also calculate the excitation spectrum in terms of Majorana fermions, and find that the excitation gap in the non-Abelian phase is fragile against thermal fluctuations of the Z_2 fluxes, while that in the Abelian phase remains open.

1. Introduction

The quantum spin models that can be solved exactly have often provided a clear insight into quantum many-body physics [1–3]. For instance, the Affleck-Kennedy-Lieb-Tasaki model [4] and the Shastry-Sutherland model [5] have been intensively studied for establishing the valence-bond solid picture in one and two dimensions, respectively. The Kitaev model is one of such solvable models, which is composed of $S = 1/2$ spins with bond-dependent interactions on a two-dimensional honeycomb lattice [6]. A striking feature in the Kitaev model is that it possesses a macroscopic number of conserved quantities, which allow us to map the model onto a free Majorana fermion system coupled to Z_2 variables. The ground state is exactly shown to be a quantum spin liquid, which is one of the most intriguing subjects in strongly correlated electron systems [7]. The exact solvability is inherited in other lattices if they are tri-coordinated: indeed, many generalizations of the Kitaev model have been studied extensively in both two and three dimensions [8–16].

Among the tri-coordinate lattices, the Kitaev model on a decorated honeycomb lattice has been of great interest as the ground state is a chiral spin liquid (CSL), where the time reversal symmetry is broken without magnetic orders [6, 8, 17, 18]. This lattice is obtained from the honeycomb lattice by replacing each lattice site by a triangle [see Fig. 1(a)]. The model has two types of conserved quantities: one is on each triangle and the other is on each dodecagon. The former corresponds to the chirality degree of freedom, and hence, it plays a key role in stabilizing the CSL ground state with time reversal symmetry breaking. Interestingly, the model exhibits two types of CSLs while changing the exchange parameters: the topologically-trivial CSL with Abelian anyon excitations and the topologically-nontrivial CSL with non-Abelian anyon excitations [8]. While the exact solution is limited for zero temperature (T), the thermodynamic properties were also investigated numerically [19, 20]. There is a finite-T phase transition between the CSLs and paramagnet, and the nature of the phase transition is dependent on the topological nature of the CSLs. Nevertheless, thermodynamic properties of CSLs remain elusive. In particular, the magnetic properties and the dynamics at finite T have not been studied in detail thus far. These are important not only for illuminating the nature of exotic CSL states but also examining the stability of anyon excitations that are useful in quantum computation.

In this paper, we investigate the spin correlation and excitation spectrum at finite T in the Kitaev model on the decorated honeycomb lattice by performing the quantum Monte Carlo (QMC) simulation in the Majorana fermion representation [16]. We find that the nearest-neighbor (NN) spin correlation develops around high-T crossovers yielded by the entropy release of the itinerant Majorana fermions. Moreover, computing the density of states (DOS) of the Majorana fermions, we show a close relationship between the crossover temperatures and the T evolution of the Majorana spectrum. We also reveal that the low-energy gap in the topologically-nontrivial CSL is smeared out above the critical temperature by thermal fluctuations of Z_2 fluxes associated with the localized Majorana fermions. In contrast, the gap in the topologically-trivial CSL remains robust against thermal fluctuations.

2. Model

The Kitaev model on the decorated honeycomb lattice is defined by the Hamiltonian [6, 8]:

$$\mathcal{H} = -J \sum_{\gamma=x,y,z} \sum_{\langle jk \rangle_\gamma} \sigma_j^\gamma \sigma_k^\gamma - J' \sum_{\gamma'=x,y,z} \sum_{\langle jk \rangle'_{\gamma'}} \sigma_j^{\gamma'} \sigma_k^{\gamma'}, \tag{1}$$

where σ_j^γ is the $\gamma(= x, y, z)$ component of the Pauli matrices representing an $S = 1/2$ spin localized on site j; $\langle jk \rangle_\gamma$ represents a NN γ bond within a triangle with the exchange constant J and $\langle jk \rangle'_{\gamma'}$ represents a NN γ' bond connecting triangles with the exchange constant J' [see Fig. 1(a)]. We parameterize the exchange constants by introducing the parameter α as $J = \cos\alpha$ and $J' = \sin\alpha$ ($0 < \alpha < \pi/2$, i.e., $J > 0$ and $J' > 0$).

By mapping the quantum spins σ_j^γ onto the fermions by the Jordan-Wigner transformation and representing the fermions by two types of Majorana fermions c_j and $\bar{c}_j$ [21–23], the Kitaev model in Eq. (1) is rewritten into the form of

$$\mathcal{H} = iJ \sum_{(jk)_x} c_j c_k - iJ \sum_{(jk)_y} c_j c_k - iJ \sum_{(jk)_z} \eta_r c_j c_k + iJ' \sum_{(jk)'_x} c_j c_k - iJ' \sum_{(jk)'_y} c_j c_k - iJ' \sum_{(jk)'_z} \eta_r c_j c_k, \tag{2}$$

where $\eta_r = i\bar{c}_j\bar{c}_k$ (r is the label of the bond). Here, $\{\eta_r\}$ are the Z_2 conserved quantities taking ± 1, as $\eta_r^2 = 1$ and they commute with each other and the Hamiltonian. The summations with $(jk)_\gamma$ and $(jk)'_{\gamma'}$ are taken for the NN pairs on the intra-triangle γ and inter-triangle γ' bonds with $j < k$, respectively. Thus, the Kitaev model is described by a quadratic Hamiltonian in terms of Majorana fermions, in which the itinerant Majorana fermions c are coupled with the Z_2 variables η composed of the localized Majorana fermions $\bar{c}$. Note that this is a faithful representation of an $S = 1/2$ spin without any extension of the Hilbert space, in contrast to

Journal of Physics: Conference Series **683** (2016) 012037 doi:10.1088/1742-6596/683/1/012037

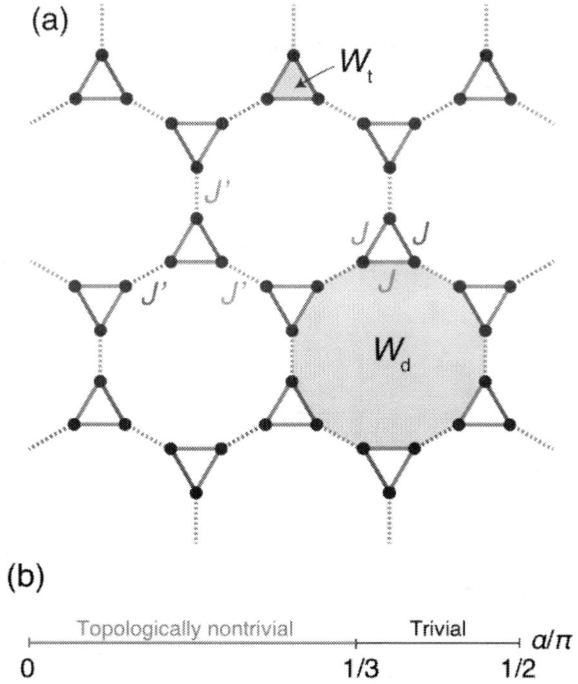

Figure 1. (a) Schematic picture of the Kitaev model in Eq. (1) on a decorated honeycomb lattice. The solid (dotted) blue, green, and red bonds represent the intra(inter)-triangle x, y, and z bonds with the exchange constant J (J'), respectively. $W_{\rm t}$ and $W_{\rm d}$ represent the conserved quantities (fluxes) defined on each triangle and dodecagon, respectively. (b) Ground state phase diagram as a function of α; we take $J = \cos\alpha$ and $J' = \sin\alpha$. There are two CSL phases with topologically trivial and nontrivial Majorana fermion bands, which meet at the critical point $\alpha_c/\pi = 1/3$.

the original Kitaev construction in Ref. [6]. This model has two types of conserved quantities, $W_{\rm t}$ and $W_{\rm d}$, defined on triangular and dodecagon plaquettes, respectively, both of which are termed fluxes. They are defined by the product of the Pauli matrices on each plaquette as $W_{\rm t} = \prod_{j=1}^{3} \sigma_j^{\gamma_j}$ and $W_{\rm d} = \prod_{j=1}^{12} \sigma_j^{\gamma_j}$, where γ_j represents the label of the bond that is not included in the plaquette among the three bonds connected to the site j. Equivalently, the fluxes are also written by the product of η_r included in the corresponding plaquette.

The ground state of the decorated-honeycomb Kitaev model is given by all $W_{\rm t} = W_{\rm d} = +1$ (zero flux state), namely, all η_r being $+1$ [8]. This is a free Majorana fermion system on a uniform background of the Z_2 variables. The spin correlation $\langle \sigma_j^{\gamma} \sigma_k^{\gamma} \rangle$ is nonzero only for the NN bonds: the ground state is a QSL with extremely short-range spin correlations [24]. As the conserved quantity $W_{\rm t}$ changes its sign by the time reversal operation, the ground state with all $W_{\rm t} = +1$ is a CSL with time reversal symmetry breaking [6, 8]. Interestingly, the current model exhibits two kinds of CSLs: a topologically-trivial CSL with Abelian anyon excitations and a topologically-nontrivial CSL with non-Abelian anyon excitations [8]. The ground state phase diagram as a function of α is presented in Fig. 1(b). There is a critical point at $\alpha_c = \pi/3$, which separates the topologically-trivial CSL for $\alpha > \alpha_c$ and the topologically-nontrivial CSL for $\alpha < \alpha_c$. Both CSLs have a nonzero gap in the excitation spectrum, while it goes to zero continuously approaching the critical point from both sides, as shown in Fig. 4. In the non-

Abelian CSL region for $\alpha < \alpha_c$, the system exhibits a chiral edge mode in the gap reflecting the topological nature [8].

3. Method

In order to investigate finite-T properties in the decorated-honeycomb Kitaev model, we perform the QMC simulation in the Majorana fermion representation [16]. As described in Eq. (2), the Kitaev model is regarded as a noninteracting Majorana fermion system coupled to the classical Z_2 variables η_r defined on the z bonds. The partition function is given by

$$Z = \mathrm{Tr}_{\{c_j\}} \mathrm{Tr}_{\{\eta_r\}} \exp\left[-\beta \mathcal{H}\right] = \sum_{\{\eta_r = \pm 1\}} \exp\left[-\beta F_f(\{\eta_r\})\right], \tag{3}$$

where $F_f(\{\eta_r\}) = -\beta^{-1} \ln \mathrm{Tr}_{\{c_j\}} \exp\left[-\beta \mathcal{H}\right]$ is the free energy of the free Majorana fermion system for a given configuration of $\{\eta_r\}$ [see Eq. (5)] and $\beta = 1/T$ is the inverse temperature (we set the Boltzmann constant to unity). In the Monte Carlo (MC) simulation, sequence of configurations of $\{\eta_r\}$ is generated by the Markov chain so as to reproduce the distribution of $\exp\left[-\beta F_f(\{\eta_r\})\right]$. The free energy F_f is computed by using the exact diagonalization method as shown below [see Eq. (5)]. The calculations are done for the $6L^2$-site clusters up to $L = 10$ with a twisted boundary condition so that the x and y bonds comprise a single one-dimensional chain. We adopted the replica exchange technique [25] with 16 replicas and performed 40000 MC steps for measurement after 10000 MC steps for thermalization in each replica.

For a given configuration of the Z_2 variables $\{\eta_r\}$, the Hamiltonian in Eq. (2) is diagonalized as

$$\mathcal{H}(\{\eta_r\}) = \sum_\lambda E_\lambda(\{\eta_r\}) \left(f_\lambda^\dagger f_\lambda - \frac{1}{2}\right), \tag{4}$$

where E_λ (≥ 0) is the one-particle energy of a fermion described by the creation (annihilation) operator $f_\lambda^\dagger$ (f_λ). The free energy of the fermion system is calculated as

$$F_f(\{\eta_r\}) = \beta^{-1} \sum_\lambda \ln \left[2 \cosh \frac{\beta E_\lambda(\{\eta_r\})}{2}\right]. \tag{5}$$

The specific heat per site is evaluated by

$$C_v = \frac{\beta^2}{6L^2} \left(\langle E_f^2 \rangle - \langle E_f \rangle^2 - \left\langle \frac{\partial E_f}{\partial \beta} \right\rangle \right), \tag{6}$$

where E_f is the energy of the fermion system given by

$$E_f(\{\eta_r\}) = -\sum_\lambda \frac{E_\lambda(\{\eta_r\})}{2} \tanh \frac{\beta E_\lambda(\{\eta_r\})}{2}. \tag{7}$$

The entropy per site is calculated from the the specific heat as

$$S = \ln 2 - \int_T^{T_m} \frac{C_v}{T'} dT', \tag{8}$$

where we choose $T_m = 10$ ($\gg J, J'$). We also introduce the DOS of the Majorana fermion system at finite T, which is defined by

$$D(\omega) = \frac{1}{3L^2} \sum_\lambda \langle \delta(\omega - E_\lambda(\{\eta_r\})) \rangle. \tag{9}$$

Journal of Physics: Conference Series **683** (2016) 012037 doi:10.1088/1742-6596/683/1/012037

Here, we take into account the effect of temperature only through thermal fluctuations of the Z_2 variables but not in the Fermi distribution function. Note that the first two terms in Eq. (6) are the contribution from the Z_2 variables and the third term comes from the itinerant Majorana fermions. The latter contribution is also written by using the DOS as

$$\tilde{C}_v = -\frac{\beta^2}{6L^2}\left\langle\frac{\partial E_f}{\partial\beta}\right\rangle = \frac{1}{2}\int_0^\infty D(\omega)\left[\frac{\beta\omega}{2\cosh(\beta\omega/2)}\right]^2 d\omega. \quad (10)$$

We also calculate the spin correlations for the intra- and inter-triangle NN bonds, which are given by

$$S_{\rm NN}^{\rm intra} = \frac{1}{6L^2}\sum_{\gamma=x,y,z}\sum_{\langle ij\rangle_\gamma}\langle\sigma_i^\gamma\sigma_j^\gamma\rangle = \frac{1}{6L^2}\left(-i\sum_{(jk)_x}\langle c_jc_k\rangle + i\sum_{(jk)_y}\langle c_jc_k\rangle + i\sum_{(jk)_z}\langle\eta_r c_jc_k\rangle\right) \quad (11)$$

and

$$S_{\rm NN}^{\rm inter} = \frac{1}{3L^2}\sum_{\gamma'=x,y,z}\sum_{\langle ij\rangle'_{\gamma'}}\langle\sigma_i^{\gamma'}\sigma_j^{\gamma'}\rangle = \frac{1}{3L^2}\left(-i\sum_{(jk)'_x}\langle c_jc_k\rangle + i\sum_{(jk)'_y}\langle c_jc_k\rangle + i\sum_{(jk)'_z}\langle\eta_r c_jc_k\rangle\right), \quad (12)$$

respectively.

4. Results

Figure 2(a) shows the specific heat for several size clusters at $\alpha/\pi = 0.15$, where the system is in the topologically-nontrivial CSL region. There is a sharp peak growing with the system size at $T_c \simeq 0.007$. This indicates that the phase transition occurs at T_c, which is associated with the time reversal symmetry breaking: the CSL phase is stabilized below T_c [19]. Through this phase transition, a half of the entropy S is released as shown in Fig. 2(d). The entropy release comes from the coherent development of both $W_{\rm t}$ and $W_{\rm d}$ [19]: the sum of the numbers of $W_{\rm t}$ and $W_{\rm d}$ is a half of the number of spins. Above T_c, however, the specific heat shows two broad peak structures at $T_{\rm H}^* \simeq 0.25$ and $T^* \simeq 1.2$. These are almost independent of the system size, and hence, correspond to crossovers [19]. (We follow the notations for the crossover temperatures in Ref. [19].) As shown in Fig. 2(d), the entropy is released by $\frac{1}{3}\ln 2$ and $\frac{2}{3}\ln 2$ around $T_{\rm H}^*$ and T^*, respectively.

In order to reveal the origin of the entropy releases, we calculate the spin correlations for the NN bonds, $S_{\rm NN}^{\rm intra}$ and $S_{\rm NN}^{\rm inter}$. Figure 2(g) shows $S_{\rm NN}^{\rm intra}$ and $S_{\rm NN}^{\rm inter}$ as functions of T at $\alpha/\pi = 0.15$. $S_{\rm NN}^{\rm intra}$ develops rapidly around T^*, and saturates to the $T = 0$ value with decreasing T. On the other hand, $S_{\rm NN}^{\rm inter}$ shows successive growth around $T_{\rm H}^*$ and T^* before the saturation. These contrastive T dependences are understood by the difference of the exchange constants J and J'. As J is larger than J' at $\alpha/\pi = 0.15$, the intra-triangle correlation $S_{\rm NN}^{\rm intra}$ develops at a higher $T \sim T^*$, while the inter-triangle one $S_{\rm NN}^{\rm inter}$ grows at a lower $T \sim T_{\rm H}^*$. The growth of $S_{\rm NN}^{\rm inter}$ around T^* is induced by the development of $S_{\rm NN}^{\rm intra}$. Although the spin correlations show large development through the crossovers, they exhibit a slight change in the vicinity of T_c, where thermal fluctuations of the Z_2 variables $\{\eta_r\}$ are enhanced. This result indicates that the NN spin correlations are not affected by the Z_2 fluctuations significantly. In other words, as the NN spin correlations are nothing but the kinetic energy of the itinerant Majorana fermions as shown in Eqs. (11) and (12), the kinetic motion of the Majorana fermions is not disturbed by the fluctuations of the localized Majorana fermions. This behavior is common in other parameter region, as shown in Figs. 2(h) and 2(i).

Next, we focus on the case of $\alpha/\pi = 0.3$, which is still in the topologically-nontrivial CSL region but closer to the critical point at $\alpha_c/\pi = 1/3$. Figure 2(b) shows the T dependence of

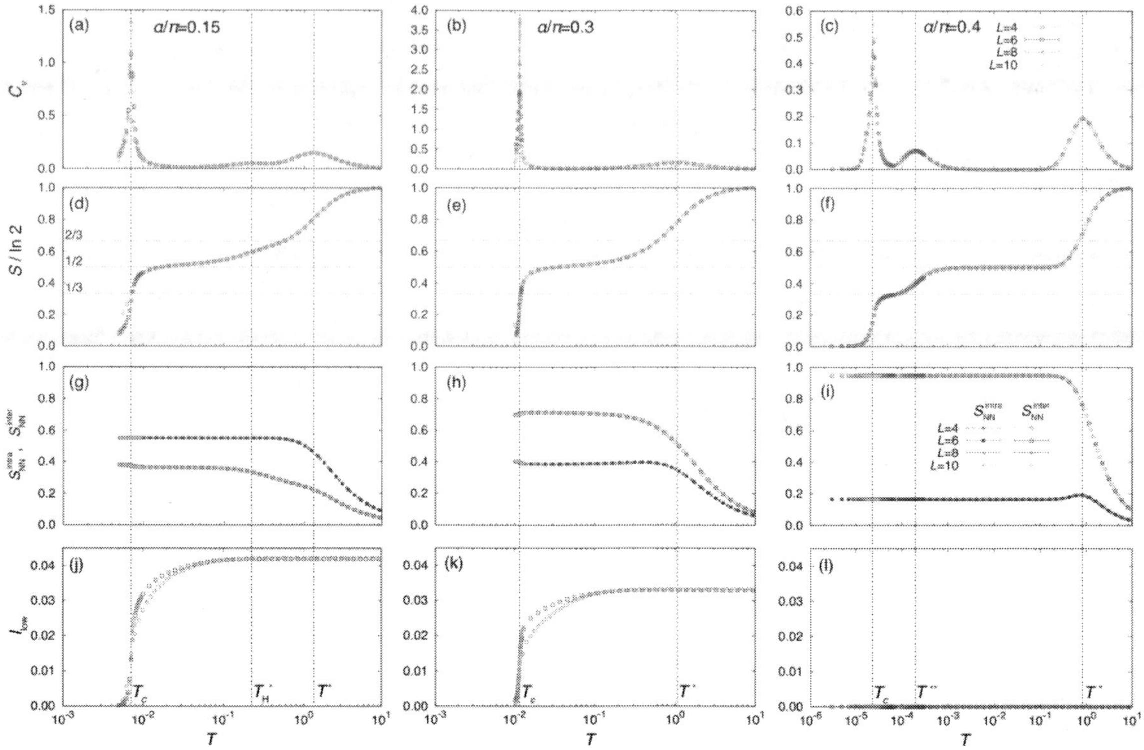

Figure 2. (a)–(c) Temperature dependences of the specific heat at (a) $\alpha/\pi = 0.15$, (b) $\alpha/\pi = 0.3$, and (c) $\alpha/\pi = 0.4$ for several cluster sizes [see Eq. (6)]. (d)–(f) The corresponding data of the entropy per site normalized by $\ln 2$ [see Eq. (8)]. (g)–(i) The NN spin correlations $S_{\rm NN}^{\rm intra}$ on the intra-triangle γ bonds with the exchange constant J and $S_{\rm NN}^{\rm inter}$ on the inter-triangle γ' bonds with the exchange constant J' [see Eqs. (11) and (12)]. (j)–(l) The low-energy weight of the DOS between 0 and $\Delta_0/2$, $I_{\rm low}$, where Δ_0 is the excitation gap at $T = 0$ [see Eq. (13)]. The vertical lines represent the transition temperature T_c and the crossover temperatures $T_{\rm H}^*$, T^*, and T^{**}.

the specific heat. The result shows a sharp peak at $T_c \simeq 0.01$ and a broad peak at $T^* \simeq 1$. The T dependence of the entropy in Fig. 2(e) indicates that the entropy is released by halves successively at T_c and T^*. This result implies that the two crossovers $T_{\rm H}^*$ and T^* at $\alpha/\pi = 0.15$ merges into a single crossover at T^* with increasing α. This is also consistent with the behavior of the NN spin correlations: $S_{\rm NN}^{\rm intra}$ and $S_{\rm NN}^{\rm inter}$ grow together around T^*, as shown in Fig. 2(h). The difference between the T dependences of the NN spin correlations at $\alpha/\pi = 0.15$ and 0.3 is understood from the DOS of the Majorana fermions, as discussed later.

With further increasing α, the ground state changes from the topologically-nontrivial CSL to the topologically-trivial CSL at $\alpha_c/\pi = 1/3$. Figure 2(c) shows the specific heat in the topologically-trivial CSL region at $\alpha/\pi = 0.4$. Similar to the result at $\alpha/\pi = 0.15$ in Fig. 2(a), there appear again two broad peaks associated with crossovers, in addition to a sharp peak corresponding to the phase transition. However, at $\alpha/\pi = 0.4$, a half of the entropy is released around the high-T crossover at $T \sim 0.8$, as shown in Fig. 2(f). Therefore, this crossover corresponds to that at T^* for $\alpha/\pi = 0.3$. Meanwhile, the low-T crossover at $T^{**} \sim 0.0002$ does not correspond to that at $T_{\rm H}^*$ for $\alpha/\pi = 0.15$. This crossover is associated with the coherent

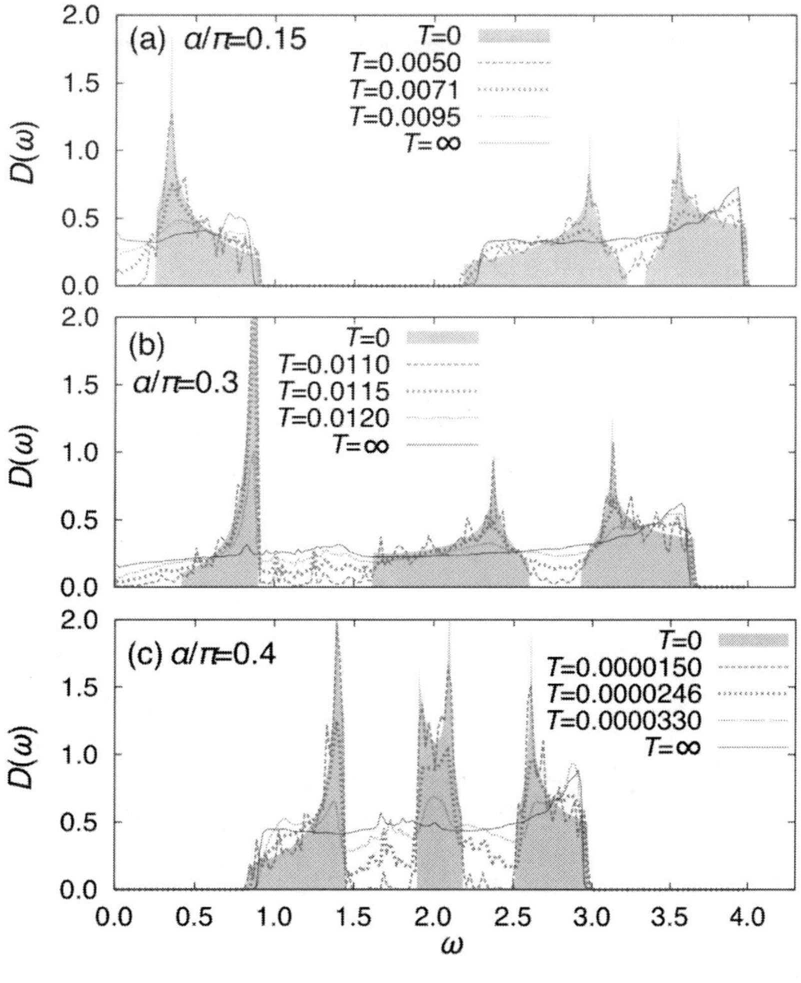

Figure 3. The DOS of the itinerant Majorana fermions, $D(\omega)$, at (a) $\alpha/\pi = 0.15$, (b) $\alpha/\pi = 0.3$, and (c) $\alpha/\pi = 0.4$ for several temperatures around T_c, which are obtained by QMC for $L = 10$ [see Eq. (9)]. For comparison, the results at $T = 0$ and $T = \infty$ are also plotted. The DOS at $T = \infty$ are obtained from 1000 random configurations of $\{\eta_r\}$ in the $L = 20$ cluster.

growth of the conserved quantity W_{d} [19, 20]. Figure 2(i) shows the T dependences of the NN spin correlations $S_{\mathrm{NN}}^{\mathrm{intra}}$ and $S_{\mathrm{NN}}^{\mathrm{inter}}$. Both of them grow with decreasing T around T^*, while $S_{\mathrm{NN}}^{\mathrm{intra}}$ shows a hump structure. However, they show no significant change at T^{**} as well as T_c.

As mentioned above, the spin correlation is given by the kinetic energy of the itinerant Majorana fermions in the current system. To gain further insight into the magnetic properties, we calculate the T evolution of the DOS of the itinerant Majorana fermions. Figure 3 shows the T dependence of the DOS $D(\omega)$ at $\alpha/\pi = 0.15$, 0.3, and 0.4. At $T = 0$, the spectrum is split into three parts for all the cases, corresponding to the three bands of the itinerant Majorana fermions. In the case of $\alpha/\pi = 0.15$, the lowest-energy band is largely separated from the two higher-energy bands at $T = 0$, as shown in Fig. 3(a). With increasing T, the gap between the two higher-energy bands is filled around T_c, and they merge into a single band above T_c. On the other hand, the gap between the lowest-energy band and the higher-energy one remains robust even at $T = \infty$. In contrast, at $\alpha/\pi = 0.3$ and 0.4, all the three Majorana fermion bands merge into a single one above T_c with increasing T, as shown in Figs. 3(b) and 3(c).

These overall evolutions of the DOS play a crucial role in the thermodynamics in the paramagnetic state above T_c. In the case of $\alpha/\pi = 0.15$, there appear two bands above T_c. This explains why the system exhibits two crossovers above T_c as follows. Suppose the DOS is approximated by a constant where it is nonzero in Fig. 3(a) for $T > T_c$, namely, $D(\omega) = D_{\text{low}}$ for $0 < \omega \lesssim 0.8$ and $D(\omega) = D_{\text{high}}$ for $2.3 \lesssim \omega \lesssim 4$, the contributions from D_{low} and D_{high} to the specific heat through Eq. (10) shows a maximum at $T \sim 0.2$ and $T \sim 1.3$, respectively. These temperatures agree well with T_{H}^{*} and T^{*}, respectively. Thus, the two well-separated bands above T_c explain the two crossovers, where the NN spin correlations $S_{\text{NN}}^{\text{inter}}$ and $S_{\text{NN}}^{\text{intra}}$ are saturated. On the other hand, at $\alpha/\pi = 0.3$ and 0.4, there is only a single band above T_c. This leads to a single crossover at T^{*}, where $S_{\text{NN}}^{\text{inter}}$ and $S_{\text{NN}}^{\text{intra}}$ develop simultaneously with decreasing T.

Next, we discuss the low-energy behavior in the DOS. In the cases of $\alpha/\pi = 0.15$ and 0.3, the energy gap to the lowest excited state is filled with increasing T in the vicinity of T_c, as shown in Figs. 3(a) and 3(b), respectively. On the other hand, as shown in Fig. 3(c), the low-energy excitation gap remains open at $\alpha/\pi = 0.4$. In order to see the T dependence of the gap more clearly, we calculate the low-energy weight of the DOS, I_{low}, which is defined by

$$I_{\text{low}} = \int_0^{\Delta_0/2} D(\omega)d\omega, \tag{13}$$

where Δ_0 represents the excitation gap at $T = 0$. Figure 2(j) shows the T dependence of I_{low} at $\alpha/\pi = 0.15$. The low-energy weight abruptly increases at T_c with increasing T. This behavior is also seen in the case of $\alpha/\pi = 0.3$ as shown in Fig. 2(k). However, as presented in Fig. 2(l), I_{low} always vanishes even at high T in the case of $\alpha/\pi = 0.4$. These results indicate that the excitation gap is filled by introducing thermal fluctuations of the Z_2 variables in the topologically-nontrivial CSL region. Namely, the low-energy gap in the non-Abelian CSL phase is fragile against thermal fluctuations of the Z_2 variables, whereas the gap in the Abelian CSL phase is robust. The gap closing in the former takes place at T_c. Therefore, in the non-Abelian CSL phase below T_c, the topological number of the Majorana fermion bands is well-defined and topological transports such as the thermal Hall effect will be observed [19]. Note that, in the present calculations, it is difficult to distinguish whether the gap closing is due to the thermal fluctuation of W_{t} or that of W_{d} because the associated entropies are released simultaneously at T_c in the cases of $\alpha/\pi = 0.15$ and 0.3. It is an intriguing issue to clarify which conserved quantity is relevant to the gap closing in the non-Abelian CSL.

Finally, in order to clarify the α dependence of the excitation gap in the paramagnetic phase above T_c, we calculate the gap in the high-T limit. This is obtained from 1000 random configurations of $\{\eta_r\}$ in the $L = 20$ cluster. Figure 4 shows the gap estimated from the DOS at $T = \infty$, Δ_∞, as a function of α, together with the value at $T = 0$, Δ_0 [8]. As mentioned before, Δ_0 vanishes at $\alpha_c/\pi = 1/3$ associated with the change of the topological nature in the Majorana fermion band. At $T = \infty$, the excitation gap disappears in the non-Abelian CSL region for $\alpha \lesssim \alpha_c$, but it does not change largely in the Abelian CSL region for $\alpha \gtrsim \alpha_c$. (We note that the gapless-gapped boundary at $T = \infty$ is slightly smaller than α_c.) This contrastive behavior is one of striking characteristics reflecting the topology of the Majorana fermion band even in the paramagnetic state above T_c. It is left for a future work to clarify the origin of the behavior.

5. Summary

To summarize, we have investigated finite-T properties in the Kitaev model on the decorated honeycomb lattice, with focusing on magnetic properties and dynamics for the Majorana fermion

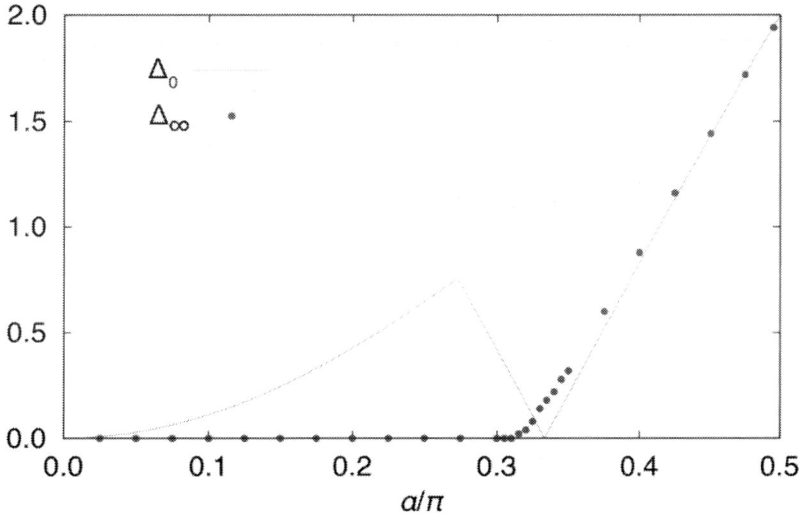

Figure 4. The excitation gaps in the itinerant Majorana fermion spectrum as functions of α: Δ_0 and Δ_∞ are the values at $T = 0$ and $T = \infty$, respectively.

system. We found that the spin correlations increase with decreasing T around the crossovers in the paramagnetic state associated with the entropy release of the itinerant Majorana fermions. The NN correlations in the intra- and inter-triangle bonds develop at different T in the small α region, whereas they develop simultaneously when α becomes larger. We also found that the DOS of the itinerant Majorana fermions changes largely at T_c, and the spectrum for $T > T_c$ gives an explanation for the crossovers with the changes of magnetic properties. Furthermore, we revealed that the excitation gap vanishes above T_c in the non-Abelian CSL region, while it remains open in the Abelian CSL region.

Acknowledgments

This work is supported by Grant-in-Aid for Scientific Research No. 15K13533, the Strategic Programs for Innovative Research (SPIRE), ME T, and the Computational Materials Science Initiative (CMSI), Japan. Parts of the numerical calculations are performed in the supercomputing systems in ISSP, the University of Tokyo.

References

[1] Bethe H A 1931 *Z. Phys.* **71** 205
[2] Hulthén L 1938 *Ark. Mat. Astron. Fys. A* **26** 1
[3] Baxter R J 1982 *Exactly Solved Models in Statistical Mechanics* (New York: Academic)
[4] Affleck I, Kennedy T, Lieb E H and Tasaki H 1987 *Phys. Rev. Lett.* **59** 799–802
[5] Shastry B S and Sutherland B 1981 *Physica B+C* **108** 1069–1070
[6] Kitaev A 2006 *Ann. Phys. (Amsterdam)* **321** 2–111
[7] Balents L 2010 *Nature* **464** 199–208
[8] Yao H and Kivelson S A 2007 *Phys. Rev. Lett.* **99** 247203
[9] Mandal S and Surendran N 2009 *Phys. Rev. B* **79** 024426
[10] Yang S, Zhou D L and Sun C P 2007 *Phys. Rev. B* **76** 180404
[11] Hermanns M and Trebst S 2014 *Phys. Rev. B* **89** 235102
[12] Kimchi I, Analytis J G and Vishwanath A 2014 *Phys. Rev. B* **90** 205126
[13] Lee E K H, Schaffer R, Bhattacharjee S and Kim Y B 2014 *Phys. Rev. B* **89** 045117
[14] Schaffer R, Lee E K H, Lu Y M and Kim Y B 2015 *Phys. Rev. Lett.* **114** 116803

[15] Nasu J, Kaji T, Matsuura K, Udagawa M and Motome Y 2014 *Phys. Rev. B* **89** 115125
[16] Nasu J, Udagawa M and Motome Y 2014 *Phys. Rev. Lett.* **113** 197205
[17] Dusuel S, Schmidt K P, Vidal J and Zaffino R L 2008 *Phys. Rev. B* **78** 125102
[18] Chung S B, Yao H, Hughes T L and Kim E A 2010 *Phys. Rev. B* **81** 060403
[19] Nasu J and Motome Y 2015 *Phys. Rev. Lett.* **115** 087203
[20] Nasu J and Motome Y unpublished *preprint* arXiv:1506.05678
[21] Chen H D and Hu J 2007 *Phys. Rev. B* **76** 193101
[22] Feng X Y, Zhang G M and Xiang T 2007 *Phys. Rev. Lett.* **98** 087204
[23] Chen H D and Nussinov Z 2008 *J. Phys. A* **41** 075001
[24] Baskaran G, Mandal S and Shankar R 2007 *Phys. Rev. Lett.* **98** 247201
[25] Hukushima K and Nemoto K 1996 *J. Phys. Soc. Jpn.* **65** 1604–1608

Journal of Physics: Conference Series **683** (2016) 012038 doi:10.1088/1742-6596/683/1/012038

Effect of monomer-monomer interactions on the phase diagrams of the S = 1/2 distorted diamond type quantum spin chain

Kiyomi Okamoto*[1], Takashi Tonegawa†[2,3] and Tôru Sakai‡[4,5]

[1]School of Arts and Sciences, College of Engineering, Shibaura Institute of Technology, Minuma-ku, Saitama-shi, Saitama, 337-8570, Japan
[2]Professor Emeritus, Kobe University, Nada-ku, Kobe-shi, Hyogo, 657-8501, Japan
[3]Department of Physical Science, School of Science, Osaka Prefecture University, Naka-ku, Sakai-shi, Osaka, 599-8531, Japan
[4]Graduate School of Material Science, University of Hyogo, Kamigori-cho, Ako-gun, Hyogo, 678-1297, Japan
[5]Japan Atomic Energy Agency, SPring-8, Sayo-cho, Sayo-gun, Hyogo, 679-5148, Japan

E-mail: *nnn2411@yahoo.co.jp, †tone0115@vivid.ocn.ne.jp, ‡sakai@spring8.or.jp

Abstract. By use of mainly the exact diagonalization and the level spectroscopy method, we investigate the ground-state phase diagrams of the $S = 1/2$ distorted diamond type quantum spin chain with the monomer-monomer interactions and/or ferromagnetic interactions for the zero magnetic field case, as well as the $M = M_s/3$ case and the $M = (2/3)M_s$ case, where M is the total magnetization and M_s is the saturation magnetization. The magnetization plateau at $M = M_s/3$ vanishes in the region where the ferromagnetic interaction is rather strong. The monomer-monomer interaction remarkably stabilizes the magnetization plateau at $M = (2/3)M_s$.

1. Introduction

In recent years, the frustration effects on low-dimensional quantum spin systems have been attracting increasing attention. The $S = 1/2$ distorted diamond type quantum spin chain [1], sketched in Fig.1, is known as an interesting model due to the frustration, the quantum nature of spins, the low dimensionality and the trimer nature. The Hamiltonian for this model is described by

$$\begin{aligned}\mathcal{H} &= J_1 \sum_j (\boldsymbol{S}_{3j\;\;1}\;\;\boldsymbol{S}_{3j} + \boldsymbol{S}_{3j}\;\;\boldsymbol{S}_{3j+1}) + J_2 \sum_j \boldsymbol{S}_{3j+1}\;\;\boldsymbol{S}_{3j+2} \\ &\quad + J_3 \sum_j (\boldsymbol{S}_{3j\;\;2}\;\;\boldsymbol{S}_{3j} + \boldsymbol{S}_{3j}\;\;\boldsymbol{S}_{3j+2}) + J_{\rm m} \sum_j \boldsymbol{S}_{3j}\;\;\boldsymbol{S}_{3(j+1)}\quad H \sum_j S_j^z \end{aligned} \qquad (1)$$

where $\boldsymbol{S}_i = (S_i^x, S_i^y, S_i^z)$ is the spin-1/2 operator at the ith site, and H denotes the magnetic filed along the z direction. Hereafter we often use the normalized coupling constants $\tilde{J}_i \quad J_i/J_1$ for $i = 1, 2, 3, {\rm m}$. In the original version of this model [1], it was supposed that all of J_1, J_2 and J_3 are antiferromagnetic and $J_{\rm m} = 0$.

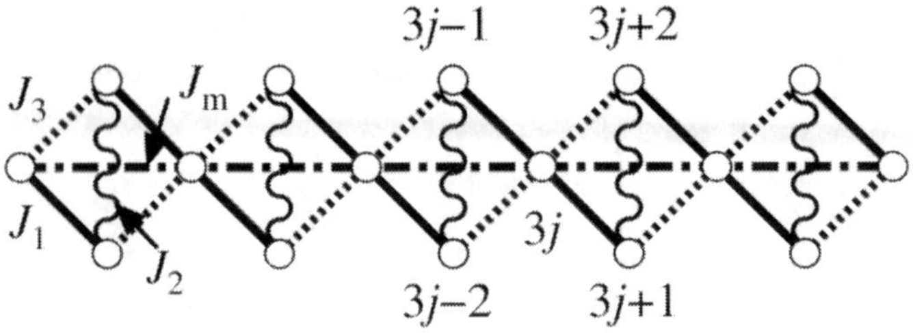

Figure 1. The distorted diamond type quantum spin chain. Open circles denote $S = 1/2$ spins, and various types of lines the interactions between spins.

Okamoto *et al.* [1–3], and Honecker and Läuchli [4] discussed the ground-state properties for the above case. The ground-state phase diagrams for the zero magnetic field case, the $M = M_s/3$ case and the $M = (2/3)M_s$ case were obtained by Okamoto *et al.* [1–3], where M is the magnetization defined by $M \quad \sum_i S_i^z$ which commutes with the total Hamiltonian $\mathcal{H}$, and M_s is the saturation magnetization. The zero magnetic field phase diagram consists of three phases, the Tomonaga-Luttinger liquid (TLL) phase, the dimerized phase and the ferrimagnetic phase. For the $M_s/3$ case, the magnetization plateau always exist except for the plateau mechanism changing line. The $(2/3)M_s$ magnetization plateau exists in a narrow region.

After the above pioneering works, Kikuchi *el al.* (including present three authors) [5, 6] reported that the magnetic properties of natural mineral azurite $Cu_3(CO_3)_2(OH)_2$ can be well described by the original distorted diamond type chain model, and estimated $J_1 : J_2 : J_3 = 1 : 1.25 : 0.45$. Gu and Su [7, 8] stated that the double peak behavior of the susceptibility $\chi(T)$ can be well fitted by the parameter set $J_1 : J_2 : J_{3z} = 1 : 1.9 : \quad 0.3$ and $J_{3x}/J_{3z} = J_{3y}/J_{3z} = 1.7$. Here, the quantities J_{3x}, J_{3y} and J_{3z} are defined by the anisotropic J_3 coupling $J_{3x}(S^x_{3j \quad 2}S^x_{3j} + S^x_{3j}S^x_{3j+2}) + J_{3y}(S^y_{3j \quad 2}S^y_{3j} + S^y_{3j}S^y_{3j+2}) + J_{3z}(S^z_{3j \quad 2}S^z_{3j} + S^z_{3j}S^z_{3j+2})$. Namely, Gu and Su claimed that the coupling J_3 is ferromagnetic and rather anisotropic, while other couplings are antiferromagnetic and isotropic. We did not agree their parameter set [9] because the angles of the Cu O Cu super exchange paths of J_1 and J_3 are very similar to each other (113.7 and 113.5 , respectively), which means that J_1 and J_3 cannot differ in sign. After the papers of Gu and Su, some papers [10, 11] support $J_3 > 0$ (antiferromagnetic) whereas others [12, 13] $J_3 < 0$ (ferromagnetic).

This controversy was resolved by Jeschke *et al.* [14] who pointed out the importance of the monomer-monomer interaction J_m by the first-principle calculation. They showed that the magnetization curve $M(H)$ and the susceptibility $\chi(T)$ can be well fitted by $J_1 : J_2 : J_3 : J_m = 1 : 2.13 : 0.45 : 0.30$. Honecker *et al.* [15] showed that the inelastic neutron scattering result [13] was well explained by the above parameter set including J_m. However, the phase diagrams in the presence of J_m have not been discussed until now.

Very recently, experimental results on new distorted diamond type chain materials have been reported, which are summarized in Table 1. In some cases the existences of J_m and ferromagnetic interactions are expected. The nonexistence of the $M_s/3$ plateau in (d) is also important, because the unit cell of the present model consists of three $S = 1/2$ spins, which often brings about the $M_s/3$ plateau by the condition of Oshikawa, Yamanaka and Affleck [28]. As already stated, we [1–3] obtained the phase diagrams of the present model for $J_1, J_2, J_3 \geq 0$ (antiferromagnetic) and $J_m = 0$ case. Considering the above situations, the phase diagrams with J_m and/or ferromagnetic interactions are urgently necessary.

In this paper, we investigate the phase diagram of the present model with J_m and/or ferromagnetic interactions by use of the exact diagonalization (ED) and the level spectroscopy (LS) method [23–26]. The phase diagrams are presented in §2, and the discussion is given in §3.

Table 1. Summary of experimental results on azurite (a) and new distorted diamond type chain materials (b)-(g). The materials for (c) is α $[Cu_3(OH)_2(CH_3CO_2)_2(H_2O)_4](C_6H_5SO_3)_2$ and that for (f) is $[Cu_3(OH)_2(CH_3CO_2)_2(OH)_2(H_2O)_2](2,6$ naphthalenedisulfonate). In (b) "A" denotes adipic acid $HOOC(CH_2)_4COOH$. In (g), the combinations of "A" and "M" are $[A, M] = [K, Ga]$, $[Rb, Al]$, $[Rb, Ga]$, $[Cs, Al]$ and $[Cs, Ga]$. The "ferromagnetic interactions" means whether there are ferromagnetic interactions within four couplings J_1, J_2, J_3 and J_m or not. All of these materials except for (f) show gapless excitation behaviors at low temperatures, which strongly suggests the TLL ground state. The spin gap behavior was found in (f), which seems to be due to the dimerization (namely, the fact that the unit cell consists of six $S = 1/2$ spins) coming from the crystal structure.

	ref.	material(s)	J_m	ferromagnetic interactions	$M_s/3$ plateau	$(2/3)M_s$ plateau
(a)	[5,6]	$Cu_3(CO_3)_2(OH)_2$	yes	no	yes (wide)	no
(b)	[16,17]	$Cu_3(A)_2(OH)_2(H_2O)_4$	?	maybe	yes (narrow)	no
(c)	[18]	see caption	?	maybe	?	?
(d)	[19,20]	$K_3Cu_3AlO_2(SO_4)_4$	maybe	no	no	?
(e)	[21]	$Cu_3(CrO_4)_2(OH)_2(C_5H_5N)_2$	?	maybe	yes (wide)	?
(f)	[22]	see caption	?	maybe	?	?
(g)	[20]	$A_3Cu_3MO_2(SO_4)_4$	maybe	no	?	?

2. Ground-state phase diagram

To determine the phase diagrams, we numerically performed the ED for finite systems up to 30 spins by use of the Lanczös algorithm under the periodic and twisted boundary conditions.

2.1. zero magnetic field case

It is expected that there are two phases, the TLL phase and the dimerized phase, for the $M = 0$ case. The former has the gapless excitation and the latter the gapped excitation. The phase boundary between these two phases is of the Berezinskii-Kosterlitz-Thouless (BKT) type, which is rather difficult to numerically detect from the ED data. We use the LS method [23, 24] which is a very powerful tool to detect the BKT transition point from the ED data. The details of the LS method for the present case are fully explained in [1].

Figure 2 shows the phase diagrams with $\tilde{J}_m = 0$ and $\tilde{J}_m = 0.3$ at zero magnetic field. The shape of the phase diagram with $\tilde{J}_m = 0.3$ was slightly modified from that with $\tilde{J}_m = 0.0$.

2.2. $M = M_s/3$ case

We determined the phase boundaries between the type A plateau region, the type B plateau region and the no plateau region from the ED data by use of the LS method [3,25–27]. The phase diagrams for the $M = M_s/3$ case with $\tilde{J}_m = 0.0$ and $\tilde{J}_m = 0.3$, as well as the spin configurations of the plateau states, are given in Fig.3. The appearance of the no $M_s/3$ plateau region in the lower-left part should be noticed. The addition of $\tilde{J}_m = 0.3$ slightly modifies the whole shape of the phase diagram.

Journal of Physics: Conference Series **683** (2016) 012038 doi:10.1088/1742-6596/683/1/012038

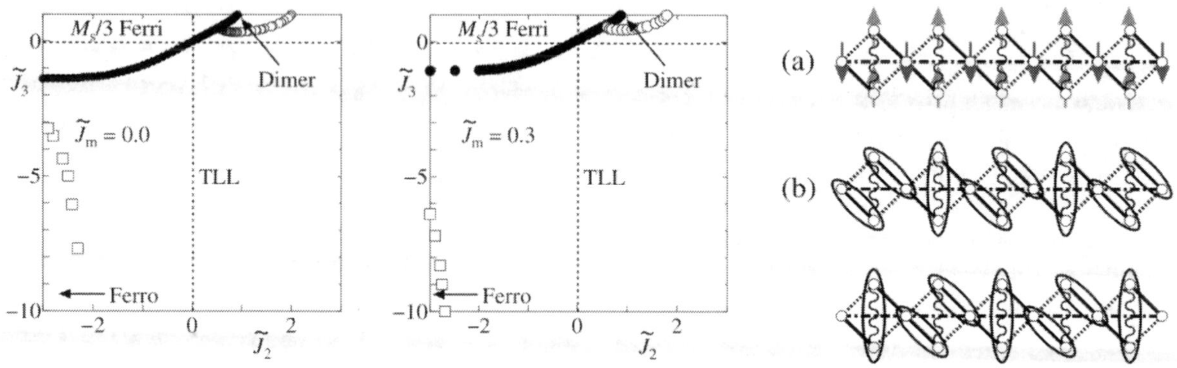

Figure 2. The left and center panels are the ground-state phase diagrams for the zero magnetic field case with $\tilde{J}_{\rm m} = 0.0$ and $\tilde{J}_{\rm m} = 0.3$, respectively. The TLL-dimer boundary (○) is of the BKT type, while other boundaries (□ and ●) are of the first order. The right panel shows the spin configurations for (a) the ferrimagnetic state and (b) the dimerized state, where ellipses denote the singlet pairs. The dimerized state is doubly degenerate.

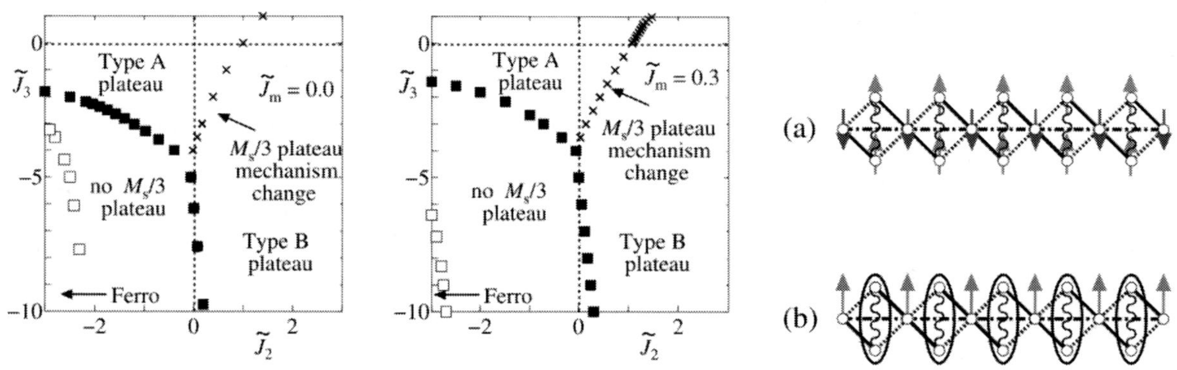

Figure 3. The left and center panels are the ground-state phase diagrams for the $M = M_{\rm s}/3$ case with $\tilde{J}_{\rm m} = 0.0$ and $\tilde{J}_{\rm m} = 0.3$, respectively. The boundary (□) between the ferromagnetic phase and the no $M_{\rm s}/3$ plateau phase is of the first order, and that (■) between the $M_{\rm s}/3$ plateau phase to the no $M_{\rm s}/3$ plateau phase is of the BKT type. The plateau mechanism changing line ( ) is of the Gaussian type. The right panel shows the spin configurations for (a) the type A plateau and (b) the type B plateau, where ellipses denote the singlet pairs. The spin configuration of the type A plateau is the same as that of the ferrimagnetic state under zero magnetic field in the right panel of Fig.2.

2.3. $M = (2/3)M_{\rm s}$ *case*

Since the unit cell of the present model consists of three $S = 1/2$ spins, the $M = (2/3)M_{\rm s}$ magnetization plateau is expected associated with spontaneous symmetry breaking of the translational invariance [28]. The LS method for the $M = (2/3)M_{\rm s}$ case was explained in [3]. Figure 4 shows the phase diagrams for the $M = (2/3)M_{\rm s}$ case with $\tilde{J}_{\rm m} = 0.0$, $\tilde{J}_{\rm m} = 0.1$ and $\tilde{J}_{\rm m} = 0.3$. We can see that the addition of $\tilde{J}_{\rm m}$ drastically extends the plateau region towards the upper left direction.

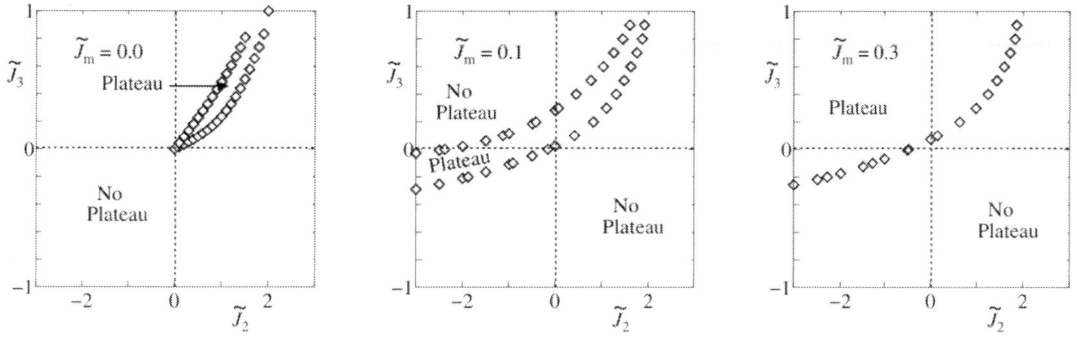

Figure 4. Ground-state phase diagrams for the $M = (2/3)M_s$ case with $\tilde{J}_m = 0.0$ (left), $\tilde{J}_m = 0.1$ (center) and $\tilde{J}_m = 0.3$ (right). The boundaries ($\Diamond$) between the plateau phase and the no plateau phase are of the BKT type.

Figure 5. The spin configurations of the $M = (2/3)M_s$ plateau states. These states are doubly degenerate because the translational invariance is spontaneously broken.

3. Discussion

We have numerically obtained the phase diagrams of the $S = 1/2$ distorted diamond type quantum spin chain with the monomer-monomer interactions and/or the ferromagnetic interactions. The remarkable nature of the phase diagrams for the $M = M_s/3$ case is the existence of the no $M = M_s/3$ plateau region in the lower-left part. Since the unit cell of the present model consists of three $S = 1/2$ spins, the appearance of the $M = M_s/3$ magnetization plateau is naturally expected [28]. However, strong quantum fluctuation can destroy this kind of magnetization plateau. In fact, in the $S = 1/2$ ferromagnetic-ferromagnetic-antiferromagnetic (F-F-AF) trimerized chain [29], the $M = M_s/3$ magnetization plateau vanishes for $|J_F/J_{AF}| > 15.4$ [27], where J_F and J_{AF} are the ferromagnetic and antiferromagnetic coupling constants, respectively. The relation between the present model and the F-F-AF trimerized chain will be discussed elsewhere.

As stated in (d) of Table 1, the $M = M_s/3$ plateau does not exist in the magnetization curve of $K_3Cu_3AlO_2(SO_4)_4$. However, unfortunately, it seems to be difficult to explain this fact from Fig.3, because the no $M = M_s/3$ plateau region appears when J_3 is strongly ferromagnetic. The reason for the absence of the $M = M_s/3$ plateau in this material is a future problem.

The effect of J_3 on the $M = (2/3)M_s$ magnetization plateau is drastic, as can be seen from Fig.4. It is easy to see that the antiferromagnetic J_m stabilizes the $M = (2/3)M_s$ magnetization plateau from the spin configuration of Fig.5. In fact, the first order perturbation theory with respect to $\tilde{J}_2$, $\tilde{J}_3$ and $\tilde{J}_m$ around the point $\tilde{J}_2 = \tilde{J}_3 = \tilde{J}_m$ leads to the plateau condition

$$\frac{5}{32}\tilde{J}_2 + \frac{1}{4}\tilde{J}_m < \tilde{J}_3 < \frac{7}{16}\tilde{J}_2 + \frac{5}{2}\tilde{J}_m \qquad (2)$$

where the lhs and rhs give the lower right and upper left boundaries, respectively. The coefficient of $\tilde{J}_m$ in the rhs is ten times as large as that in the lhs, which semi-quantitatively explains the fact that the plateau region strongly extends towards the upper left direction as $\tilde{J}_m$ increases.

As ca be seen from Fig.5, the $M = (2/3)M_s$ plateau is attributed to the two-fold degeneracy of the wave function [28]. Other types of degeneracy such as three-fold, four-fold, etc. may bring about other plateaus. However, in this work, we could not find any evidence for the plateaus other than $M = M_s/3$ and $M = (2/3)M_s$ plateaus.

We have numerically calculated the magnetization curve by the ED at several points on the phase diagrams. All of their behaviors are consistent with the phase diagrams obtained by the LS method. The details of the magnetization process will be published elsewhere.

Acknowledgments
We thank Hikomitsu Kikuchi and Masayoshi Fujihala for informing us their experimental results prior to publications. For the numerical diagonalization, we used the package program TITPACK ver.2 developed by Hidetoshi Nishimori, to whom we are grateful. For computer facilities, we also thank the Supercomputer Center, Institute of Solid State Physics, University of Tokyo, and the Computer Room, Yukawa Institute of Theoretical Physics, Kyoto University.

References

[1] Okamoto K, Tonegawa T, Takahashi Y and Kaburagi M 1999 *J. Phys.: Cond. Mat.* **11** 10485
[2] Tonegawa T, Okamoto K, Hikihara T, Takahashi T and Kaburagi M 2000 *J. Phys. Soc. Jpn.* **69** Suppl. A 332
[3] Okamoto K, Tonegawa T, Takahashi Y and Kaburagi M 2003 *J. Phys.: Cond. Mat.* **15** 5979
[4] Honecker A and Läuchli A 2001 *Phys. Rev.* B **63** 174407
[5] Kikuchi H, Fujii Y, Chiba M, Mitsudo S, Idehara T, Tonegawa T, Okamoto K, Sakai T, Kuwai T and Ohta H 2005 *Phys. Rev. Lett.* **94** 227201
[6] Kikuchi H, Fujii Y, Chiba M, Mitsudo S, Idehara T, Tonegawa T, Okamoto K, Sakai T, Kuwai T, Kindo K, Matsuo A, Higemoto W, Nishiyama K, Horović M and Bertheir C 2005 *Prog. Theor. Phys.* Suppl. No.159, 1
[7] Gu B and Su G 2006 *Phys. Rev. Lett.* **97** 089701
[8] Gu B and Su G 2007 *Phys. Rev.* B **75** 174437
[9] Kikuchi H, Fujii Y, Chiba M, Mitsudo S, Idehara T, Tonegawa T, Okamoto K, Sakai T, Kuwai T and Ohta H 2006 *Phys. Rev. Lett.* **97** 089702
[10] Mikeska H-J and Luckmann C 2008 *Phys. Rev.* B **77** 054405
[11] Kang J, Lee C, Kremer R K and Whangbo M-H 2009 *J. Phys.: Cond. Mat.* **21** 392201
[12] Li Y-C 2007 *J. Appl. Phys.* **102** 113907
[13] Rule K C, Wolter A U B, Su Süllow, Tennant D A, Brühl A, Köhler S, Wolf B, Lang M and Schreuer J 2008 *Phys. Rev. Lett.* **100** 1172202
[14] Jeschke H, Opahle I, Kandpal H, Valentí R, Das H, Saha-Dasgupta T, Janson O, Rosner H, Brühl A, Wolf B, Lang M, Richter J, Hu S, Wang X, Peters R, Pruschke T and Honecker A 2011 *Phys. Rev. Lett.* **106** 217201
[15] Honecker A, Hu S, Peters R and Richter J 2011 *J. Phys.: Cond. Mat.* **23** 164211
[16] Kikuchi H, Fujii Y, Matsuo A, Kindo K 2013 JPS 68th Annual Meeting 26aPS-87 and private communications
[17] Asano Y, Kikuchi H, Fujii Y, Matsuo A and Kindo K 2013 JPS 2013 Autumn Meeting 28aKF-6 and private communications
[18] Yoneyama S, Kodama T, Kikuchi K, Fujii Y 2014 Kikuchi H and Fujita W 2014 CrystEngComm **16** 10385
[19] Fujihala M, Koorikawa H, Mitsuda S, Hagihala M, Morodomi H, Kawae T, Matsuo A, and Kindo K 2015 *J. Phys. Soc. Jpn.* **84** 073702
[20] Koorikawa H, Fujihala M, Mitsuda S, Sagayama H, Kumai R, Murakami Y,Nakamura D and Takeyama S 2015 JPS 2015 Autumn Meeting 16aPS-46 and private communications
[21] Asano T, Ichimura S, Inagaki Y, Kawae T, Matsuo, Kindo K, Matsumoto M and Kikuchi H 2015 JPS 2015 Autumn Meeting 16aPS-38 and private communications
[22] Fujita W, Fujii Y, Kikuchi H 2015 JPS 2015 Autumn Meeting 16aPS-39 and private communications
[23] Okamoto K and Nomura K 1992 *Phys. Lett.* A **169** 433
[24] Nomura K and Okamoto K 1994 *J. Phys. A: Math. Gen.* **27** 5773
[25] Kitazawa A 1997 *J. Phys. A: Math. Gen.* **30** L285
[26] Nomura K and Kitazawa A 1998 *J. Phys. A: Math. Gen.*, **31** 7341
[27] Kitazawa A and Okamoto K 1999 *J. Phys.: Cond. Mat.* **11** 9765
[28] Oshikawa M, Yamanaka M and Affleck I 1997 *Phys. Rev. Lett.* **78** 1984
[29] Hida K 1994 *J. Phys. Soc. Jpn.* **63** 2359

TMU International Symposium 2015 IOP Publishing
Journal of Physics: Conference Series **683** (2016) 012039 doi:10.1088/1742-6596/683/1/012039

Ground-state phase diagram of an anisotropic S=1/2 ladder with alternating rung interactions

Takashi Tonegawa[1,2], **Kiyomi Okamoto**[3], **Toshiya Hikihara**[4] **and Tôru Sakai**[5,6]

[1]Professor Emeritus, Kobe University, Kobe 657-8501, Japan
[2]Department of Physical Science, Osaka Prefecture University, Sakai, 599-8531, Japan
[3]College of Engineering, Shibaura Institute of Technology, Saitama, 337-8570, Japan
[4]Faculty of Science and Technology, Gunma University, Kiryu, 376-8515, Japan
[5]Graduate School of Material Science, University of Hyogo, Hyogo 678-1297, Japan
[6]Japan Atomic Energy Agency, SPring-8, Hyogo 679-5148, Japan

E-mail: tone0115@vivid.ocn.ne.jp

Abstract. Employing mainly numerical methods, we explore the ground-state phase diagram of an anisotropic $S=1/2$ ladder, in which leg interactions are uniform and isotropic, while rung interactions are alternating and have a common Ising-type anisotropy. We determine the phase diagram in the case where $J_{\rm leg}=0.2$ (antiferromagnetic), $J_{\rm rung}=-1.0$ (ferromagnetic) and $|J'_{\rm rung}|\leq 1.0$, the first one being the magnitude of the leg interaction and the second and third ones those of the rung interactions, which are alternating. It is emphasized that the system has a frustration when $J'_{\rm rung}$ is positive. We find that, in the frustrated region, the Haldane state appears as the ground state even when the Ising character of rung interactions is strong. This appearance of the Haldane phase is contrary to the ordinary situation, and it is called the inversion phenomenon concerning the interaction anisotropy. We also find that an incommensurate state becomes the ground state in a portion of the Haldane phase region.

1. Introduction

The frustration effect on the ground-state properties of low-dimensional quantum spin systems with competing interactions has long been a subject of active research. According to a significant amount of theoretical and experimental effort which has been devoted so far, it is now widely known that an interplay between two phenomena of great interest, frustration and quantum fluctuation, leads to various exotic ground states. A typical and long-established example of these ground states is the dimer state accompanying spontaneous translational symmetry breaking in an $S=1/2$ zigzag chain in which antiferromagnetic nearest-neighbor (nn) and next-nearest-neighbor (nnn) interactions are competing with each other [1–3].

The effect of the frustration on the ground-state properties of an $S=1/2$ two-leg ladder system has been extensively studied in the cases where additional leg nnn and/or diagonal interactions are competing with the leg nn and rung interactions [4, 5]. In the present paper, as another example of the frustrated $S=1/2$ two-leg ladder systems, we discuss the case where the rung interactions are alternating [6, 7] and aim at exploring its ground-state phase diagram. We express the Hamiltonian describing this system in the following form:

$$\mathcal{H} = J_{\rm leg} \sum\nolimits_{j=1}^{L} \{\vec{S}_{j,a}\cdot\vec{S}_{j+1,a} + \vec{S}_{j,b}\cdot\vec{S}_{j+1,b}\}$$

$$+ J_{\rm rung} \sum_{j=1}^{L/2} \{\gamma(S_{2j-1,a}^x S_{2j-1,b}^x + S_{2j-1,a}^y S_{2j-1,b}^y) + S_{2j-1,a}^z S_{2j-1,b}^z\} \qquad (1)$$
$$+ J'_{\rm rung} \sum_{j=1}^{L/2} \{\gamma(S_{2j,a}^x S_{2j,b}^x + S_{2j,a}^y S_{2j,b}^y) + S_{2j,a}^z S_{2j,b}^z\},$$

where $\vec{S}_{j,l} = (S_{j,l}^x, S_{j,l}^y, S_{j,l}^z)$ is the $S=1/2$ operator acting at the (j,l) site assigned by rung j and leg $l(=a$ or $b)$; $J_{\rm leg}$ denotes the magnitude of the isotropic leg interaction; $J_{\rm rung}$ and $J'_{\rm rung}$ denote those of the two kinds of anisotropic rung interactions which are alternating, the XXZ-type anisotropy being controlled by the parameter γ in common with both interactions; L is the total number of spins in each leg, which is assumed to be even. It is emphasized that this system has a frustration when $J_{\rm rung}J'_{\rm rung} < 0$ irrespective of the sign of $J_{\rm leg}$.

Although materials corresponding to the present model have been neither yet found nor synthesized so far, we believe that it is a physically realistic model. In fact, for example, Yamaguchi *et al.* [8, 9] have recently demonstrated the modulation of magnetic interactions in spin ladder systems by using verdazyl-radical crystals. The flexibility of molecular arrangements in such organic-radical materials is expected to realize spin ladder systems with alternating rung interactions.

In the following discussions, we confine ourselves to the case where $J_{\rm rung}$ is ferromagnetic, and we put $J_{\rm rung} = -1$, choosing $|J_{\rm rung}|$ as the unit of energy. We also consider, for simplicity, only the case where $J_{\rm leg} = 0.2$, $|J'_{\rm rung}| \le 1$ and $0 \le \gamma < 1$, that is, we assume that $J_{\rm leg}$ is antiferromagnetic and relatively weak, and that the anisotropy of the rung interactions is of the Ising-type. Determining the ground-state phase diagram on the γ versus $J'_{\rm rung}$ plane, we mainly employ the numerical methods such as the exact-diagonalization (ED) method and the density-matrix renormalization-group (DMRG) method [10, 11] with the help of physical considerations as well as already-known results for some special cases.

2. Special cases

We discuss here some special cases for which the ground states have already been clarified or for which they are reasonably anticipated by physical considerations.

2.1. Case where $J_{\rm rung} = J'_{\rm rung}\ (= -1)$

In this case where $0 < J_{\rm leg} \ll |J_{\rm rung}| = |J'_{\rm rung}|$, by the use of the degenerate perturbation theory, the present system can be mapped onto the $S=1$ chain in the following way. The eigenstates of an isolated $J_{\rm rung}$- or $J'_{\rm rung}$-rung are given by $\phi_j^{(1,+)} = \alpha_{j,a}\alpha_{j,b}$, $\phi_j^{(1,0)} = (\alpha_{j,a}\beta_{j,b} + \beta_{j,a}\alpha_{j,b})/\sqrt{2}$, $\phi_j^{(1,-)} = \beta_{j,a}\beta_{j,b}$ and $\phi_j^{(0,0)} = (\alpha_{j,a}\beta_{j,b} - \beta_{j,a}\alpha_{j,b})/\sqrt{2}$, where $\alpha_{j,l}$ denotes the $S_{j,l}^z = +1/2$ state and $\beta_{j,l}$ the $S_{j,l}^z = -1/2$ state. The corresponding energies are, respectively, $E^{(1,+)} = -1/4$, $E^{(1,0)} = (1-2\gamma)/4$, $E^{(1,-)} = -1/4$ and $E^{(0,0)} = (1+2\gamma)/4$, for all j's. Then, it is easy to see that when γ is sufficiently large, the state $\phi_j^{(0,0)}$ can be neglected. We introduce the pseudo $S=1$ operator $\vec{T}_j$ for rung j, and make the $T_j^z = +1$, 0 and -1 states correspond to $\phi_j^{(1,+)}$, $\phi_j^{(1,0)}$ and $\phi_j^{(1,-)}$, respectively. The relation $\vec{T}_j = \vec{S}_{j,a} + \vec{S}_{j,b}$ holds, as is readily shown by comparing the matrix elements of both operators $\vec{T}_j$ and $\vec{S}_{j,l}$ with respect to $\phi_j^{(1,+)}$, $\phi_j^{(1,0)}$ and $\phi_j^{(1,-)}$. Thus, the Hamiltonian (1) for the $S=1/2$ operator $\vec{S}_{j,l}$ can be mapped onto the effective Hamiltonian $\mathcal{H}_{\rm eff}$ for the $S=1$ operator $\vec{T}_j$, which is given by

$$\mathcal{H}_{\rm eff} = 0.1 \sum_{j=1}^{L} \vec{T}_j \cdot \vec{T}_{j+1} + D \sum_{j=1}^{L} (T_j^z)^2, \qquad D = (\gamma - 1)/2, \qquad (2)$$

where the on-site anisotropy (D-) term comes from the difference between $E^{(1,+)} = E^{(1,-)}$ and $E^{(1,0)}$.

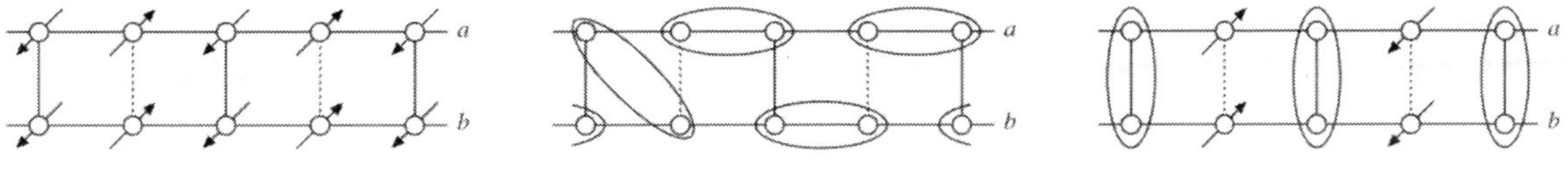

Figure 1. Schematic pictures of the AFstN (left panel), Haldane (center panel) and F-SD (right panel) phases. Each of the small open circles denotes an $S=1/2$ spin. Two $S=1/2$ spins in each ellipse form a singlet dimer. An $S=1/2$ spin with an upward arrow and that with a downward arrow are, respectively, in the $S^z=+1/2$ (α) and $S^z=-1/2$ (β) states.

It has already been clarified by Chen *et al.* [13] that in the $S=1$ chain governed by the effective Hamiltonian (2), the phase transition from the Néel to the Haldane state takes place at $D\sim -0.04$ as D increases. This suggests that in the present $S=1/2$ ladder, the ground state is the antiferromagnetic stripe Néel (AFstN) state, sketched in the left panel of Fig. 1, or the Haldane state, sketched in the center panel of Fig. 1, depending upon whether $\gamma<\gamma_{\rm c}^{\rm (AFstN,H)}$ or $\gamma_{\rm c}^{\rm (AFstN,H)}<\gamma<1$ with $\gamma_{\rm c}^{\rm (AFstN,H)}\sim 0.9$. (It is noted that the Néel state in the $S=1$ chain corresponds to the AFstN state in the $S=1/2$ ladder.)

2.2. Case where $|J_{\rm rung}|(=1)\gg J'_{\rm rung}>0$
In this case, the procedure discussed in the previous subsection can be applied to the $J_{\rm rung}$-rung only. Thus, the Hamiltonian (1) can be mapped onto an anisotropic '$S=1$'-'$S=1/2$' diamond chain. Hida and Takano [12] have studied the ground state of this chain. By using their results, we may conclude that, when $\gamma\to 1.0$, the ground state in this case is the Haldane state if $J'_{\rm rung}\lesssim 0.26$.

2.3. Case where $J'_{\rm rung}\sim -J_{\rm rung}(=1)$
In this case where $0<J_{\rm leg}\ll|J_{\rm rung}|\sim|J'_{\rm rung}|$, the lowest-energy state of an isolated $J'_{\rm rung}$-rung is the singlet dimer state $\phi_{2j}^{(0,0)}$, while that of the ferromagnetic $J_{\rm rung}$-rung is one of the ferromagnetic states $\phi_{2j-1}^{(1,+)}$ and $\phi_{2j-1}^{(1,-)}$. The effective interaction between the neighboring $J_{\rm rung}$-rungs through the in-between $J'_{\rm rung}$-rung can be obtained by carrying out a third-order perturbation calculation. The resulting effective interaction is considered to be antiferromagnetic judging from the numerical result discussed below (see the right panel of Fig. 4). Therefore, the $J_{\rm rung}$-rung in the $\phi_{2j-1}^{(1,+)}$ (or $\phi_{2j-1}^{(1,-)}$) state and that in the $\phi_{2j+1}^{(1,-)}$ ($\phi_{2j+1}^{(1,+)}$) state arrange alternatively. Thus, the ground state in the present case is anticipated to be the 'ferromagnetic'-'singlet dimer' (F-SD) state sketched in the right panel of Fig. 1.

3. Ground-state phase diagram

We denote, respectively, by $E_0^{\rm (p)}(L,M)$ and $E_1^{\rm (p)}(L,M)$, the lowest and second-lowest energy eigenvalues of the Hamiltonian (1) under periodic boundary conditions within the subspace characterized by L and the total magnetization $M\equiv\sum_{j=1}^{L}(S_{j,a}^z+S_{j,b}^z)$. It is noted that $E_0^{\rm (p)}(L,0)$ always gives the ground-state energy for the finite-L system. Then, the excitation energy gap $\Delta_{00}(L)$ within the $M=0$ subspace is given by

$$\Delta_{00}(L)=E_1^{\rm (p)}(L,0)-E_0^{\rm (p)}(L,0)\,. \tag{3}$$

We also define the site magnetization $m_{j,l}(L)$ as

$$m_{j,l}(L)=\langle S_{j,l}^z\rangle_L\,, \tag{4}$$

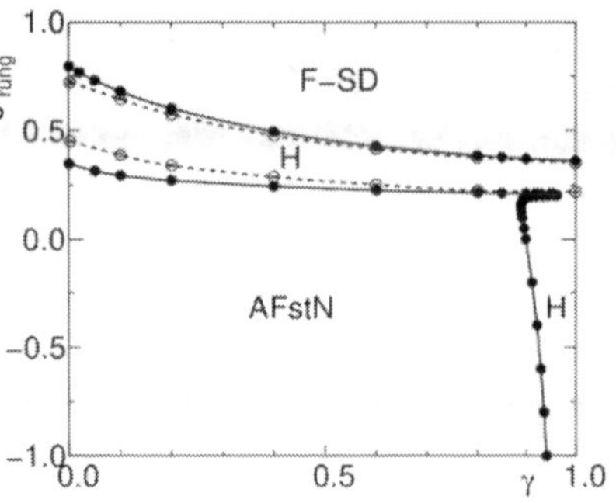

Figure 2. Ground-state phase diagram on the γ versus $J'_{\rm rung}$ plane. Here, F-SD, H and AFstN stand, respectively, for 'ferromagnetic'-'singlet dimer', Haldane and antiferromagnetic stripe Néel. The solid lines are the second-order (2D Ising) phase boundary lines. The dotted lines, being the Lifshitz lines, separate the commensurate and incommensurate regions; the region between these lines is the incommensurate one.

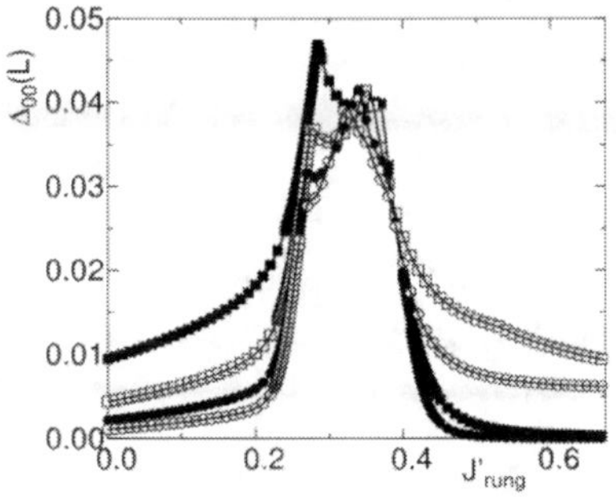

Figure 3. $J'_{\rm rung}$-dependence of the excitation energy gap $\Delta_{00}(L)$ for $L=8$ (closed squares), 10 (open squares), 12 (closed circles) and 14 (open circles) along the $\gamma=0.6$ line.

where $\langle\cdots\rangle_L$ denotes the ground-state expectation value for the system with L rungs.

Figure 2 shows our final result for the ground-state phase diagram on the γ versus $J'_{\rm rung}$ plane. It consists of the F-SD, Haldane and AFstN phases. The solid lines are the 2D Ising phase boundary line between the F-SD and Haldane phases and that between the Haldane and AFstN phases. On the other hand, the dotted lines, which are often called the Lifshitz lines, separate the commensurate and incommensurate regions; the latter region is between them. The most striking feature of this phase diagram is the fact that an incommensurate region appears within the Haldane phase region, which can be attributed to the frustration effect. The fact that the Haldane state appears as the ground state even at the $\gamma\to0$ limit is also interesting. This is because the Haldane state is known to become the ground state mainly in the case of the XY-type anisotropy in the $S=1$ XXZ chain [13]. This phenomenon is called the inversion phenomenon concerning the interaction anisotropy [14–17].

Let us turn to a discussion on the determination of the phase diagram. In Fig. 3 we plot the $J'_{\rm rung}$-dependence of the excitation energy gap $\Delta_{00}(L)$ for $\gamma=0.6$, calculated by the ED method. This figure demonstrates that the ground state is doubly degenerate when

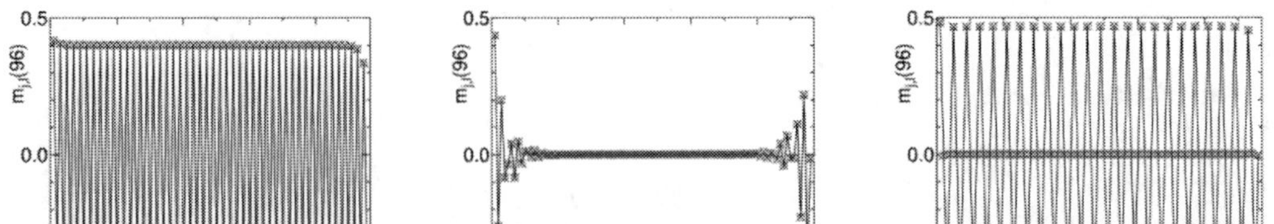

Figure 4. Plots of the site magnetizations $m_{j,l}(96)$ $(l=a,\ b)$ versus j with γ fixed at $\gamma=0.6$; the left, center and right panels are, respectively, for $J'_{\rm rung}=0.1$, 0.3 and 0.5. The red circles and the blue crosses show, respectively, $m_{j,a}(96)$ and $m_{j,b}(96)$.

$-1 \leq J'_{\rm rung} \lesssim 0.22$ and when $0.43 \lesssim J'_{\rm rung} \leq 1$, while it is unique when $0.22 \lesssim J'_{\rm rung} \lesssim 0.43$. From these results together with the physical considerations discussed in section 2, we may expect that, when $-1 \leq J'_{\rm rung} \lesssim 0.22$, $0.22 \lesssim J'_{\rm rung} \lesssim 0.43$ and $0.43 \lesssim J'_{\rm rung} \leq 1$, the ground states are, respectively, the AFstN, Haldane and F-SD states. In order to ascertain these expectations, we have performed DMRG calculations for the finite-size system with $2L = 192$ spins under open boundary conditions to evaluate the site magnetization $m_{j,l}(96)$ for various values of $J'_{\rm rung}$ [18]. As examples, the results for the cases of $J'_{\rm rung} = 0.1$, 0.3 and 0.5 are depicted in Fig. 4. We see from this figure that in all cases $m_{j,a}(96)$ and $m_{j,b}(96)$ are almost equal to each other. In the first case, $m_{2j-1,a}(96) \simeq -m_{2j,a}(96)$, which shows that the ground state in this case is the AFstN state. In the second case, the edge states clearly exist; this is one of the most representative features of the Haldane state [19–21]. Finally in the third case, $m_{4j-3,a}(96) \simeq -m_{4j-1,b}(96)$ and $m_{2j,a}(96) \simeq 0$, which show that the ground state is the F-SD state.

Both of the phase transition between the AFstN and Haldane states and that between the Haldane and F-SD states are of the 2D Ising type, since the Z_2 symmetry is broken in the AFstN and F-SD states while it is not broken in the Haldane state. It is well known that the phenomenological renormalization group (PRG) method [22] is a useful one to determine the phase boundary line for this phase transition. The PRG equations for the (AFstN,Haldane)-transition and for the (Haldane,F-SD)-transition are, respectively, given by

$$L\,\Delta_{00}(L) = (L+2)\,\Delta_{00}(L+2)\,, \qquad L\,\Delta_{00}(L) = (L+4)\,\Delta_{00}(L+4)\,. \tag{5}$$

This is because the periods along each leg are 2 and 4 in the AFstN and F-SD states, respectively. Solving numerically these PRG equations for a given value of γ (or $J'_{\rm rung}$), we have computed the finite-size (AFstN,Haldane)-transition point $J'^{\rm (AFstN,H)}_{\rm rung,c}(L)$ ($\gamma_{\rm c}^{\rm (AFstN,H)}(L)$) for $L = 6$, 8, 10, 12, and the finite-size (Haldane,F-SD)-transition point $J'^{\rm (H,F-SD)}_{\rm rung,c}(L)$ ($\gamma_{\rm c}^{\rm (H,F-SD)}(L)$) for $L = 4$, 8, 12. We have extrapolated these finite-size data to the thermodynamic ($L \to \infty$) limit by fitting them for the former and for the latter to quadratic functions of $(L+1)^{-2}$ and of $(L+2)^{-2}$, respectively. Some examples of the results are $J'^{\rm (AFstN,H)}_{\rm rung,c}(\infty) = 0.225 \pm 0.001$ and $J'^{\rm (H,F-SD)}_{\rm rung,c}(\infty) = 0.428 \pm 0.001$ for $\gamma = 0.6$, and also $\gamma_{\rm c}^{\rm (AFstN,H)}(\infty) = 0.942 \pm 0.002$ for $J'_{\rm rung} = -1.0$. It is noted that the above value of $\gamma_{\rm c}^{\rm (AFstN,H)}(\infty)$ for $J'_{\rm rung} = -1.0$ is in fairly good agreement with the corresponding value ~ 0.9 obtained by mapping the present Hamiltonian (1) onto the effective Hamiltonian (2) for the $S = 1$ chain (see subsection 2.1). We have carried out the same procedure for various values of γ and $J'_{\rm rung}$ to obtain the 2D Ising phase boundary lines shown by the solid lines in Fig. 2.

The Lifshitz line which separates the commensurate and incommensurate regions can be estimated by examining the Fourier transform of the site magnetization $m_{j,l}(L)$ [23,24]. When we adopt open boundary conditions, the present $S = 1/2$ rung-alternating ladder system has no inversion symmetry with respect to its center position, that is, $m_{j,l}(L) \neq m_{L+1-j,l}(L)$. Therefore, it is important, especially in the region of the Haldane phase, to discuss two kinds of the Fourier transforms, $\mathcal{S}^z_q(L;{\rm left})$ and $\mathcal{S}^z_q(L;{\rm right})$, defined by

$$\mathcal{S}^z_q(L;{\rm left}) = \frac{1}{\sqrt{L/2}} \sum\nolimits_{j=1}^{L/2} \exp(iqj)\, m_j(L)\,, \quad \mathcal{S}^z_q(L;{\rm right}) = \frac{1}{\sqrt{L/2}} \sum\nolimits_{j=1}^{L/2} \exp(iqj)\, m_{L+1-j}(L) \tag{6}$$

with $m_j(L) = m_{j,a}(L) + m_{j,b}(L)$, in order to avoid the mismatch of the edge states of both edges. In the above equations q is the wave number.

In Fig. 5 the q-dependences of the squared modulus of $\mathcal{S}^z_q(192;{\rm left})$, which have been calculated by using the DMRG results for $m_{j,l}(192)$ for the finite-size system with $2L = 384$ spins, are shown for the cases of $J'_{\rm rung} = 0.1$, 0.3, 0.37 and 0.5. It is noted that the q-dependences of $|\mathcal{S}^z_q(192;{\rm right})|^2$ show almost the same behavior. Thus, the values $q^z_0(192;{\rm left})$

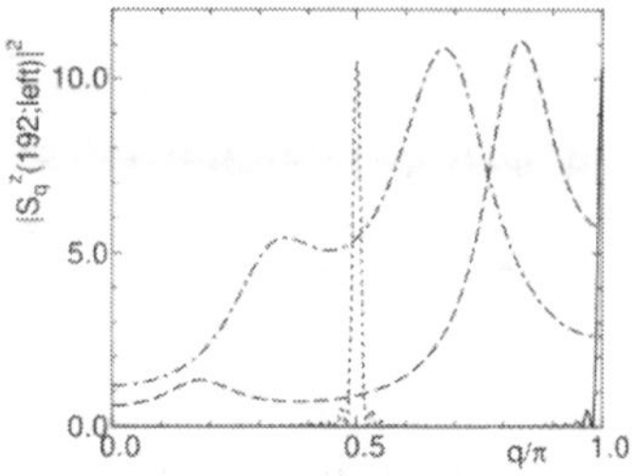

Figure 5. Squared modulus of the Fourier transform of the site magnetization, $|S_q^z(192;\text{left})|^2$, plotted versus q/π with γ fixed at $\gamma=0.6$. The solid, dashed, dot-dashed and dotted lines show, respectively, the result for $J'_{\text{rung}}=0.1$ (divided by 3), that for $J'_{\text{rung}}=0.3$ (multiplied by 460), that for $J'_{\text{rung}}=0.37$ (multiplied by 900) and that for $J'_{\text{rung}}=0.5$.

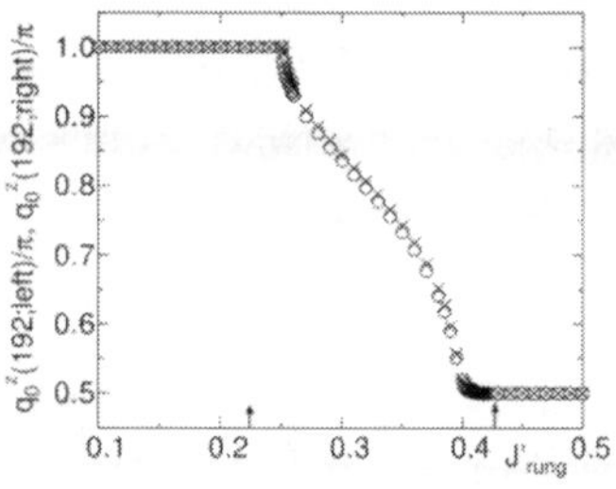

Figure 6. Plots of $q_0^z(192;\text{left})$ (circles) and $q_0^z(192;\text{right})$ (crosses) versus J'_{rung} with γ fixed at $\gamma=0.6$. The two arrows on the abscissa indicate the (AFstN,Haldane)-transition and (Haldane,F-SD)-transition points.

and $q_0^z(192;\text{right})$ of q which give, respectively, the maximum values of $|S_q^z(192;\text{left})|^2$ and $|S_q^z(192;\text{right})|^2$ are almost equal to each other. Figure 6 depicts the plots of $q_0^z(192;\text{left})$ and $q_0^z(192;\text{right})$ versus J'_{rung} for $\gamma=0.6$. From this figure we clearly see that, there are two Lifshitz points $J'^{(1)}_{\text{rung,Lifshitz}}(192)$ and $J'^{(2)}_{\text{rung,Lifshitz}}(192)$, and in the region of $J'^{(1)}_{\text{rung,Lifshitz}}(192) \le J'_{\text{rung}} \le J'^{(2)}_{\text{rung,Lifshitz}}(192)$, the system has the incommensurate character in the sense that both of $q_0^z(192;\text{left})$ and $q_0^z(192;\text{right})$ are larger than $\pi/2$ and smaller than π, where $J'^{(1)}_{\text{rung,Lifshitz}}(192)=0.251 \pm 0.001$ and $J'^{(2)}_{\text{rung,Lifshitz}}(192)=0.415 \pm 0.001$. These values of the Lifshitz points yield good approximations for the corresponding $L\to\infty$ ones, since our calculations show that the finite-size values for $L=192$, 96 and 72 agree with each other within numerical errors. We have performed the same procedure for various values of γ, and obtained the Lifshitz lines shown by the dotted lines in Fig. 2. It is noted that, in the region between these two Lifshitz lines, the ground state of the present system has an incommensurate character.

4. Summary

We have numerically determined, with the help of some physical considerations, the ground-state phase diagram of an anisotropic rung-alternating $S=1/2$ ladder, which is described by the Hamiltonian (1), in the case where $J_{\text{leg}}=0.2$, $J_{\text{rung}}=-1$, $|J'_{\text{rung}}|\le 1$ and $0\le\gamma<1$. The obtained phase diagram on the γ versus J'_{rung} plane is shown in Fig. 2. We hope that the present research stimulates future experimental studies on related subjects, which include the synthesization of spin ladder systems with alternating rung interactions.

Acknowledgments

We would like to express our sincere thanks to Professors U Schollwöck, T Vekua, K Hida, J Richter and H Yamaguchi for their invaluable discussions and comments. We also thank the Supercomputer Center, Institute for Solid State Physics, University of Tokyo and the Computer Room, Yukawa Institute for Theoretical Physics, Kyoto University for computational facilities. Finally, T.H. acknowledges the support by JSPS KAKENHI Grant Nunmber 15K05198.

References

[1] Majumdar C K and Ghosh D K 1969 *J. Math. Phys.* **10** 1399

[2] Tonegawa T and Harada I 1987 *J. Phys. Soc. Jpn.* **56** 2153
[3] Okamoto K and Nomura K 1992 *Phys. Lett.* A **169** 433
[4] Lavarélo A, Guillaume G and Laflorencie N 2011 *Phys. Rev.* B **84** 144407 and references therein
[5] Vekua T and Honecker A 2006 *Phys. Rev.* B **73** 214427 and references therein
[6] Japaridze G I and Pogosyan E 2006 *J. Phys.: Condens. Matter* **18** 9297
[7] Amiri F, Sun G, Mikeska H-J and Vekua T 2015 arXiv:1509.04062
[8] Yamaguchi H, Iwase K, Ono T, Shimokawa T, Nakano H, Shimura Y, Kase N, Kittaka S, Sakakibara T, Kawakami T, and Hosokoshi Y 2013 *Phys. Rev. Lett.* **110** 157205
[9] Yamaguchi H, Miyagai H, Shimokawa T, Iwase K, Ono T, Kono Y, Kase N, Araki K, Kittaka S, Sakakibara T, Kawakami T, Okunishi K and Hosokoshi Y 2014 *J. Phys. Soc. Jpn.* **83** 033707
[10] White S R 1992 *Phys. Rev. Lett.* **69** 2863
[11] White S R 1993 *Phys. Rev.* B **48** 10345
[12] Hida K and Takano K 2011 *J. Phys. Soc. Jpn.* **80** 104710
[13] Chen W, Hida K and Sanctuary B C 2003 *Phys. Rev.* B **67** 104401
[14] Okamoto K and Ichikawa Y 2002 *J. Phys. Chem. Solids* **63** 1575
[15] Okamoto K 2002 *Prog. Theor. Phys.* Suppl. No.145 208
[16] Tokuno A and Okamoto K 2005 *J. Phys. Soc. Jpn.* **74** Suppl. 157
[17] Okamoto K 2014 *JPS Conf. Proc.* **1** 012031
[18] We note that, fundamentally speaking, the site magnetization $m_{j,l}(L)$ must be zero in the ground state of the finite-L system. However, when the ground states are degenerate (or the energy gap is exponentially small), the DMRG calculation for a finite system often selects one of them: In our calculation for the $M = 0$ subspace, we have obtained one of the magnetic ordered states in the F-SD and AFstN phases and one of the ground states in which free effective spins at open edges are polarized in opposite directions in the Haldane phase.
[19] Miyashita S and Yamamoto S 1993 *Phys. Rev.* B **48** 913
[20] White S R 1993 *Phys. Rev.* B **48** 10345
[21] Yamamoto S and Miyashita S 1994 *Phys. Rev.* B **50** 6277
[22] Nightingale M P 1976 *Physica* A **83** 561
[23] Hikihara T, Kecke L, Momoi T and Furusaki A 2008 *Phys. Rev.* B **78** 144404
[24] Hikihara T, Momoi T, Furusaki A and Kawamura H 2010 *Phys. Rev.* B **81** 224433

Journal of Physics: Conference Series **683** (2016) 012040 doi:10.1088/1742-6596/683/1/012040

Intersite electron correlations in a Hubbard model on inhomogeneous lattices

Nayuta Takemori[1], Akihisa Koga[1] and Hartmut Hafermann[2]

[1]Department of Physics, Tokyo Institute of Technology, Meguro, Tokyo 152-8551, Japan
[2]Mathematical and Algorithmic Sciences Lab, France Research Center, Huawei Technologies Co. Ltd., 92100 Boulogne-Billancourt, France

E-mail: takemori@stat.phys.titech.ac.jp

Abstract. We study intersite electron correlations in the half-filled Hubbard model on square lattices with periodic and open boundary conditions by means of a real-space dual fermion approach. By calculating renormalization factors, we clarify that nearest-neighbor intersite correlations already significantly reduce the critical interaction. The Mott transition occurs at $U/t \sim 6.4$, where U is the interaction strength and t is the hopping integral. This value is consistent with quantum Monte Carlo results. It shows the importance of short-range intersite correlations, which are taken into account in the framework of the real-space dual fermion approach.

1. Introduction

The successful development of dynamical mean field theory (DMFT) [1] enables us to investigate a wide range of heavy fermion systems. In its real-space formulation (RDMFT), it has successfully been applied to inhomogeneous systems such as quasiperiodic systems [2, 3, 4], fermionic cold atoms [5, 6, 7, 8], electron systems on surfaces [9], interfaces [10, 11] and topological insulating systems [12]. However, this method does not take into account intersite correlation effects, which should be important at very low temperatures. Spatial correlations in homogeneous systems have therefore been treated by means of extensions of DMFT, e.g., cellular DMFT (CDMFT) [13], dynamical cluster approximation (DCA) [14], or diagrammatic extensions of DMFT [15, 16, 17, 18, 19]. Even though a generalization of CDMFT to inhomogeneous systems exists (the so-called I-CDMFT [20]), it is in general not clear how to treat inhomogeneous systems by means of CDMFT or DCA, due to the absence of translational symmetry. Diagrammatic extensions of DMFT, however, are naturally well suited to treat such systems [21]. Here we present a real-space formulation of the dual fermion approach (RDF), which includes spatial correlations beyond RDMFT. It allows us to study intersite electron correlations in inhomogeneous lattices. To clarify how intersite electron correlations affect Mott physics, we study the Hubbard model on square lattices with periodic and open boundary conditions. By calculating renormalization factors, we discuss intersite correlations for the Mott transitions in these systems.

This paper is organized as follows. In Sec. 2, we introduce the model Hamiltonian. In Sec. 3, we explain our theoretical approach. We discuss how the intersite electron correlations affect the Mott transition point in Sec. 4. A brief summary is given in Sec. 5.

2. Model

We consider the single-band Hubbard model on the square lattice [22], which is given by the following Hamiltonian

$$H = -t\sum_{\langle i,j\rangle}(c^\dagger_{i\sigma}c_{j\sigma} + h.c.) + U\sum_i n_{i\uparrow}n_{i\downarrow}, \tag{1}$$

where the summation is over nearest-neighbors $\langle i,j\rangle$, $c^\dagger_{i\sigma}(c_{i\sigma})$ is a creation (annihilation) operator of an electron at the ith site with spin $\sigma(=\uparrow,\downarrow)$ and $n_{i\sigma} = c^\dagger_{i\sigma}c_{i\sigma}$. t is the transfer integral between sites and U is the Coulomb interaction. We discuss Mott transitions in the half-filled system at finite temperatures, setting the chemical potential to $\mu = U/2$.

3. Method

In this paper, we use the dual fermion approach, which has been developed recently [15, 17, 18, 19]. In the method, auxiliary fermions, so-called dual fermions (f, f^*) are introduced via a Hubbard-Stratonovich transformation, and intersite correlations are treated by means of diagrammatic expansions. The diagrammatic expansion is formalized by the dual propagator $[\hat{G}^{\rm d}_{\omega\sigma}]_{ij} = -\langle f_{i\omega\sigma}f^*_{j\omega\sigma}\rangle$, where ω denotes fermionic Matsubara frequency $i\omega_n = (2n+1)\pi/\beta$.

Since the dual fermion approach has originally been formulated in momentum space, inhomogeneous systems cannot be treated in this framework. Therefore, we introduce the real-space formulation of this method. We note that it is equivalent to the conventional dual fermion approach for the regular lattice if the same diagrams are taken into account. All the assumptions made in its derivation are the same as in the conventional dual fermion approach. We therefore only give an outline of the derivation and highlight the differences to the conventional dual fermion approach. For details of the derivation, we refer the interested reader to the above references for further details.

In the real-space formulation, the concerned system quantities are represented as $N \times N$ matrices (denoted by a hat), where N is the total number of sites. For example, the system's bare dual Green's function is denoted $[\hat{G}^{\rm d,0}_{\rm syst}]_\omega = -\hat{g}_\omega[\hat{g}_\omega + (\hat{\Delta}_\omega - \hat{t})^{-1}]^{-1}\hat{g}_\omega$, where $\hat{\Delta}_\omega$ denotes the hybridization function, $\hat{g}_\omega$ is the impurity Green's function and $[\hat{t}]_{ij} = t\delta_{\langle i,j\rangle}$. Correlations between any two sites are straightforwardly included through the corresponding diagram connecting these two sites. By considering first-order and second-order diagrams (see Fig. 1), the dual self-energy $\hat{\Sigma}^{\rm d}_\omega$ can be written

$$\hat{\Sigma}^{\rm d}_{ij,\omega} = \hat{\Sigma}^{\rm d(1)}_{i,\omega}\delta_{ij} + \hat{\Sigma}^{\rm d(2)}_{ij,\omega}, \tag{2}$$

where

$$[\hat{\Sigma}^{{\rm d},(1)\,\sigma}_{\nu}]_{ii} = -T\sum_{\nu'}\sum_{\sigma'}\gamma^{\sigma\sigma'\,i}_{\nu\nu',\omega=0}[\hat{G}^{{\rm d}\,\sigma'}_{\nu'}]_{ii} \tag{3}$$

and

$$\begin{aligned}[\hat{\Sigma}^{{\rm d},(2)\,\sigma}_{\nu}]_{ij} = & -\tfrac{1}{2}T^2\sum_{\nu'}\sum_{\omega}\sum_{\sigma'}\gamma^{\sigma'\sigma\,i}_{\nu'\nu\omega}[\hat{G}^{{\rm d}\,\sigma}_{\nu+\omega}]_{ij}[\hat{G}^{{\rm d}\,\sigma'}_{\nu'+\omega}]_{ji}[\hat{G}^{{\rm d}\,\sigma'}_{\nu'}]_{ij}\gamma^{\sigma\sigma'\,j}_{\nu\nu'\omega} \\ & -\tfrac{1}{2}T^2\sum_{\nu'}\sum_{\omega}\gamma^{\bar{\sigma}\sigma\sigma\bar{\sigma}\,i}_{\nu'\nu\omega}[\hat{G}^{{\rm d}\,\bar{\sigma}}_{\nu+\omega}]_{ij}[\hat{G}^{{\rm d}\,\bar{\sigma}}_{\nu'+\omega}]_{ji}[\hat{G}^{{\rm d}\,\sigma}_{\nu'}]_{ij}\gamma^{\bar{\sigma}\sigma\sigma\bar{\sigma}\,j}_{\nu\nu'\omega}.\end{aligned} \tag{4}$$

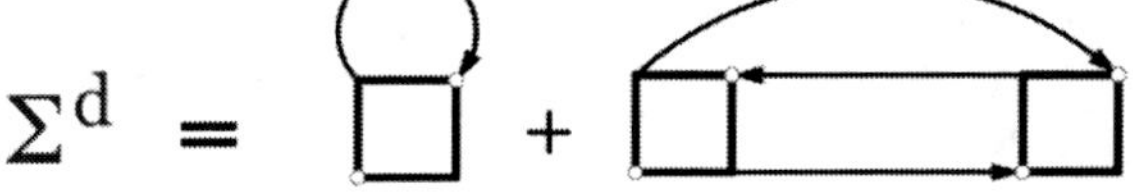

Figure 1. Diagrams for the dual self-energy up to second-order.

Here, we use shorthand notation for the vertex function

$$\gamma_{\nu'\nu\omega}^{\sigma\sigma' i} \equiv \gamma_{\nu'\nu\omega}^{\sigma\sigma\sigma'\sigma' i}. \tag{5}$$

We note that in the paper, we only consider intersite correlations between nearest neighbor sites ($\hat{\Sigma}_{ij}^{d(2)} \sim \hat{\Sigma}_{i,j}^{d(2)}\delta_{\langle i,j\rangle}$), which we expect to be dominant (correlations of larger distances are straightforwardly included). Therefore, our method does not reduce to the conventional dual fermion approach even in the homogeneous system. Nevertheless, we can discuss the effect of intersite correlations beyond DMFT. The system's self-energy is given as

$$\hat{\Sigma}_{\rm syst} = \hat{\Sigma}_{\rm imp}^{i} + [(\hat{1} + \hat{\Sigma}^{\rm d}\hat{g}_{\rm imp})^{-1}\hat{\Sigma}^{\rm d}]. \tag{6}$$

The self-consistency condition for the real-space dual fermion approach is the same as that in the conventional dual fermion approach,

$$\hat{G}^{\rm d}{}_{ii} = 0. \tag{7}$$

Since the off-diagonal component of the dual self-energy $\hat{\Sigma}^{\rm d}$ originates from the vertex function, the real-space dual fermion approach reduces to RDMFT in the weak coupling limit, or when no diagrammatic corrections are taken into account. In these cases, the system self-energy is diagonal and the dual self-energy is equal to zero.

In order to discuss Mott transitions at finite temperatures, we calculate the following quantities,

$$z_i = \left[1 - {\rm Im}\hat{\Sigma}_{\rm syst}^{ii}(i\omega_0)/\omega_0\right]^{-1}. \tag{8}$$

They can be regarded as renormalization factors at finite temperature. In the following, we use the transfer integral t as the unit of energy. We treat the two-dimensional square lattice with 4×4 sites with periodic and open boundary conditions to discuss the effect of intersite correlations.

4. Results

We consider electron correlations in the system with periodic boundary condition (see Fig. 2(a)). To clarify how intersite correlations are taken into account in the real-space dual fermion approach, we calculate the local Green's function at $U = 4.0$. For comparison, we also show results obtained by RDMFT and the conventional dual fermion approach in Fig. 2(b). These three results are in good agreement with each other in the intermediate energy region and show the same high-energy asymptotic behavior. On the other hand, they deviate in the low energy region. We find that our results are located between those obtained by RDMFT and dual fermion methods. This is because our real-space dual fermion method takes into account some (i.e., those between nearest neighbor sites) but not all intersite correlations beyond the simple RDMFT. These correlations lower the spectral weight at the Fermi level. The real-space dual fermion result must approach the conventional dual fermion result when considering longer-range correlations. This is currently under consideration. We note that even though the effect of the intersite correlations on the Green's function is small at these parameters, we will see below that their effect can be significant.

Figure 3 shows the interaction dependence of renormalization factors in the system with periodic boundary conditions. In the noninteracting case $U = 0$, the normal metallic state is realized with $z_i = 1$. By introducing the interaction, the renormalization factor at each site monotonically decreases, similarly as in the conventional Hubbard model [23]. The small difference in the low-energy region in local Green's function obtained by RDMFT and real-space dual fermion approach, which is discussed already, leads to the small supression in

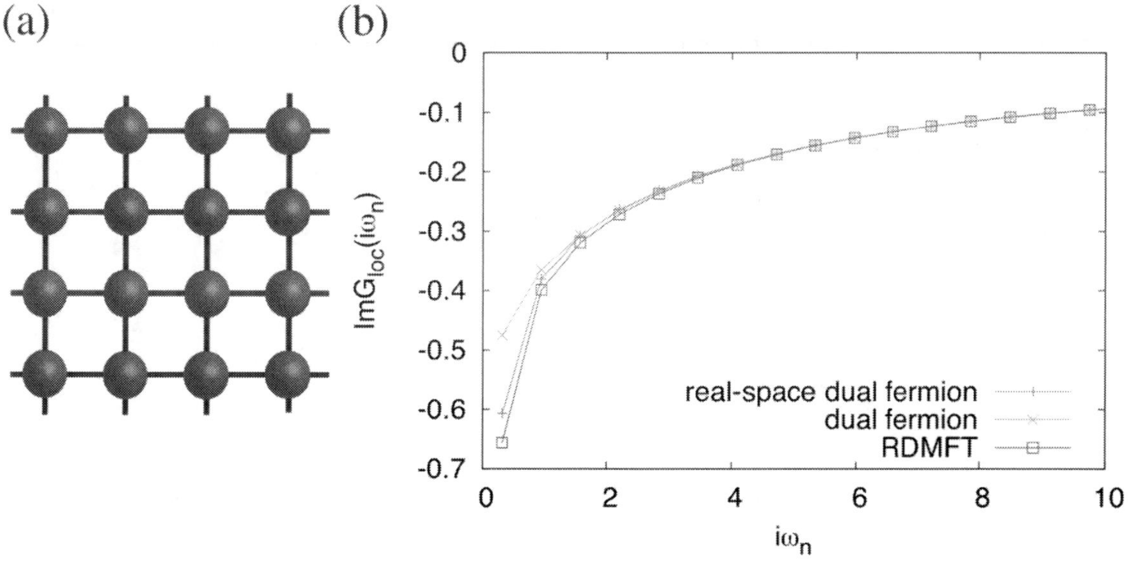

Figure 2. (a) Square lattice with periodic boundary condition.
(b)The imaginary part of the local Green's function $G_{\rm loc}(i\omega_n)$ for $U = 4.0$ at $T = 0.1$ obtained by RDMFT, real-space dual fermion approach and dual fermion approach. The unit of energy is set to be t.

renormalization factors for real-space dual fermion approach compared to RDMFT at $U = 4.0$. At last, the first-order phase transition occurs to the Mott insulating state at $U_c^{\rm periodic}$. The critical interaction U_c is deduced as $U_c^{\rm periodic} = 9.4$ (RDMFT) and 6.4 (real-space dual fermion approach). In the homogeneous system, the transition point $U_c^{\rm periodic} \sim 6.5$ has been obtained precisely by the dual fermion approach [19], $U_c^{\rm periodic} = 6.5$ by DCA [24] and $U_c^{\rm periodic} \sim 6$ by CDMFT [25] and QMC simulation [26]. Since the dominant correlations are between nearest-neighbors, finite size effects are expected to be small. We conclude that the real-space dual fermion approach properly takes intersite correlations into account and gives us an improved picture of the Mott transition.

Next, we apply the approach to the Hubbard model with open boundary conditions (see Fig. 4(a)) as an example of inhomogeneous lattices. In this system with 4×4 sites, there exist three independent sites: the corner, edge, and bulk, as shown in Fig. 4(a). Figure 4(b) shows the interaction dependence of renormalization factors for these sites. It is found that these quantities are split into three lines in an interacting case. This fact indicates that a site-dependent renormalization appears. On the other hand, the Mott transition occurs simultaneously for all sites. Due to the open boundary conditions the critical value is lower and the transition point is deduced as $U_c^{\rm open} = 8.6$ (RDMFT; not shown) and 5.6 (real-space dual fermion approach), where the renormalization factors have jump singularities. By comparing with the RDMFT results, we find a suppression of renormalization factors, which is similar to the homogeneous system. Therefore, we conclude that real-space dual fermion approach can take intersite correlations into account correctly even in inhomogeneous systems.

5. Summary

In summary, we have investigated the half-filled Hubbard model on square lattices with periodic and open boundary conditions by means of a real-space dual fermion (RDF) approach. By calculating the local Green's function, we have studied how nonlocal correlations affect the local

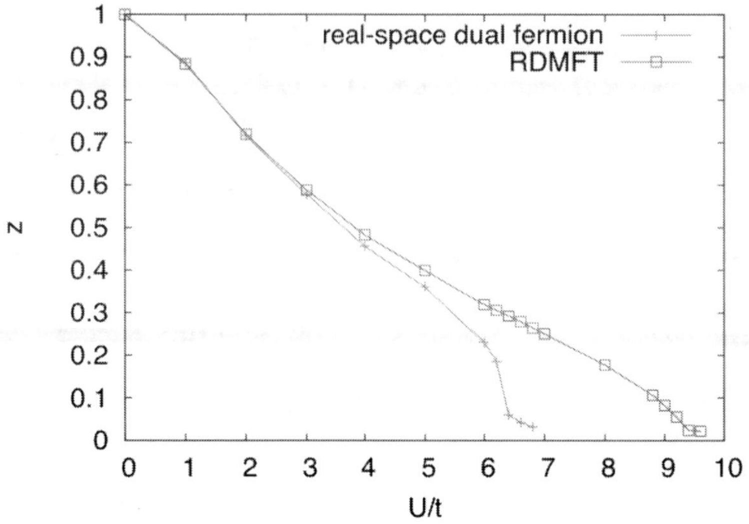

Figure 3. Renormalization factors in the system with periodic boundary condition obtained by means of the real-space dual fermion and RDMFT when $T = 0.1$. The unit of energy is set to be t.

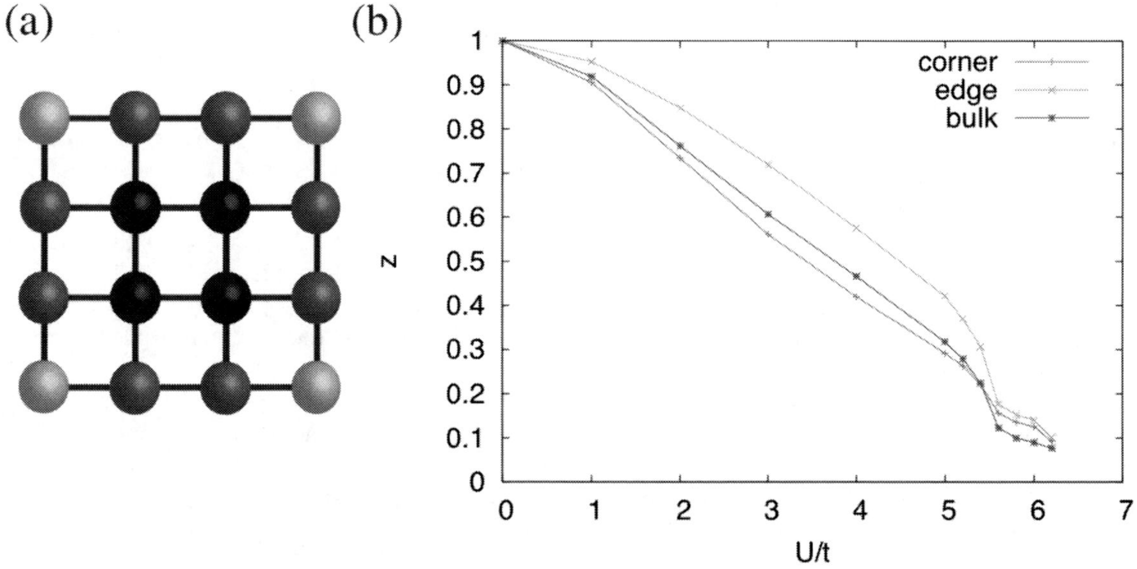

Figure 4. (a)Square lattice with open boundary condition. Three independent sites are depicted in different gray scales.
(b)Renormalization factors in the system with open boundary condition obtained by means of the real-space dual fermion when $T = 0.1$. The unit of energy is set to be t.

quantities. Moreover, we have discussed the Mott transition in this system to clarify that the real-space dual fermion approach successfully captures the suppression of the renormalization factor. This leads to a Mott transition point with a relatively low critical Coulomb interaction compared to RDMFT. The Mott transition point is close to that obtained by means of the quantum Monte Carlo method in the homogeneous system. Therefore, we can say that our real-space dual fermion method therefore takes intersite correlations properly into account, and the nearest-neighbor correlations are dominant in the system. This new method allows to study intersite electron correlation effects in various other inhomogeneous systems, such as cold atoms in a trapping potential, nanosystems, impurities in metal hosts, interfaces and surfaces, topological insulators and quasiperiodic lattices, which is now under consideration.

Acknowledgments
The authors would like to thank J. Otsuki and P. Werner for valuable discussions. This work was partly supported by JSPS KAKENHI Grant No. 15J12110 (N. T.) and 25800193 (A. K.). Part of the computations was carried out on TSUBAME2.0 at Global Scientific Information and the Computing Center of Tokyo Institute of Technology and at the Supercomputer Center at the Institute for Solid State Physics, the University of Tokyo. The simulations have been performed using some ALPS libraries [27].

[1] Georges A, Kotliar G, Krauth W and Rozenberg M J 1996 *Rev. Mod. Phys.* **68**(1) 13–125
[2] Takemori N and Koga A 2015 *J. Phys. Soc. Jpn.* **84** 023701
[3] Takemori N and Koga A 2015 *J. Phys.: Conf. Ser.* **592** 012038
[4] Takemura S, Takemori N and Koga A 2015 *Phys. Rev. B* **91**(16) 165114
[5] Helmes R W, Costi T A and Rosch A 2008 *Phys. Rev. Lett.* **100**(5) 056403
[6] Snoek M, Titvinidze I, Tőke C, Byczuk K and Hofstetter W 2008 *New J. Phys.* **10** 093008
[7] Koga A, Higashiyama T, Inaba K, Suga S and Kawakami N 2008 *J. Phys. Soc. Jpn.* **77** 073602
[8] Koga A, Higashiyama T, Inaba K, Suga S and Kawakami N 2009 *Phys. Rev. A* **79**(1) 013607
[9] Potthoff M and Nolting W 1999 *Phys. Rev. B* **59**(4) 2549–2555
[10] Okamoto S and Millis A J 2004 *Phys. Rev. B* **70**(24) 241104
[11] Okamoto S and Millis A 2004 *Nature* **428** 630–633
[12] Tada Y, Peters R, Oshikawa M, Koga A, Kawakami N and Fujimoto S 2012 *Phys. Rev. B* **85**(16) 165138
[13] Kotliar G, Savrasov S Y, Pálsson G and Biroli G 2001 *Phys. Rev. Lett.* **87**(18) 186401
[14] Hettler M H, Tahvildar-Zadeh A N, Jarrell M, Pruschke T and Krishnamurthy H R 1998 *Phys. Rev. B* **58**(12) R7475–R7479
[15] Kusunose H 2006 *J. Phys. Soc. Jpn.* **75** 054713
[16] Toschi A, Katanin A A and Held K 2007 *Phys. Rev. B* **75**(4) 045118
[17] Rubtsov A N, Katsnelson M I and Lichtenstein A I 2008 *Phys. Rev. B* **77**(3) 033101
[18] Otsuki J, Hafermann H and Lichtenstein A I 2014 *Phys. Rev. B* **90**(23) 235132
[19] Hafermann H 2010 *Numerical Approaches to Spatial Correlations in Strongly Interacting Fermion Systems* (Cuvillier Verlag, Göttingen)
[20] Charlebois M, Sénéchal D, Gagnon A M and Tremblay A M S 2015 *Phys. Rev. B* **91**(3) 035132
[21] Valli A, Sangiovanni G, Gunnarsson O, Toschi A and Held K 2010 *Phys. Rev. Lett.* **104**(24) 246402
[22] Imada M, Fujimori A and Tokura Y 1998 *Rev. Mod. Phys.* **70**(4) 1039–1263
[23] Rozenberg M J, Chitra R and Kotliar G 1999 *Phys. Rev. Lett.* **83**(17) 3498–3501
[24] Werner P private communication
[25] Park H, Haule K and Kotliar G 2008 *Phys. Rev. Lett.* **101**(18) 186403
[26] Vekić M and White S R 1993 *Phys. Rev. B* **47**(2) 1160–1163
[27] Bauer B, Carr L D, Evertz H G, Feiguin A, Freire J, Fuchs S, Gamper L, Gukelberger J, Gull E, Guertler S, Hehn A, Igarashi R, Isakov S V, Koop D, Ma P N, Mates P, Matsuo H, Parcollet O, Pawłowski G, Picon J D, Pollet L, Santos E, Scarola V W, Schollwöck U, Silva C, Surer B, Todo S, Trebst S, Troyer M, Wall M L, Werner P and Wessel S 2011 *J. Stat. Mech. Theor. Exp.* **2011** P05001

Journal of Physics: Conference Series **683** (2016) 012041 doi:10.1088/1742-6596/683/1/012041

DMFT Study for Valence Fluctuations in the Extended Periodic Anderson Model

Ryu Shinzaki, Joji Nasu, and Akihisa Koga

Department of Physics, Tokyo Institute of Technology, Tokyo 152-8551, Japan

E-mail: ryushin@stat.phys.titech.ac.jp

Abstract. We study valence fluctuations at finite temperatures in the extended periodic Anderson model, where the Coulomb interaction between conduction and localized f-electrons is taken into account, using dynamical mean-field theory combined with the continuous-time quantum Monte Carlo (CT-QMC) method. The valence transition with the hysteresis is clearly found, indicating the first-order phase transition between the Kondo and mixed-valence states. We demonstrate that spin correlation rapidly develops when the system approaches the valence transition point. The comparison of the impurity solvers, the CT-QMC, non-crossing approximation, and one-crossing approximation, is also addressed.

1. Introduction

Heavy fermion systems with rare-earth or actinide elements have attracted considerable attention. One of the interesting phenomena in the systems is the valence transition, where the valence in rare-earth ions abruptly changes due to applied field and/or pressure. Recently, it has been suggested that the superconductivity in the rare-earth compound $CeCu_2Si_2$ [1] and quantum critical behavior in the quasicrystal $Au_{51}Al_{34}Yb_{15}$ [2] are related to valence fluctuations. These stimulate further theoretical and experimental investigations on valence fluctuations in the heavy-fermion systems [3–8].

Valence fluctuations and valence transitions have been studied theoretically in terms of the extended periodic Anderson model (EPAM) [9–15]. The possibility of the valence transition in this model has been suggested by the slave boson mean-field theory [12], density matrix renormalization group [10], and dynamical mean-field theory (DMFT) [11,15]. Although a jump singularity in the curve of the f-electron number has been reported in several papers [10–15], the hysteresis has not been found clearly. Therefore, it is instructive to examine carefully this problem by means of unbiased numerical methods.

Motivated by this, we study the EPAM at finite temperatures, combining DMFT with a continuous-time quantum Monte Carlo (CT-QMC) method [27–29], which is one of the most powerful numerical techniques. We clarify that the first-order valence transition occurs with a hysteresis in the curve of the valence. We also compare the CT-QMC results with those obtained by the non-crossing approximation (NCA) [24–26] and the one-crossing approximation (OCA) [26], and discuss the advantage of the CT-QMC method.

This paper is organized as follows. In Sec. 2, we introduce the model Hamiltonian and summarize our numerical method briefly. In Sec. 3, we demonstrate the valence transition at a finite temperature in the framework of DMFT with the CT-QMC method. We also discuss spin correlation in the vicinity of the valence transition. Furthermore, quantitative discussions for the valence transition in the EPAM are given in Sec. 4, by comparing with the DMFT results obtained by means of the CT-QMC, NCA, and OCA impurity solvers. In the final section, we state our conclusions.

2. Model and Method

We study valence fluctuations at finite temperatures in the EPAM, which is described by the following Hamiltonian:

$$\begin{aligned} H_{\mathrm{EPAM}} &= -t \sum_{\langle i,j\rangle,\sigma} c_{i\sigma}^{\dagger} c_{j\sigma} + V \sum_{i,\sigma} (c_{i\sigma}^{\dagger} f_{i\sigma} + \mathrm{h.c.}) + \epsilon_f \sum_{i,\sigma} n_{i\sigma}^{f} \\ &\quad + U_{ff} \sum_{i} n_{i\uparrow}^{f} n_{i\downarrow}^{f} + U_{cf} \sum_{i,\sigma,\sigma'} n_{i\sigma}^{c} n_{i\sigma'}^{f}, \end{aligned} \tag{1}$$

where $c_{i\sigma}$ ($f_{i\sigma}$) is the annihilation operator of a conduction electron (f-electron) with spin $\sigma(=\uparrow,\downarrow)$. $n_{i\sigma}^c(= c_{i\sigma}^\dagger c_{i\sigma})$ and $n_{i\sigma}^f(= f_{i\sigma}^\dagger f_{i\sigma})$ are the number operators of the conduction and the f- electrons at ith site, respectively. t is the hopping integral of the conduction electrons between the nearest-neighbor sites, V is the hybridization between the conduction band and the f-orbitals, and ϵ_f is the energy level of the f-orbitals. U_{ff} is the repulsive interaction between the f-electrons and U_{cf} is the repulsive interaction between the conduction and the f- electrons. We here consider the Bethe lattice with a large coordination number $z \to \infty$, which has a semi-elliptic density of states with the bandwidth $2D$.

In order to study low temperature properties in the EPAM [17, 18], we make use of DMFT [19–22]. In the framework of DMFT, the lattice model is mapped to an effective impurity model, where local electron correlations are taken into account precisely. This method is formally exact in the infinite dimensions and has successfully been used even in three dimensions. The site-diagonal lattice Green's function is required to be equal to the Green's function of the effective impurity model. This condition leads to the self-consistent equation [23], as

$$[\hat{\mathscr{G}}_\sigma^{-1}(\omega)]_{cc} = \mathrm{i}\omega - \mu - \left(\frac{D}{2}\right)^2 [\hat{G}_{\mathrm{imp}\sigma}(\omega)]_{cc}, \tag{2}$$

where μ is the chemical potential, $\hat{\mathscr{G}}_\sigma(\omega)$ $[\hat{G}_{\mathrm{imp}\sigma}(\omega)]$ is the noninteracting (full) impurity Green's function with spin σ. To solve the effective impurity model, we make use of the CT-QMC method based on the hybridization expansion [27–29]. In this method, the partition function is described by the expansion in powers of the impurity-bath mixing, which allows us to evaluate physical quantities quantitatively in terms of the Monte Carlo procedure. This is contrast to other biased methods such as the NCA and OCA methods. In the paper, we set $U_{ff}/D = 20$ and $U_{cf}/D = 8$, and use the half bandwidth D as the unit of energy.

3. First-order valence transition

Now, we study the EPAM at finite temperatures to clarify how valence fluctuations develop in the system. To characterize the Kondo and mixed-valence states, we first examine the particle number when the chemical potential is varied. In Fig. 1, we show the total particle number per site $\langle n_{\mathrm{tot}}\rangle$ and the f-electron number per site $\langle n_f\rangle$ when $T/D = 0.04$ and $V/D = 0.13$. In the case of $\epsilon_f/D = -7.8$, the total particle number increases monotonically with increase of the chemical potential as shown in Fig. 1(a) whereas the f-electron number little changes within $7.7 < \mu/D < 8.0$ as shown in Fig. 1(d). This indicates that the Kondo state is stabilized with localized electrons in the f-orbitals. On the other hand, we find in Figs. 1(c) and 1(f) that both $\langle n_{\mathrm{tot}}\rangle$ and $\langle n_f\rangle$ are smoothly changed in the case $\epsilon_f/D = -7.45$. This implies that

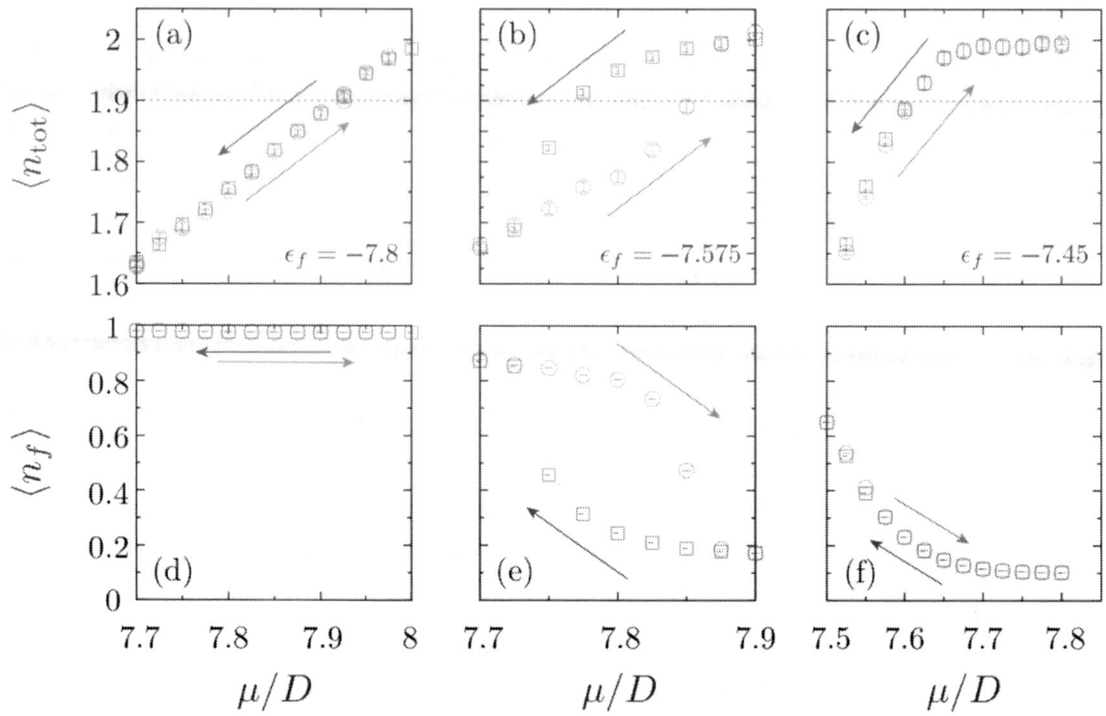

Figure 1. The total electron number (upper panels) and f-electron number (lower panels) as functions of the chemical potential in the system with U_{cf}/D=8, V/D=0.13, and $T/D = 0.04$ when $\epsilon_f/D = -7.8$(left), -7.575(middle), and -7.45(right). The squares (circles) are the data obtained by changing μ from the higher (lower) side.

the mixed-valence state is realized when $7.5 < \mu/D < 7.8$. In the intermediate case with $\epsilon_f = -7.575$, remarkably, the hysteresis appears in the total number of electrons and the f-electron number, as shown in Figs. 1(b) and 1(e). These results indicate the existence of the first-order valence transition in the system with $\epsilon_f/D = -7.575$. In fact, if one considers the system with $\langle n_{\rm tot}\rangle \sim 1.9$, two solutions are obtained: the Kondo state with $\langle n_f\rangle \sim 1$ and the mixed-valence state with $\langle n_f\rangle \ll 1$.

By performing similar calculations in the system with $\langle n_{\rm tot}\rangle = 1.9$, we discuss how the f-electron number depends on the energy level of the f-orbital. Figure 2(a) shows the f-electron number at $T/D = 0.04$. When $\epsilon_f/D = -8.0$, the f-electron number is almost one, and the Kondo state with the localized f-electrons is realized. With increase of ϵ_f, $\langle n_f\rangle$ decreases and the first-order valence transition with a jump singularity occurs at $\epsilon_f \simeq -7.575$. Beyond the transition point, the mixed valence state is stabilized with $\langle n_f\rangle \sim 0.2$. As discussed above, we have found two solutions in the case $\epsilon_f/D = -7.575$. Therefore, the first-order phase transition with the hysteresis indeed occurs between the Kondo and mixed-valence states.

In the vicinity of the valence transition point, the singularity appears in other physical quantities. Fig. 2(b) shows the spin correlation between the conduction and f- electrons $\langle \mathbf{S}_c\cdot\mathbf{S}_f\rangle$. It is found that the $c-f$ spin correlation is enhanced when the system approaches the valence transition point. Then, the jump singularity appears at the first-order transition point. In order to clarify how such singularities in the valence and the spin correlation develop with decrease of temperature, we show in Fig. 3 the temperature dependence of the quantities in the system with $U_{cf}/D = 8.0$ and $V/D = 0.15$. When $T/D = 1.0$, the valence is smoothly varied and

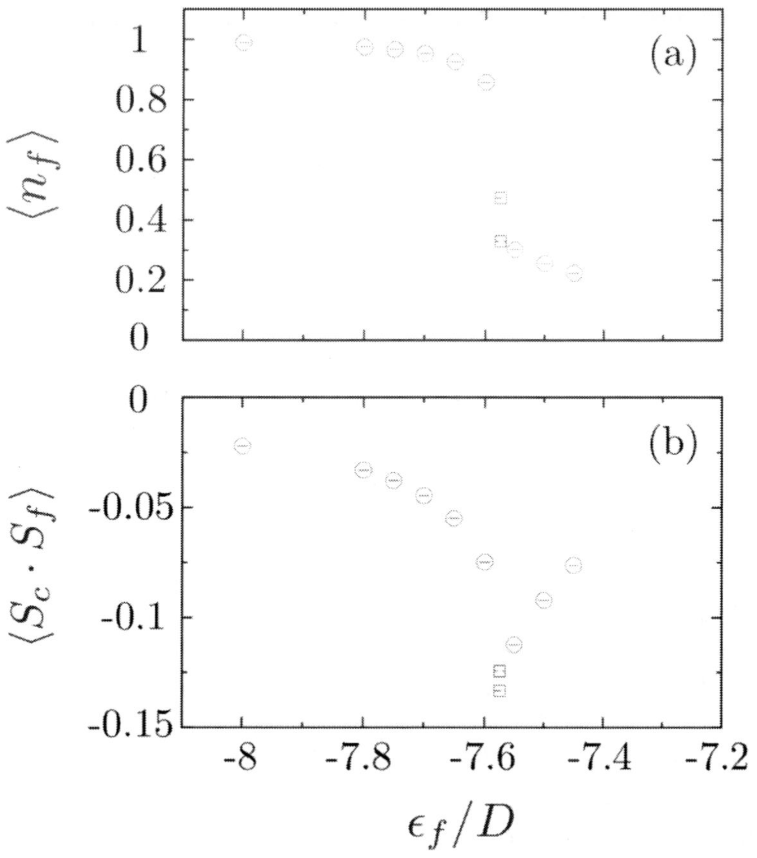

Figure 2. f-level dependence of the f-electron number (a) and $c-f$ spin correlation (b) when T/D=0.04, $U_{cf}/D = 8$ and $V/D = 0.13$. The squares represent the two solutions at $\epsilon_f/D = -7.575$.

the spin correlation is little changed with increase of the energy of the f-level, as shown in Fig. 3(d). This implies that the high temperature paramagnetic state is realized. The decrease of the temperature leads to the rapid decrease of the valence, as shown in Figs. 3(b) and 3(c). However, we could not find the jump singularity in the valence and thereby a crossover between the Kondo and mixed-valence states occurs. In these cases, interesting behavior appears in the spin correlation, as shown in Figs. 3(d), 3(e), and 3(f). At $T/D = 1.0$, the spin correlation is almost zero and the Kondo singlet state does not appear. With decrease of temperature, the spin correlation is enhanced around the crossover region at $T/D = 0.33$ and 0.1, as shown in Figs. 3(e) and 3(f). This indicates that the valence fluctuations enhance the spin correlation between the conduction and the localized f- electrons. Further decrease of temperature induces the valence transition at a certain ϵ_f with the cusp singularity in the $c-f$ spin correlation. Similar behavior in the magnetization process has been reported thus far [10, 12, 16]. Therefore, the spin correlation is one of the appropriate quantities to discuss the nature of the valence transitions.

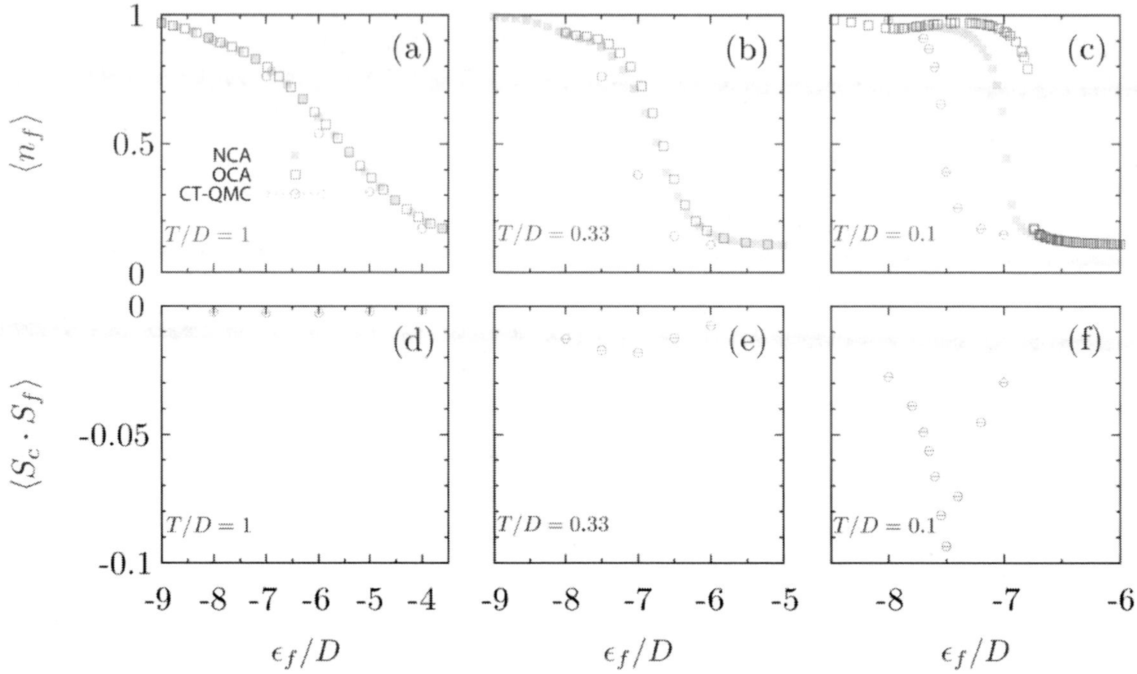

Figure 3. The f-electron number (upper panels) and $c-f$ spin correlation (lower panels) in the system with $U_{cf}/D = 8.0$ and $V/D = 0.15$ when $T/D = 1.0$(left), 0.33(middle), and 0.1(right). The open circles, filled squares, and open squares are the results obtained by means of the CT-QMC, NCA, and OCA solvers, respectively.

4. Comparison with NCA and OCA results

In this section, we compare the CT-QMC method with the NCA and OCA methods. To this end, we show in the upper panels of Fig. 3 the f-electron number obtained by means of the NCA and OCA solvers. At $T/D = 1.0$, the system is gradually changed from the Kondo like state to the mixed-valence state, as discussed above. In this case, we could not find a big difference in the curves obtained by means of the CT-QMC, NCA, and OCA methods. Therefore, the NCA and OCA methods are appropriate solvers in the case. With decrease of temperature, slightly different behavior appears, as shown in Figs. 3 (b) and 3(c). We find that the valence crossover points obtained by the NCA and OCA methods are shifted, comparing with that obtained by the CT-QMC method. In addition, when $T/D = 0.1$, a jump singularity in $\langle n_f \rangle$ appears around $\epsilon_f/D \sim 6.75$, which is obtained by means of the OCA method. This valence transition does not appear when the CT-QMC numerical solver is used. Therefore, we conclude that the OCA method as well as the NCA method overestimates valence fluctuations at low temperatures.

5. Summary

We have studied the extended periodic Anderson model, combining dynamical mean-field theory with the continuous-time quantum Monte Carlo method. By calculating the f-electron number, we have clarified that, at a low temperature, the first-order valence transition occurs between the Kondo and the mixed-valence states together with the hysteresis in the curve of the valence. We have also found that the $c-f$ spin correlation rapidly develops when the system approaches the valence transition point. In addition, The comparison of the impurity solvers such as the CT-QMC, NCA, and OCA has been discussed.

6. Acknowledgements

This work was partly supported by the Grant-in-Aid for Scientific Research from JSPS, KAKENHI No. 25800193. (A.K.) Some of the computations in this work have been carried out by using the facilities of the Supercomputer Center, the Institute for Solid State Physics, the University of Tokyo. The simulations have been performed by using some of ALPS libraries [30].

References

[1] Thomas F. et al. 1992 *J. Phys.* C **8** L51
[2] Deguchi K, Matsukawa S, Sato N K, Hattori T, Ishida K, Takakura H and Ishimasa T 2012 *Nat. Mater.* **11** 1013.
[3] Holmes T A, Jaccard D and Miyake K 2004 *Phys. Rev.* **69** 024508
[4] Sugibayashi T, Saiga Y and Hirashima S D 2008 *J. Phys. Soc. Jpn.* **77** 024716
[5] Watanabe S and Miyake K 2013 *J. Phys. Soc. Jpn.* **82** 083704
[6] Matsukawa S, Tanaka K, Nakayama M, Deguchi K, Imura K, Takakura H, Kashimoto S, Ishimasa T and Sato K N 2014 *J. Phys. Soc. Jpn.* **83** 034705
[7] Takemura S, Takemori N and Koga A 2015 *Phys. Rev.* B **91** 165114
[8] Andrade C E, Jagannathan A, Miranda E, Vojta M and Dobrosavljević 2015 *Phys. Rev. Lett.* **115** 036403
[9] Onishi Y and Miyake K 2000 *J. Phys. Soc. Jpn.* **69** 3955
[10] Watanabe S, Imada M and Miyake K 2006 *J. Phys. Soc. Jpn.* **75** 043710
[11] Saiga Y, Sugibayashi T and Hirashima S D 2008 *J. Phys. Soc. Jpn.* **77** 114710
[12] Watanabe S, Tsuruta A, Miyake K and Flouquet J 2008 *Phys. Rev. Lett.* **100** 236401
[13] Kubo K 2008 *J. Phys. Soc. Jpn.* **80** 114711
[14] Hagymási I, Itai K and Sólyom 2013 *Phys. Rev.* B **87** 125146
[15] Kojima Y and Koga A 2014 *JPS Conf. Proc.* **1** 012106
[16] Miyake K and Watanabe S 2014 *J. Phys. Soc. Jpn.* **83** 061006
[17] Yoshida T, Ohashi T and Kawakami N 2011 *J. Phys. Soc. Jpn.* **80** 064710
[18] Yoshida T and Kawakami N 2012 *Phys. Rev.* B **85** 235148.
[19] Pruschke T, Jarrell M, Freericks K J 1995 *Adv. Phys.* **44** 187
[20] Georges A, Kotliar G, Krauth W and Rozenberg J M 1996 *Rev. Mod. Phys.* **68** 13
[21] Müller-Hartmann E 1989 *Z. Phys.* B **74** 507
[22] Metzner W and Vollhardt D 1989 *Phys. Rev. Lett.* **62** 324
[23] Schork T, Blawid S and Igarashi J 1999 *Phys. Rev.* B **59** 9888
[24] Y. Kuramoto 1983 *Z. Phys.* B **53** 37
[25] Y. Kuramoto, Zeitschrift fü 1986 *Physik B Condensed Matter* **65** 29
[26] Eckstein M and Werner P 2010 *Phys. Rev.* B **82** 115115.
[27] Werner P, Comanac A, de' Medici L, Troyer M and Millis J A 2006 *Phys. Rev. Lett.* **97** 076405
[28] Werner P and Millis J A 2006 *Phys. Rev.* B **74** 155107
[29] Gull E, Millis J A, Lichtenstein I A, Rubtsov N A, Troyer M and Werner P 2011 *Rev. Mod. Phys.* **83** 349
[30] Albuquerque F A, Alet F, Corboz P, Dayal P, Feiguin A, Fuchs S, Gamper L, Gull E, G'urtler S, Honecker A, Igarashi R, K'orner M, Kozhevnikov A, L'auchli A, Manmana R S, Matsumoto M, McCulloch P I, Michel F, Noack M R, Pawłowski G, Pollet L, Pruschke T, Schollw'ock U, Todo S, Trebst S, Troyer M, Werner P and Wessel S 2007 *J. Mag. Mag. Mat.* **310** 1187

Valence Imbalance of Manganese Ions between Surface and Bulk Enhanced by Fermi-Surface Structure in Layered Manganites

Ryosuke Yamamura and Takashi Hotta

Department of Physics, Tokyo Metropolitan University, Hachioji, Tokyo 192-0397, Japan

E-mail: yamamura-ryosuke@ed.tmu.ac.jp

Abstract. To investigate valence imbalance phenomena between the surface and bulk in layered manganites, we analyze an e_g-orbital degenerate double-exchange model with surfaces for two types of t_{2g} spin structures. We reconfirm that the surface-induced Friedel oscillations occur in the charge structure and that the number of Mn^{4+} ions on the surface is larger than that in the bulk. This tendency is found to be more significant, when the e_g-electron system has Fermi-surface structures with better nesting properties. The behavior of the Friedel oscillations depends on the nesting properties of the Fermi-surface curves along the direction of the oscillations, and not on the dimension of the system itself. We believe that these results will be useful for the development of high-efficiency cathodes in Li-ion batteries and catalysts.

1. Introduction

The development of rechargeable batteries has been an important milestone in the quest for clean sources of energy. Among batteries, Li-ion batteries with $LiCoO_2$ as the cathode material have significant high energy density and found remarkable success in application to portable electronic devices [1, 2]. However, since $LiCoO_2$ suffers from high cost, toxicity, and safety issues, the material is not appropriate for use in electric vehicles. In recent decades, substitutional materials for $LiCoO_2$ have been investigated, and some manganites are preferred for Li-ion batteries since they have low cost, no toxicity, and better safety characteristics. In fact, $LiMn_2O_4$ with a spinel structure has been successfully employed as the cathode in electric vehicles [3]. $LiMnO_2$ and related materials including other transition metal elements with a layered structure are promising cathode materials [4, 5]. Manganites such as $CaMn_4O_5$ and MnO_2 are also expected to act as catalysts in photosystems and artificial photosynthesis [6, 7, 8].

For application to cathodes and catalysts, *surface* electron states play important roles because manganites work as cathodes and catalysts in chemical reactions involving a change of the valence of surface manganese ions mainly from Mn^{4+} to Mn^{3+}. Furthermore, in the case of cathodes, if Mn^{3+} ions are present on the surface, manganese dissolution from the lattice into the electrolyte occurs owing to the disproportion of Mn^{3+} ions into Mn^{2+} and Mn^{4+} ions [9]. This reaction is one of the most important factors causing the capacity fading of cathodes in Li-ion batteries. Thus, for the material to achieve high efficiency as a cathode or catalyst, it is desirable to retain the Mn^{3+} ions in the bulk while the Mn^{4+} ions reside on the surface. Previous studies approached this problem using first-principles calculations of the charge structure on the surface of manganites. For λ-MnO_2, which is obtained by removing Li ions from $LiMn_2O_4$, a portion of Mn^{4+} ions transition to Mn^{3+} ions because of surface reconstruction [10]. First-principles calculations have enabled us to consider surface reconstruction or so, but such calculations are limited to small clusters. Therefore, studies of the microscopic mechanisms causing the valence imbalance of manganese ions between the surface and bulk are meaningful when trying to understand increasingly large clusters.

To further the above research, we have investigated the effect of the surface on the electronic state in layered manganites using a $1000 \times 1000 \times 100$ lattice on the basis of an orbital-degenerate double-exchange model with surfaces [11]. In Ref. [11], in order to clarify the difference in the electron numbers between the surface and bulk, we have concentrated on e_g electron states for some fixed t_{2g} spin patterns. As typical examples, we have chosen two types of t_{2g} spin structures: the C-type antiferromagnetic (AF) and sheet-like AF phases. These phases were chosen because they have been found in the bulk for some values of J, when we include the effect of Jahn-Teller distortions in the $2 \times 2 \times 2$ and $4 \times 4 \times 4$ lattices [12, 13]. In those cases, we have confirmed that the number of Mn^{4+} ions on the surface is always larger than that in the bulk, which is understood by surface-induced Friedel oscillations. Note that the behavior of Friedel oscillations is different between two t_{2g} spin structures. Thus, it is necessary to carefully investigate the behavior of these oscillations for future applications to cathodes and catalysts.

In this paper, we investigate the behavior of surface-induced Friedel oscillations in layered manganites using an orbital-degenerate double-exchange model with surface effects for other types of t_{2g} spin structures. It is found that the number of Mn^{4+} ions on the surface is always larger than that in the bulk owing to surface-induced Friedel oscillations. Furthermore, irrespective of the type of t_{2g} spin structures, we find that the Friedel oscillations are enhanced when the Fermi-surface curves with better nesting properties exist. Namely, the dimension of the area of ferromagnetic (FM) t_{2g} spin structure is not essential to the enhancement of Friedel oscillations. We believe that these results will contribute to the development of highly-efficient cathodes and catalysts. For the calculations in this paper, we use $k_B = \hbar = 1$.

2. Model and Method

Herein we introduce the orbital-degenerate double-exchange model [14]. This model has been widely used to successfully investigate the physical properties in layered manganites, such as colossal magneto-resistance phenomena [15] and various types of charge and/or orbital orders in the bulk [12, 13, 16, 17]. To simplify the model, we consider a widely-used approximation that invokes infinite Hund coupling between e_g and t_{2g} electron spins. In this limit, the e_g electron spin perfectly aligns along the t_{2g} spin direction, reducing the spin degrees of freedom of e_g electrons. Namely, the e_g electrons move only in the FM region of t_{2g} spins, while in the AF region, e_g electrons are localized. The model Hamiltonian for this system is given by

$$H = -\sum_{\boldsymbol{i}\boldsymbol{a}\gamma\gamma'} D_{\boldsymbol{i},\boldsymbol{i}+\boldsymbol{a}} t^{\boldsymbol{a}}_{\gamma\gamma'} d^{\dagger}_{\boldsymbol{i}\gamma} d_{\boldsymbol{i}+\boldsymbol{a}\gamma'} + J \sum_{\langle \boldsymbol{i},\boldsymbol{j} \rangle} S_{z\boldsymbol{i}} S_{z\boldsymbol{j}}, \tag{1}$$

where $d_{\boldsymbol{i}a}$ ($d_{\boldsymbol{i}b}$) is the annihilation operator for a spinless e_g electron in the $d_{x^2-y^2}$ ($d_{3z^2-r^2}$) orbitals at site $\boldsymbol{i}$, $\boldsymbol{a}$ is the vector connecting nearest-neighbor sites, $D_{\boldsymbol{i},\boldsymbol{i}+\boldsymbol{a}} = (1 + S_{z\boldsymbol{i}} S_{z\boldsymbol{i}+\boldsymbol{a}})/2$, $S_{z\boldsymbol{i}}$ is an Ising-like t_{2g} spin at site $\boldsymbol{i}$ with $S_{z\boldsymbol{i}} = \pm 1$, and $t^{\boldsymbol{a}}_{\gamma\gamma'}$ denotes the nearest-neighbor hopping amplitude between γ and γ' orbitals along the $\boldsymbol{a}$ direction with $t^{\boldsymbol{x}}_{aa} = -\sqrt{3}t^{\boldsymbol{x}}_{ab} = -\sqrt{3}t^{\boldsymbol{x}}_{ba} = 3t^{\boldsymbol{x}}_{bb} = 3t^{\boldsymbol{z}}_{bb}/4$, $t^{\boldsymbol{y}}_{aa} = \sqrt{3}t^{\boldsymbol{y}}_{ab} = \sqrt{3}t^{\boldsymbol{y}}_{ba} = 3t^{\boldsymbol{y}}_{bb} = 3t^{\boldsymbol{z}}_{bb}/4$, and $t^{\boldsymbol{z}}_{aa} = t^{\boldsymbol{z}}_{ab} = t^{\boldsymbol{z}}_{ba} = 0$ [20]. We set the energy, $t^{\boldsymbol{z}}_{bb} = 1$. We note that the hopping amplitudes between different orbitals of a spinless e_g-electron system change sign between along the x-and y- directions, $t^{\boldsymbol{x}}_{ab} = t^{\boldsymbol{x}}_{ba} = -t^{\boldsymbol{y}}_{ab} = -t^{\boldsymbol{y}}_{ba}$. The second term in Eq. (1), J, is the AF coupling between the nearest-neighbor t_{2g} spins and $\langle \boldsymbol{i}, \boldsymbol{j} \rangle$ denotes the nearest-neighbor site pair. In this paper, we do not include explicitly the Coulomb

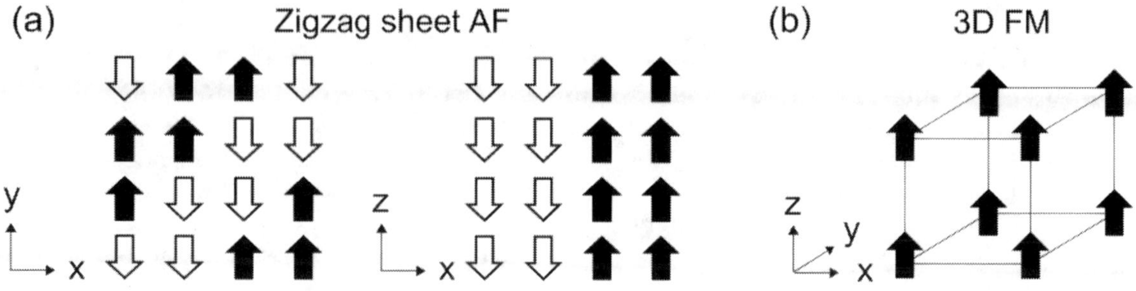

Figure 1. Schematic views of spin structures cited in this paper. Solid and open arrows indicate up and down t_{2g} spins, respectively. (a) The spin structure of the zigzag sheet AF state. (b) The spin structure in the FM state.

interaction terms, but the effect is partially considered by working with a large Hund coupling. Namely, we consider a manganite with large bandwidth. We define x as the hole doping and n as the electron number of e_g electrons per site with a relation of $x = 1 - n$. Thus, x indicates the average number of Mn^{4+} ions per site.

We diagonalize the Hamiltonian in a $1000 \times 1000 \times 100$ lattice: one layer is composed of a 1000×1000 square lattice with periodic boundary conditions along the x- and y-axes. To introduce surfaces to the model, we consider a stack of 100 layers with open boundary condition (OBC) along the z-axis. In this paper, we introduce the OBC in the sense of $\psi(0) = \psi(101) = 0$, where $\psi(z)$ denotes the wavefunction of a spinless e_g electron along the z-axis and z is a layer number. Namely, the first layer along the z-axis is the surface, whereas the 50-th layer denotes the bulk.

In this paper, we remark the e_g-electron properties on the background of the fixed t_{2g} spin structure. In our previous study, we investigated two t_{2g} spin structures: a C-type AF (C-AF) state and an A-type AF (A-AF) state [11]. In the C-AF state, e_g electrons can move only along the z-axis owing to infinite Hund coupling. Thus, the system is regarded as a one-dimensional (1D) chain with the OBC along the z-axis. In the A-AF state, e_g electrons can move in the x-z plane, leading to a two-dimensional (2D) sheet with edges. On the other hand, in this paper, we assume two other possible t_{2g} spin structures, zigzag sheet AF and FM states, as shown in Figs. 1(a) and 1(b), respectively. At least in a $\sqrt{8} \times \sqrt{8} \times 3$ lattice with surfaces, it is easy to show that these magnetic states actually appear as the ground states for appropriate values of J, but detailed discussion on the stability of ground-state magnetic structures in the large cluster is postponed in future. In the zigzag sheet AF state, the FM regions of t_{2g} spins form zigzag structures in the x- and y- directions, whereas t_{2g} spins are ferromagnetic in the z-direction. Thus, the system is regarded as a 2D sheet with edges, where hopping amplitudes change periodically. In the three-dimensional (3D) FM state, e_g electrons can move in all three dimensions.

By using these lattices, we evaluate $N(z)$, the number of Mn^{4+} ions per site in layer z, and investigate the behavior of the surface-induced Friedel oscillations. For several t_{2g} spin structures with different FM region dimensions, we examine the factor that determines the behavior of the oscillations. Note that $N(1)$ and $N(50)$ denote the number of Mn^{4+} ions per site on the surface and in the bulk, respectively.

3. Calculation Results

3.1. Brief review of previous results

Before proceeding to the results of our calculations, we need to briefly review our previous results [11]. We observe that $N(z)$ clearly exhibits a damped oscillation from the surface ($z = 1$) to the bulk ($z = 50$), which is understood to be the surface-induced Friedel oscillation, and results in an increase in the number of Mn^{4+} ions on the surface. In the 1D system such as the C-AF state, the formula for Friedel oscillation is expressed as [21, 22, 23, 24, 25]

$$N(z) - x \propto \cos(2k_{\mathrm{F}}z)z^{-(1+K_\rho)/2}, \tag{2}$$

where k_{F} is the Fermi momentum and K_ρ is the correlation exponent [26]. Since we do not include Coulomb interactions among e_{g} electrons, $K_\rho = 1$, which leads to z^{-1}.

On the other hand, in the 2D system such as the A-AF state, the formula for Friedel oscillations is modified to

$$N(z) - x \propto \cos(2k_{\mathrm{Fz}}z)z^{-C(x,\gamma)}, \tag{3}$$

where k_{Fz} is the z-component of the Fermi momentum with good nesting properties along the z-axis. $C(x,\gamma)$ is the damping power for the oscillations. We have confirmed that in the 2D system C depends on hole doping, x, and a relevant orbital, γ, which leads to $C(x,\gamma) \geq 1$. In other words, the envelope of the 2D Friedel oscillations decays faster than z^{-1}. At this stage, the definite expression of $C(x,\gamma)$ is not obtained. However, in general, $N(1)$ seems to be increasingly suppressed in higher dimensional systems since the influence of the surface on the system becomes relatively weaker as the dimension of the FM region increases.

3.2. Zigzag sheet AF state

First, let us consider the charge structure in the zigzag sheet AF state. In Fig. 2(a), we show $N(z)$ vs. z for $x = 0.65$. Since the zigzag sheet AF indicates a 2D-like system even though the FM region is not a plane (see Fig. 1(a)), one could naively consider $N(z)$ to be expressed by Eq. (3). However, as observed in Fig. 2(a), the envelope of the oscillations is well expressed by z^{-1}. Namely, the Friedel oscillation pattern is rather 1D-like in spite of the 2D-like FM region.

To investigate one of the factors that determines the value of $C(x,\gamma)$, let us examine the oscillations from a viewpoint of the Fermi-surface structure. In Fig. 2(b), we show two Fermi-surface curves at $x = 0.65$. Note that we depict the Fermi-surface curves considering OBC along the z-axis. When we consider the 100 layers stacked along the z-axis with OBC, k_z is given by $k_z(\ell) = \ell\pi/101$ where ℓ $(= 1, 2, \cdots, 100)$ is an integer, leading to $\delta k_z = \pi/101$. In the zigzag sheet AF state, we find that the spinless e_{g}-electron system possesses a quasi-1D Fermi-surface structure along the z-axis for a certain hole doping, x. Even in the 2D system, when the Fermi-surface curves have a quasi-1D structure along the z-axis, the envelope of the oscillations is expressed by z^{-1} as if in the 1D system. The period of the oscillations in Fig. 2(a) is given by $2 \times \pi/3$, consistent with the k_{Fz} shown in Fig. 2(b) and used in Eq. (3). Fig. 2(c) shows the Fermi-surface curves at $x = 0.7$. For $x = 0.7$, the Fermi-surface curves also retain the quasi 1D structure along the z-axis and we can confirm the value of $C(x,\gamma)$ is almost unity. The nesting properties are lost for $x < 0.65$ and $x > 0.7$, so the envelope of the oscillations decays faster than z^{-1}.

To understand the increase in $N(1)$ compared with x, we evaluate the efficiency R, given by $R = N(1)/x$ [11]. This quantity denotes to what degree the Mn^{4+} ions appear on the surface at a given hole doping x. When $N(1)$ is enhanced, we expect that the efficiency as cathodes and catalysts is increased. Thus, R has been called the efficiency . Note that R is associated with the magnitude of amplitude of the Friedel oscillations. Fig. 2(d) shows the plot of R as a function of x for the zigzag sheet AF state. We find the magnitude of R exhibits a peak between $x = 0.65$ and 0.7, for which one Fermi-surface curve has perfect nesting properties, leading to a Van Hove singularity. Note that the small oscillations in R around $x = 0.0$ are effects due to the size of the lattice. If we increase the size of the lattice, these oscillations disappear.

Thus, we can conclude that when the spinless e_{g}-electron system possesses a Fermi-surface structure with decent nesting properties along the z-axis, the value of $C(x,\gamma)$ becomes unity and

Journal of Physics: Conference Series **683** (2016) 012042 doi:10.1088/1742-6596/683/1/012042

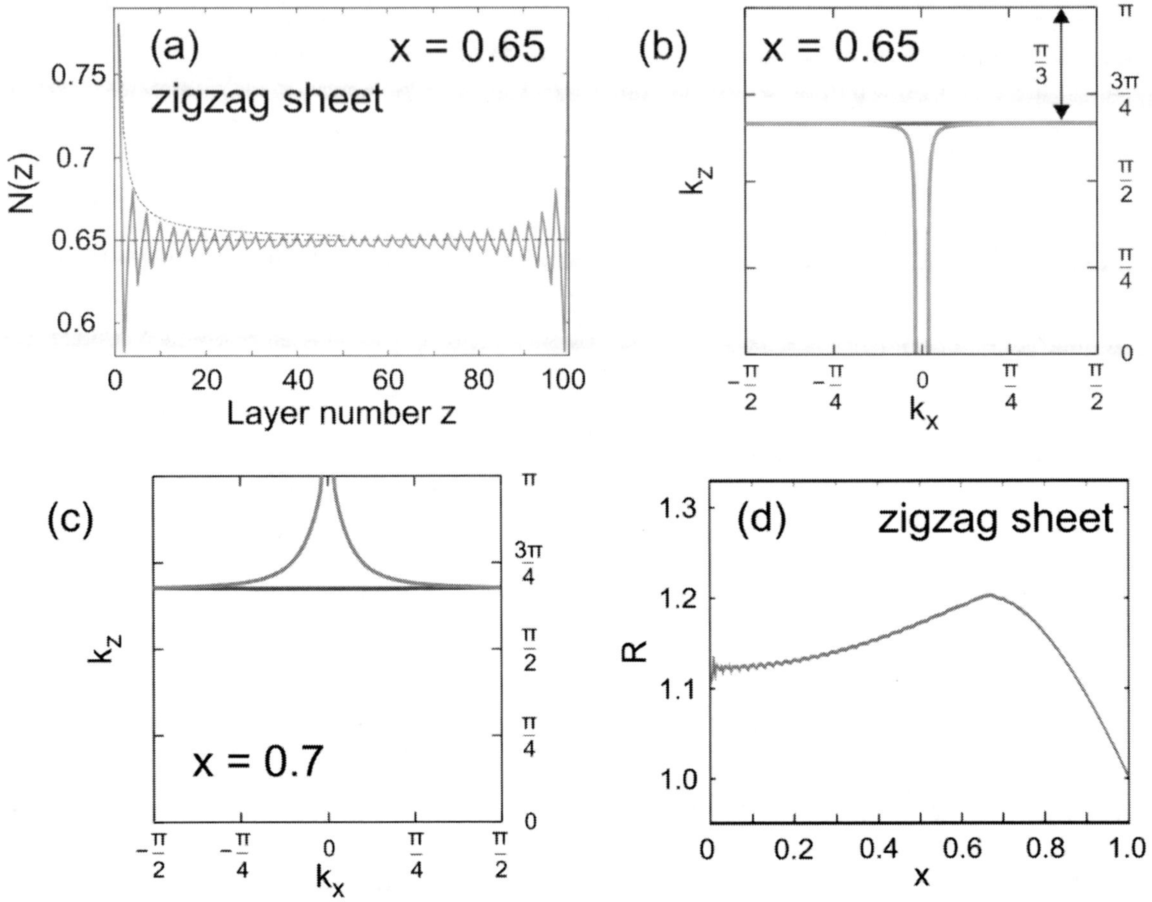

Figure 2. (a) $N(z)$ vs. z plot for $x = 0.65$ in the zigzag sheet AF state. the blue dotted curve denotes the envelope expressed by z^{-1}, and the broken horizontal line denotes the value of x. (b) Two Fermi-surface curves of the e_g-electron system for $x = 0.65$. The magnitudes of the Fermi momentum along the z-axis are both $\pi/3$, which is consistent with the period of the oscillation in (a). (c) Two Fermi-surface curves for $x = 0.7$. (d) Efficiency R vs. x. R is defined as $R = N(1)/x$.

the amplitude of the Friedel oscillations tends to be larger. Namely, the behavior of the Friedel oscillations is *not* determined by the dimension of t_{2g} spin structure itself, but the Fermi-surface structure of e_g-electron system is essentially important.

3.3. 3D FM state

Fig. 3(a) shows the plot of $N(z)$ vs. z for $x = 0.3$; surface-induced Friedel oscillations can be observed in the figure. Note that $N(50)$ is slightly smaller than x. In contrast to the case of the zigzag sheet AF state, the envelope of the oscillations decays faster than z^{-1}, since the nesting properties are weak in the 3D e_g-electron system. As observed in Fig. 3(b), R decreases monotonically in the FM state, which can be explained by the fact that the nesting properties in the 3D e_g-electron system are enhanced to some extent for smaller x.

A characteristic point for the FM phase is the appearance of an orbital arrangement on the surface, in sharp contrast to other t_{2g} spin structures. In Fig. 3(c), we plot $[n_a(z) - n_b(z)]/[1 -$

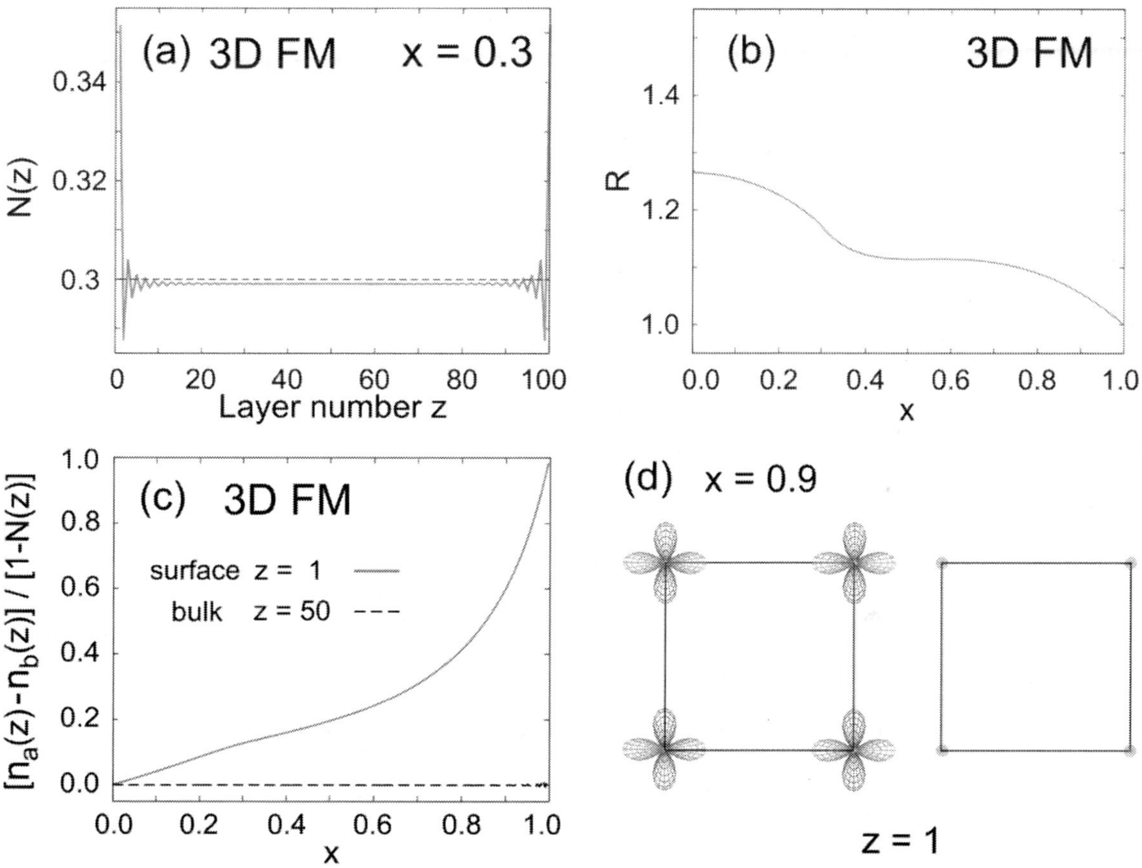

Figure 3. (a) $N(z)$ vs. z plot for $x = 0.3$ in the 3D FM state. the broken horizontal line denotes the value of x. (b) Efficiency R vs. x. (c) $[n_a(z) - n_b(z)]/[1 - N(z)]$ vs. x for $z = 1$ and 50. (d) Orbital arrangement on the surface x-y plane for $x = 0.9$. The orbital size is chosen to be proportional to the orbital density.

$N(z)]$ vs. x for $z = 1$ and 50, where $n_a(z)$ and $n_b(z)$ denote the average electron number in the $d_{x^2-y^2}$ and $d_{3z^2-r^2}$ orbitals, respectively. In the bulk, $d_{x^2-y^2}$ and $d_{3z^2-r^2}$ orbitals are occupied with equal weights because cubic symmetry is recovered in the 3D environment. On the surface, $d_{x^2-y^2}$ orbitals are selectively occupied, since e_g electrons on the surface layer can gain kinetic energy by the 2D motion in the x-y plane than the 1D motion along the z-direction. In fact, as shown in Fig. 3(d), there is a large density of $d_{x^2-y^2}$ orbitals with a ferromagnetic-arrangement on the surface, while $n_b(1)$ is relatively small. Namely, electrons on the surface continue to occupy $d_{x^2-y^2}$ and escape only from $d_{3z^2-r^2}$ orbitals, leading to the overall increase of $N(1)$.

4. Conclusion

In this paper, we have evaluated $N(z)$, the number of Mn^{4+} ions per site in layer z, by analyzing the orbital-degenerate double-exchange model with surfaces for zigzag sheet AF and 3D FM states. We have found that the surface-induced Friedel oscillations in $N(z)$ are enhanced, when the spinless e_g-electron systems possess Fermi-surface structures with improved nesting properties along the z-axis, leading to an increase of Mn^{4+} ions on the surface. For cathodes and catalysts, retaining Mn^{4+} ions on the surface and Mn^{3+} in the bulk is desirable. Subsequently, we can conclude that the t_{2g} spin structures possessing Fermi-surface structures with good nesting properties are possibly suitable for high-efficiency cathodes in Li-ion batteries and catalysts. We believe that the present results provide a fundamental understanding for further application of manganites to such systems.

Acknowledgment

We thank K. Hattori, K. Kubo, and K. Ueda for discussions and comments. The computation in this work was partly carried out using the facilities of the Supercomputer Center of Institute for Solid State Physics, University of Tokyo.

References

[1] Nagaura T and Tazawa K 1990 *Prog. Batteries Solar Cells* **9** 20
[2] Ozawa K 1994 *Solid State Ionics* **69** 212
[3] Yang H, Amiruddin S, Sun H J, and Prakash J 2006 *J. Indust. Eng. Chem* **12** 12
[4] Armstrong A R and Bruce P G 1996 *Nature* **381** 499
[5] Kang K, Meng Y S, Breger J, Grey C P, and Ceder G 2006 *Science* **311** 977
[6] Umena Y, Kawakami K, Shen J R, and Kamiya N 2011 *Nature* **473** 55
[7] Robinson D M, Go Y B, Greenblatt M, and Dismukes G C 2010 *J. Am. Chem. Soc* **132** 11467
[8] Takashima T, Hashimoto K, and Nakamura R 2012 *J. Am. Chem. Soc* **134** 18153
[9] Choi W and Manthiram A 2006 *J. Electrochem. Soc* **153** A1760
[10] Ouyang C Y, Sljivancanin Z, and Baldereschi A 2010 *J. Chem. Phys* **133** 204701
[11] Yamamura R and Hotta T *arXiv:*1510.05348 (to appear in Phys. Procedia)
[12] Hotta T, Yunoki S, Mayr M, and Dagotto E 1999 *Phys. Rev.* B **60** R15009
[13] Hotta T 2003 *Phys. Rev.* B **67** 104428
[14] Zener C 1951 *Phys. Rev.* **82** 403
[15] 2000 *Colossal Magnetoresistance Oxides* ed Y Tokura (New York: Gordon Breach)
[16] Dagotto E, Hotta T, and Moreo A 2001 *Phys. Rep.* **344** 1
[17] Dagotto E 2002 *Nanoscale Phase Separation and Colossal Magnetoresistance* (Berlin: Springer)
[18] Hotta T and Dagotto E 2004 *Colossal Magnetoresistive Manganites* ed T K Chatterji (Amsterdam: Kluwer) p 207
[19] Hotta T 2006 *Rep. Prog. Phys.* **69** 2061
[20] Slater J C and Koster G F 1954 *Phys. Rev.* **94** 1498
[21] Friedel J 1954 *Adv. Phys.* **3** 446
[22] Shibata N, Ueda K, Nishino T, and Ishii C 1996 *Phys. Rev.* B **54** 13495
[23] Fabrizio M and Gogolin A O 1995 *Phys. Rev.* B **51** 17827
[24] Egger R and Grabert H 1995 *Phys. Rev. Lett.* **75** 3505
[25] Egger R and Schoeller H 1996 *Czech. J. Phys.* **46** Suppl 54 1909
[26] Schulz H J *Phys. Rev. Lett.* **64** 2831

Institute of Physics
Dirac House, Temple Back
Bristol BS1 6BE UK

ISSN: 1742-6588
ISBN 978-1-5108-2093-7

CURRAN ASSOCIATES INC.
proceedings.com